SEPARATIONS TECHNOLOGY

Pharmaceutical and Biotechnology Applications

Wayne P. Olson, Editor

Interpharm Press, Inc.
Buffalo Grove, IL

10 9 8 7 6 5 4 3 2

ISBN: 0-935184-72-4

Interpharm Press, Inc.
1358 Busch Parkway
Buffalo Grove, IL 60089, USA

Phone: + 1 + 708 + 459-8480
Fax: + 1 + 708 + 459-6644

Contents

Preface

My interest in separations commenced with attendance at the weekly staff meetings of Dan Campbell's group at CalTech where, in 1964, I first learned of immunoaffinity chromatography. At Hyland (division of Baxter) I was using ammonium sulfate, gel filtration, and ion-exchange methods to produce research lots of fluorescent antibodies and other diagnostic reagents. It was the beginning of a lifelong interest in separations technologies.

Separations systems are at the heart of product purification and the maintenance of purity. Chemists usually think of separations as synonymous with column chromatography, but separations also embrace molecular and particulate separations by membranes, centrifugal fields, phase partitioning, and other means. Microbiologists equate separations with the removal of microorganisms from nonsterile environments by filtrative methods.

Two of the areas in which the U.S. FDA currently is very active are included in this text: the resolution of chiral (enantiomeric or racemic) drugs and the elimination of viruses in biologics. Supercritical fluid extraction and chromatography are an emerging area of interest that, to date, has been exploited only by the food industry.

As with *Aseptic Pharmaceutical Manufacturing* (1987) [*Aseptic*] and *Sterile Pharmaceutical Manufacturing* (1991) [*Sterile*], this text is intended for scientists and managers involved in the planning and execution of process, quality assurance, and regulatory affairs. The information in each chapter is of practical use and in each instance has been written by someone with hands-on experience. A number of the chapters in *Aseptic* and *Sterile* were cutting edge, as are certain chapters in this text. Much of the content of the three texts is complementary although a modest degree of overlap exists. For example, the excellent chapter on membrane methods in this volume stands alone, but also provides considerable new information on and insights to the same topic presented in *Aseptic* and in Pete Johnson's phenomenally good *Fluid Sterilization by Filtration*. Monoclonal antibodies have been dealt with in *Sterile* and in this text, in part because different methods have been employed and because the immunosorbent and monoclonal chapters in this text are complementary. This text is intended for the pharmaceutical manufacturing community at large, but the reader with particular interests in aseptic manufacturing will find much of use in this volume and is advised to peruse both *Aseptic* and *Sterile*, most of which remain current.

Wayne P. Olson
Beecher, Illinois
April 1995

Editor's Note

Work on this text started late in 1991 when I was completing my thesis at the University of Illinois at Chicago. Keeping the text, and its contributors, on track required more than a minor effort. Amy Davis, Director of Publications and Software at Interpharm Press, made it happen. This publication belongs as much to her as to me and the other contributors.

Dedication

Timothy Leahy, Ph.D., 1949–1992

This text is dedicated to Dr. Timothy Leahy, Vice-President and General Manager, Pharmaceutical Division of Millipore Corporation, Bedford, Massachusetts.

Tim earned his doctorate in microbiology while working at Millipore. His thesis provided direct proof of how membrane filters remove bacteria from suspension. Tim's subsequent work at Millipore resulted in the *Pseudomonas diminuta* challenge tests that now is used, worldwide, in defining sterilizing-grade filters. That body of work also related the bubble point pressure to microbial retention. Prior to his doctorate, he established the relationship between membrane filter bubble point and diffusion test values. These and related applications are critical to the aseptic manufacture of parenteral drugs.

There are many measures of a person, but one of the best is the quality of his or her legacy. Dr. Leahy left scientific benchmarks by which we now define and measure filtration practices; he maintained and embellished a tradition of excellence at Millipore; and he left some very fine memories.

Wayne P. Olson
Beecher, Illinois
April 1995

1

Separations in Pharmaceutical Manufacturing

Wayne P. Olson

PURE DRUGS v. CRUDE PREPARATIONS: CONVENTIONAL PHARMACEUTICALS v. FOLK MEDICINE

The current fashion among many Western nonphysicians and nonpharmacists is to extol native medicines as having great virtues, not least of which is low cost. Simple aqueous or ethanolic extracts of plants for centuries have had known efficacy in the treatment of particular disease states. However, in the absence of the pure drug and analytical methods of quantitating the drug in biological fluids, one cannot develop a reproducible dose-effect relationship; nor guarantee dose-to-dose uniformity, efficacy of the drug, or safety. In the view of most Western physicians and pharmacists, such a system is anathema although the bulk of the drugs presently in widespread use in the West had their origins in folk medicine.

The current understanding is that each drug has a receptor with which it reacts to initiate an event at the cellular level. Current regulatory philosophy at the U.S. FDA and elsewhere is that no solutes, other than the active drug and excipients, should be formulated with a drug. The target level for impurities in First World therapeutics for repeated injections is less than 1–2 parts per million (Wheelwright 1989). This level of purity often is achieved although, in the strict sense, racemates of a drug may be inactive or even counteractive, and usually are included in the dosage form. The philosophy is that a highly purified drug is formulated only with those excipients that are required to effect delivery of the drug with safety to the patient.

DEVELOPING A DRUG (APART FROM REGULATORY ISSUES)

The following is a somewhat simplistic and abbreviated approach to drug development, and indicates what must be done before and as scaleup commences. One of the earliest steps is the development of an activity assay on which subsequent chemical or physical assays are based.

Developing a Bioassay

The first step in the development of a drug is the development of a bioassay for activity. Consider an hypothetical drug, perhaps a neutralizing humanized mouse monoclonal antibody (MAb) directed against a critical enzyme of the rabies virus (RV). The initial assay might be in rabid mice. One would inject a highly uniform line of mice with varying levels of RV so as to establish a relationship for time to death as a function of RV dose. At a uniform RV dose, one then would inject the RV–injected mice with the MAb at varying levels, taking note of the survival times. If the enzyme is critical to the replication of RV, and if the MAb inactivates the enzyme, the MAb may be protective. Control mice with RV might be injected with bovine serum albumin (BSA) since BSA is unlikely to protect RV–injected mice. Because such assays are inordinately expensive, tedious and sometimes hazardous, a chemical or biochemical assay is developed that correlates with this hypothetical bioassay.

Developing a Chemical/Chromatographic/Biochemical Assay

By means of the bioassay, a chemical or biochemical assay is developed [e.g., an enzyme-linked immunoassasy (ELISA) or elution of a peak with absorbance at 260–280 nm at a particular concentration of acetonitrile in a reverse-phase gradient on a C_{18} alkylated column, i.e., RP–HPLC]. Validation of the correlation between bioactivity and the chemical assay is critical to scaleup and, during production, is critical to routine quality control of the drug.

RP–HPLC is the most popular routine drug assay method in use, and is applicable to conventional drugs (synthetics or secondary metabolites, e.g., digoxin) as well as peptides and proteins (e.g., insulin). In addition to use in scaleup and process monitoring during production, such an assay may be used in the critical studies of pharmacokinetics, wherein blood samples are recovered at various times following the dosing of an animal or human and the blood is analyzed, quantitatively, for the drug. This is necessary if one is to determine the minimum effective dose and the dosing regimen that maintains the desired concentration in vivo. Those quantities must be estimated as part of scaleup information.

In the case of the hypothetical MAb that neutralizes an enzyme contained in RV, an RP–HPLC assay (or a capillary electrophoresis assay or any of a variety of other reliable, sensitive, reproducible tests for the purified MAb) might be correlated with the bioassay. Relative to bioassays, such tests are inexpensive, rapid and simple. They do not demonstrate the biological function of the drug but identify the drug from a well-described characteristic. Such tests are important for scaleup, become a routine part of QC, and are used in process monitoring during production.

Purifying the Drug

For many conventional drugs, the cost of purification is modest. For biologics, purification is a dominant part of the overall process cost (Dwyer 1983). Ideally, purification is done with a simple, inexpensive, single step that provides (in the case of biologics such as the anti-RV MAb) an immunologically pure product in a high yield. In fact, there is mutual exclusivity between several of these. Ion-exchange chromatography (IXC) is simple, relatively inexpensive, and a high-yield step when used alone. Ion exchange in two or more iterations may provide the required level of purity but that is exceptional. Affinity methods, utilizing an immobilized ligand that binds the drug, usually result in high yield and high purity but are far more complex and expensive than IXC. Two or three relatively inexpensive steps (A-B-C) that result in a highly purified drug (> 99 percent purity) in good yield are preferred to a single affinity step that costs ten times more than A-B-C. The exception is a very expensive drug (e.g., recombinant human clotting factor VIII), for which the expense of an affinity step is readily justified.

The best approach to the least-cost method for purifying a drug is to outline the two or more schemes that seem the most practical, and to make the decision as to which method will be used based on projected cost. For example, this author sees no need for affinity purification of a monoclonal antibody (MAb) when two or (at most) three passes on one or two strong ion-exchanger(s) are equally as effective at a fraction of the cost.

Scaleup of the Manufacturing Process

If one has a product with a very high probability of therapeutic utility and a high return on investment, it may not be wise to attempt to optimize the process too early in the game. The tissue plasminogen activator (tPA) footrace was a case in point, as was the case with erythropoietin (EPO). As quickly as is practicable, one must have a process that is adequate to produce reagents for the clinical trials and, at that stage, one may not be able to alter too much the manufacturing process due to regulatory pressures. Early compromises may be necessary.

Therefore, scaleup must be considered before human clinical reagents are prepared. In a highly competitive situation, the time constraints can be very serious. Elegance and optimization may be less important than timeliness. If you are competing with another firm for IND status on a drug that falls within the provisions of orphan drug legislation, you may not have the luxury of time to optimize drug production and purification. You may be obliged to produce using, for example, mammalian cell culture in roller bottles followed by three IXC steps and ultrafiltration (UF) for buffer exchanges and concentration of a biological. The yield may be 65 percent rather than the 85 percent possible with an optimized process, but if the choice is between time for optimization of process versus obtaining orphan drug status and sole access to a market, one goes for the market.

Timelines must be established and critical path/PERT (process evaluation and research technique) scheduling applied to scaleup through to the production of human clinical reagents because PERTing promotes the most efficient use of time and resources. The beauty of classical methods, such as ion-exchange and affinity chromatography, is that they are well characterized, readily scalable, multiple vendors are in the marketplace, and costs are predictable.

SIZING A CHROMATOGRAPHIC SYSTEM

Market Projections

If the world market for a particular drug is about 1 kg, the average use level will be about 20 g/week. On the smallest scale, if the drug can be purified to homogeneity in 3 g quantities in 1 hr on a silica column of 10 cm diameter by 50 cm length, then that column can be automated and, even with a 3 hr regeneration time for returning the column to starting conditions, material for the world market can be produced in 2 months. At a purification rate of 6 g/day, the world supply could be purified on the column in 167 working days. However, each column run likely would be interpreted as a lot and QC costs would be extraordinarily high.

QC Costs and the Value of Product at Risk

Large, infrequent lots reduce QC costs, but increase the value of material put at risk in each production run. Usually, the compromise lies in a lot size that risks no less than about $250,000 and no more than $2 million.

If the dose is 50 mg of active ingredient in a single-dose vial, then 120 g represents 2,400 vials which is a small lot. If a 50 mg vial sells for $100, then the value of a lot is approximately $240,000, which is a reasonable dollar amount to put at risk for one lot. QC costs would be sustainable, but substantial as a proportion of the gross. The supply for one year would be 20,000 × 50 mg vials. The gross for this product would be $100/vial × 20,000 = $2,000,000. [For such a small return, the product might not be made. I would guess that the product would be priced at $400/vial, putting the gross at $4,000,000, still far too little for most firms.]

Dosage, Demand for Product, and QC Costs

Let us say that the drug is an organic of molecular weight (M_r) of 250 Daltons. The dose is 50 mg (as above). On the basis of the number of vials, provided the shelf life is appropriately long and the manufacturing and QC systems have been properly validated, it makes sense to manufacture 250 g of drug once per quarter rather than a 30 g column runs once per week for 34 weeks. This reduces QC costs (but not overhead) by 88 percent. However, a once-quarterly lot is vastly different in the scale of equipment and reagents required, hence the upfront investment is greater. Much more drug is put at risk per lot, but facility and space are freed for other work and products. However, if the firm is severely undercapitalized and does filling with a small filling machine in a biohazard hood, weekly or biweekly purifications may be necessary.

If the demand for product has been underestimated, which often occurs as new indications are found, doubling production is done easily with no additional costs in equipment. Similarly, a low concentration of drug in the final container or other QC test failure can be met in a timely manner since manufacturing, in this instance, need not be scheduled too closely.

Column Life

Just as the drug has a finite shelf life, so the column has a finite life. Column matrix may fragment or dissolve. For example, microporous glass is notoriously labile at pHs greater than 8; the glass goes into solution. Column matrix comprised of a cross-linked polysaccharide may compact and become essentially impermeable (e.g., Sephadex G-200).

Ligand (e.g., C_{18}) will elute from the matrix over time and with use, reducing column capacity. Some solutes will adsorb irreversibly, also reducing column capacity. Particles entering the column will reduce flow rate if the column is fed by head pressure, or increase back pressure if the column is fed by pump; fragmentation of the column matrix will have the same effect.

Much of this is predictable from bench-scale experiments with 1–10 ml beds, run for a minimum of 50 cycles. Fifty cycles for production beds are not uncommon and upwards of 80 cycles have been known. However, beyond about 20 cycles the savings achieved by prolonging the column life represent a diminishing return. Primary concerns are the quality of the feed and the cleaning, sanitization, and storage of the columns between runs. Validation of the cleaning, sanitization, and storage of production columns is most important.

Purification Schemes

An incredible variety of permutations are possible in the selection of column matrix, ligand, coupling conditions, equilibration, and elution conditions. Appropriate action is not to investigate all of these but only the most-likely-to-succeed. Fortunately, for some of the purification schemes, a wealth of information is available in the literature (e.g., *Journal of Chromatography,* which has occasional articles, symposia, and issues devoted to process scale). For the industrial scale purification of biologics, *Process Biochemistry* often is a useful source.

Commonly, one attempts to translate information on analytical separations to the preparative scale. Problems that are invisible to the researcher are magnified by orders of magnitude on a production scale to issues of system operation and product purity in the final container. The classic case is reverse-phase high performance liquid chromatography, which in most cases does not perform to requirements on an industrial scale.

Separations Based on Analytical Systems

Consider that a drug, produced under patent protection, is about to come off patent and a generic manufacturer wishes to produce the drug in bulk. The literature may contain information on quantitative analysis by RP–HPLC, employing an acetonitrile gradient. It may be argued that acetonitrile can be recovered and recycled, and that drying the product at 10 Torr should reduce the acetonitrile content to nondetectable levels.

However, with conventional HPLC, the sample volume is small and the gradient is linear or moderately concave or convex. Product elutes from the column as a peak containing a modest amount of product (Figure 1.1). For analytical columns, sample volume usually

Figure 1.1. Automated columns-in-series system for the purification of human growth hormone produced in *E. coli*, after Chapman et al. (1990).

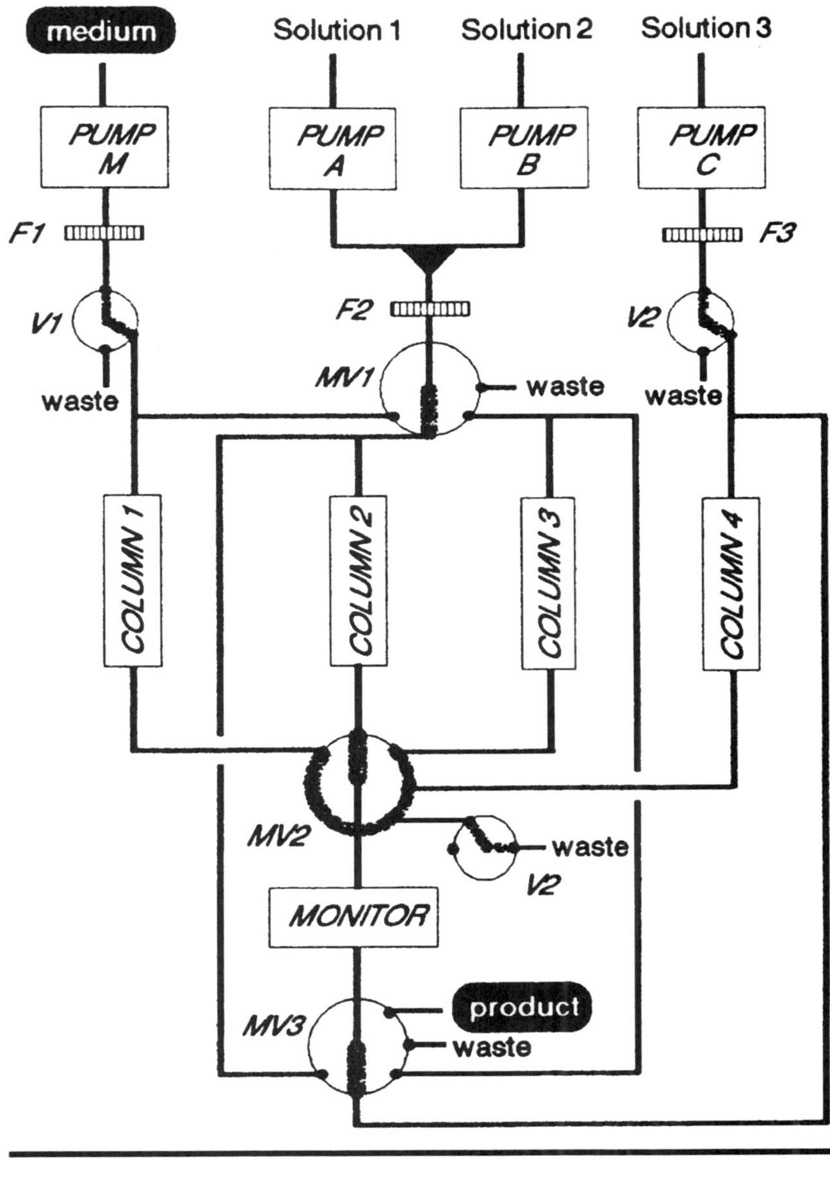

is, at most, 4 or 5 percent of the bed volume. For this reason, RP–HPLC usually is considered unsuitable for production scale.

What one wants is a column bed that is used to some substantial proportion of its capacity to bind solute. This can occur only if the column matrix or ligand binds the desired product strongly under a set of conditions that do not degrade the drug (i.e., K_d is small, e.g., 10^{-6}). Alternatively, one may bind the contaminants and collect the product directly from the column under starting condition development. This strategy can be applied with, for example, strong ion-exchangers. As with any chromatographic method, it is crucial to know the conditions at which the columns are saturated and breakthrough occurs, and to oversize the columns by 50 to 100 percent as a margin of safety.

A system for the isolation of an enzyme from native or recombinant *Escherichia coli* is shown in Figure 1.2 (Titchener-Hooker et al. 1990). Column steps would be preferred downstream. The present author would prefer, following centrifugal clarification, a rapid solvent exchange on a column of Sephadex G-25 coarse or diafiltration and concentration in an ultrafilter, followed by ion-exchange chromatography on a strong exchanger such as sulfopropyl cellulose or the equivalent.

PURIFICATION METHODS

Particle Separations

Sieving (Separation by Size)

The most common method for separation of dry particles is sieving, and a wide variety of analytical and process equipment is available. Most of the sieves are of woven wire, hence relatively well defined. Small-aperture etched sieves are manufactured by BuckBee Meers in Minneapolis/St. Paul, MN. Sieving is not used in the manufacture of injectables. Particles for injection invariably are produced from solution; such methods are amenable of tight controls and standards, and the fractionation of particles so generated should be unnecessary. Trituration and other strictly mechanical methods are more difficult to control and are quite unsuitable for the large-scale generation of submicron dispersions. Consequently, sieving usually is done with, for example, particulate drugs for dispersion in ointments (e.g., griseofulvin).

Air Elutriation (Separation by Size and Density)

On an industrial scale, particles can be separated by size and density on a column of air. For example, Pharmacia separates beads of

Figure 1.2. Effect of borate anion concentration and pH on the clarification of a yeast homogenate as a result of flocculation and centrifugation. After Titchener-Hooker et al. (1990).

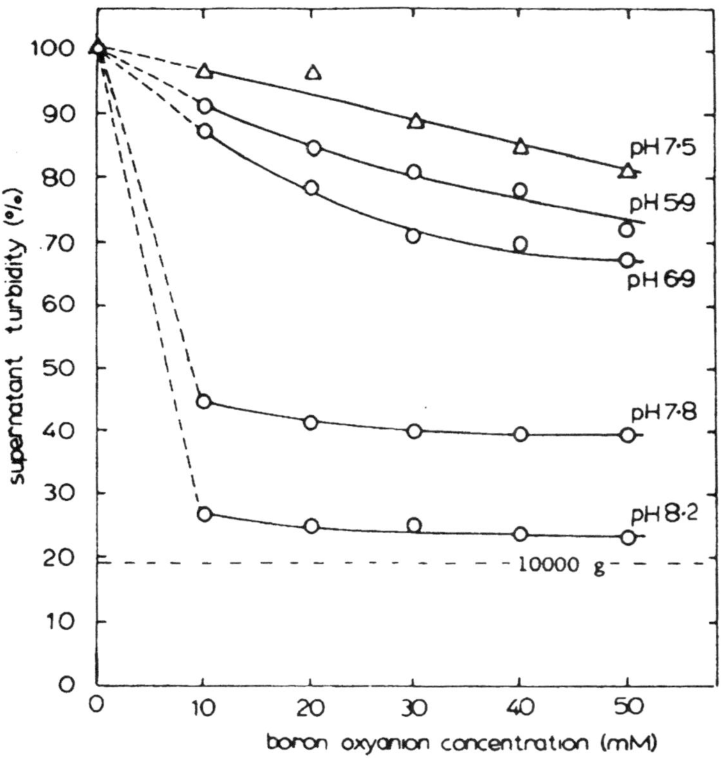

Sephadex in a batch by air elutriation. For particles of equivalent density, the larger particles move a shorter distance up the pipe. The efficiency of the separation is not particularly high, but for many purposes (e.g., the sorting of Sephadexes) is adequate, gentle, simple, and inexpensive. Some coating of formulated drug particles is done in a modification of this method: the particles, suspended by warm air, are sprayed with the various coatings in solution. However, the spraying of drug particles suspended by blown air is not air elutriation and is not a separation process.

Removal of Particles and Microbes (Microporous Filtration)

Whether from air entering a sterile tank or from a liquid (solvent or aqueous) entering a sterile batch tank, the favored method for *cold* sterilization is microporous filtration (Meltzer 1987; Olson 1987). Microporous filters are rated on the basis of their ability to retain various organisms. For example, a filter intended for pharmaceutical use is rated at a pore diameter of 0.2 μm if the filter can retain 10^7 *Pseudomonas diminuta* ATCC 19146 (grown in saline-lactose broth in static culture) per cm^2 filter area (Meltzer 1987; Olson 1987). The same filter may pass precision polystyrene latex beads of diameter greater than 0.2 μm, but the microbial retention test is considered definitive.

Liquids for injections that are not packaged aseptically also are filtered, at least to 0.45 μm, if possible. This includes large-volume parenterals (LVPs) such as sterile saline. Filtration is necessary as it has been established that a heavy particle burden in LVPs tends to occlude the capillaries of the lungs and other organs, much to the detriment of the recipient (Garvan and Gunner 1963, 1964; Hearse et al. 1985; and discussion in Groves 1993).

Asahi Chemical (Osaka, Japan) has a filtration group that does exquisite research under the guidance of Prof. Dr. Kenji Kamide. That group has developed what they call the BMM Planova 15 N filter, of cuprammonium-regenerated cellulose hollow fibers with a mean pore size of 15 nm. A 35 nm pore-size filter also is available. In deadend filtration, a French Red Cross group (Burnouf-Radosevich et al. 1994) found that single filtrations could remove more than 5.7–7.8 $\log_{10}$ of HIV, bovine viral diarrhoea virus, porcine pseudorabies virus, reovirus type 3, SV-40, Sabin type 1 poliovirus, or bovine parvovirus.

Such a small pore size represents an enormous resistance to flow. It is interesting to note that 3–4 ℓ of the clotting factor IX that had been virus-spiked were filtered, *dead-ended*, through *one* m^2 of the 15 nm hollow fiber membranes. The removal of all of the viruses by means of a membrane filter is most impressive, especially since the Sabin type 1 poliovirus has a diameter of 25–30 nm and bovine parvovirus has a diameter of 20–25 nm. The implication of the methods section in Burnouf-Radosevich is that transmembrane pressures were similar to those used with sterilizing-grade filters, in which case the flow rates must have been very slow, one meter squared of membrane surface in a hollow fiber configuation notwithstanding. However, this method and the filter require serious consideration since, like Virasorb (Millipore), this represents an additional method for virus removal from therapeutic proteins.

Particle Recovery or Particle-Free Liquid (Ultrafiltration)

Ultrafiltration (UF) is one of the simplest and most economic scalable methods for generating aqueous systems with low particle burdens. *Particle-free* is a misnomer, but ultrafiltered water or buffers or proteinaceous solutions should contain modest amounts of submicron particulate contaminants and very small concentrations of particles greater than 1 μm in diameter. A/G Technologies (Lexington, MA), Amicon (Beverly, MA), Filtron (Northborough, MA), Gelman (Ann Arbor, MI), and Millipore (Bedford, MA), amongst others, offer microfiltration and ultrafiltration systems for particle reduction.

Particle Recovery or Desludging (Centrifugation)

The best-characterized and most widely used methods for desludging a suspension to reduce particle burden are centrifugation, depth filtration (usually with a filter aid) and cross-flow filtration. Where the particles are desired and the supernatant or infranatant is to be discarded, centrifugation or cross-flow filtration with systems lacking upstream supports are preferred (e.g., in the De Danske Sukkerfabriker system, and certain of the systems available from Millipore). The use of depth filters (especially with diatoms) is not acceptable. The exception obtains where particles are to be recovered and then resolubilized. Although this can be done by centrifugation and resuspension of the paste, the plate-and-frame depth filter method cited in Fiore et al. (1980) has many advantages, provided one can backflush the filter under conditions that resolubilize the trapped particles (e.g., insolubilized proteins).

An interesting problem in particle removal is the case of the removal of cellular debris when recombinant microbes are disrupted, (e.g., in a Manton-Gaulin homogenizer or a French press). Many of the particles fall into the submicron range and a substantial proportion can be precipitated only after 20 min at 10,000 × *g* (Bonnerjea et al. 1988). However, if the suspension is made 10 m*M* in borax and taken to pH 8.2, over 99 percent of the cellular debris can be removed at low speed (equivalent to 2,000 × *g*). The effect of borate anion concentration and pH on the clarification of a yeast homogenate is shown in Titchener-Hooker et al. (1990) (see Figure 1.2). Borate anion reacts specifically with carbohydrates, but this includes glycosylated proteins. Hence, where recovery of glycosylated material in solution is desired, borate flocculation of debris will reduce yield.

Recombinant proteins that are overexpressed in *Escherichia coli* usually form insoluble particles (inclusion bodies) that have a high density (e.g., 1.3–1.4 g/cc). When the cells have been disrupted (lysozyme, homogenization, or both) and centrifuged, the inclusion bodies sediment first. A centrifugal system that is efficient (flow-through tube), very user-friendly and readily cleanable is the Sharples™ tubular centrifuge (models A-16 and A-26 as well as the larger units commonly are used) (Figure 1.3). The bowl usually is lined

Figure 1.3. A Sharples™ centrifuge.

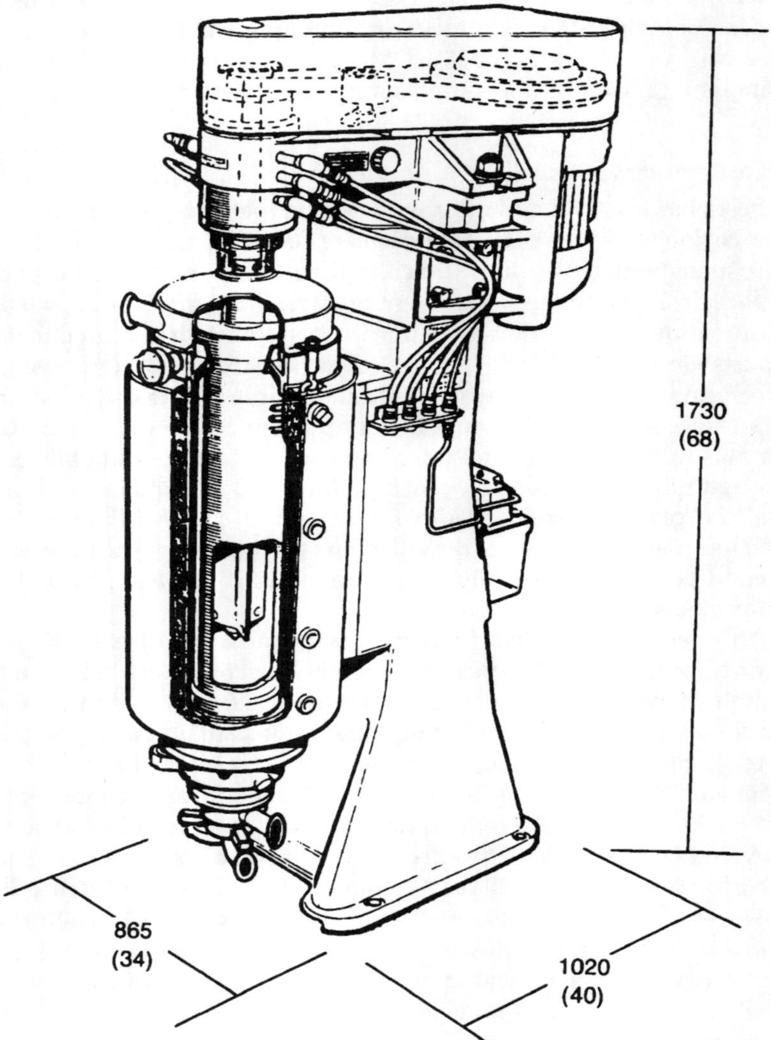

with poly(tetrafluoroethylene) or similar sheets for the rapid removal of the precipitate. Feed of the suspension is from the bottom of the upright centrifuge, rotating at about 15,000–17,000 rpm. These centrifuges develop about 15,000 × g, which is exceptionally high for a production centrifuge. However, flow rates as low as 1–5 ml/minute may be required to effect the desired separation in which case multiple centrifuges are manifolded. Other types of centrifuges also have been used for the recovery of inclusion bodies. Most non-Sharples units are more difficult to clean and develop lower centrifugal force.

Particle (Cell) Separation by Charge or Density

Fisker et al. (1990) developed a simple discontinuous density system for the purification of mononuclear and polymorphonuclear cells. For purposes of industrial manufacturing, the method is of limited interest. However, an early step in the development of monoclonal antibodies is the enrichment of hybridomas prior to plating out in a 96 well system for the identification of the desired cells. Koolwijk et al. (1988) have described a Percoll™ (Pharmacia, Uppsala, Sweden) density gradient system that, coupled with cell sorting based on an appropriate label, should improve the efficiency of hybridoma recovery.

Molecular Separations and Conversions

Distillation

Distillation has been used in the purification of ethanol, diethyl ether, and other solvents since antiquity. The primary applications for pharmaceutical manufacturing are in solvent recovery and in the preparation of water for injection (WFI). The reader is referred to Ted Meltzer's (1991) definitive chapter on WFI for a discussion.

Phase-Partitioning

A technique of co-culture of one cell line with another is emerging in in vitro fertilization. If it is adopted in recombinant fermentations, host-cell concentration and purification may become necessary.

Per-Ake Albertsson (1971) developed simple aqueous two-phase systems (ATPS) and multi-phase systems for the purification of mammalian cells. Arne Boyum et al. (1983) established systems for the purification of peripheral mammalian blood cells using Albertsson's partitioning systems.

One advantage of a purification step based on the partitioning of a drug between two largely immiscible phases is that, for many systems, the recovery can be very high. A second advantage is that such systems are readily scaled up. For example, aqueous stream A enters a Y from one pipe, organic solvent stream B enters the Y from a second pipe, the two mix in a static mixer as they pass through the stem of the Y, from which the stream may be fed to an homogenizer or into a tank where the stream may encounter a high-speed or ultrasonic mixer. Alternatively, the fluid exiting the static mixer may be processed through a Nuclepore track-etch filter (resistant to the solvent) at high pressure. These are mechanics; the theoretical basis for the selection of bi- or triphasic partitioning systems may be found in texts such as Wolf (1969). Such steps are used where there is a dramatic change in drug polarity (e.g., following the esterification of a carboxyl group). For subtle and complex separations, column methods are preferred.

ATPS can be highly suitable for preliminary steps in the purification of proteins (personal opinion). The polymer-salt or polymer-polymer (usually PEG–dextran) systems that are suitable for phase-partitioning (Albertsson 1971) can be applied to commercial processes that are automatable. The greatest impediment to use is the cost of dextran ($100–600/kg). Hydroxypropyl starch ($1.00–2.50/kg) may be employed in place of dextran, and other low-cost polymer systems also may prove suitable. Such systems are not presently in use in the industrial-scale purification of proteins for injection. The power of these systems might best be applied to separations for the preparation of difficult- or expensive-to-purify species such as DNA. For example, in a salt–PEG system, nucleic acids partition efficiently to the salt phase whereas proteins concentrate in the PEG phase and cell debris concentrates at the interface.

Hydrophobic proteins partition to the detergent-rich (coacervate) phase in ATPS employing PEG–based detergents (Terstappen et al. 1993). Since many receptors bear hydrophobic membrane anchors, one would expect that receptors would migrate to the interface of such systems. Detergent-containing two-phase systems may be particularly useful in the purification of proteins like viral capsid antigens produced in recombinant bacterial cells. Such proteins often contain major hydrophobic sequences that cause the protein to aggregate, impeding the purification process.

ATPS have not (to the author's knowledge) been used routinely in the preparation of cells for reactors. Such preparation usually is unnecessary since one commences with a clone that is grown in a fermentor and frozen in aliquots at -70°C or thereabouts for the preparation of inocula on demand. Other than on a one-time basis, cell purification is inappropriate.

Phase partitioning for the purification of organic drugs also can be done on column. The best-known supports are Sephadex LH-20 and Sephadex LH-60. The vida infra (solvent inside the beads of the column) is water-saturated solvent. After application of the sample, column development is with solvent-saturated water. The water may be buffered with an inorganic salt(s). The advantage of the column system is carefully controlled conditions and short diffusion distances. The Sephadex supports are viewed as quite expensive and are used primarily with high-priced drugs.

Crystallization

Crystallization is a powerful technique requiring time, tanks, and seed crystals. An interesting example of the purification of a peptide for use as a vaccine was the crystallization of tetanus toxoid in 1946. Usually, crystalline product, whether an organic or a protein, has a high degree of purity, often well in excess of 90 percent. For peptides or proteins such as tetanus toxoid, crystallization commences with a solution of partially purified solute that is concentrated to, for example, 5 g percent. This solution then is made supersaturated in the protein of interest by the addition of a precipitant [e.g., 15 percent (w/v) poly(ethylene glycol) (PEG)] of ca. 4 kDa, or ammonium sulfate to, for example, 20–25 percent of saturation. If the solute is a macromolecule such as a protein, the solution then may be concentrated slowly, as with an ultrafilter which does not alter the salt molarity but concentrates the protein. Crystallization is not a favored method because it is very slow.

Precipitation-Resolubilization

One of the most common steps in a purification scheme is the precipitation of an organic chemical drug, a protein, or a polysaccharide. An example of a drug that is precipitated and resolubilized in the process of purification is human serum albumin (HSA or NSA). HSA is precipitated at its isoelectric pH (pI) of about 4.8 in the cold and in the presence of ethanol, after several other fractions have been precipitated and removed by centrifugation (Cohn et al. 1946).

Dextrans used as plasma expanders (in place of human serum albumin) are fractionated by the addition (in the cold) of acetone or ethanol; precipitation is in the order of molecular weight, the highest molecular weight fractions precipitating first. The advantage of solvents of high volatility is that solvent removal is simple. The solvent in effect displaces some of the water solubilizing the drug or macromolecule and the partially dehydrated molecules self-associate, then precipitate. Charge effects may be modest but real for a protein at the pI and are negligible for uncharged polysaccharides like the dextrans. Neutral salts are held to a minimum.

Water also can be sequestered from a protein or a polysaccharide by a highly charged inorganic salt that binds considerable amounts of water (e.g., sodium or ammonium sulfate). This is salting-out (Scopes 1982). The result is that insufficient water is available to solvate the macromolecule and so it precipitates as an amorphous solid. Such precipitates containing the precipitant are quite stable in the cold, and may be stable for many hours at room temperatures.

A novel approach to the manipulation of solubility, which may merit its own category, is repeated cycles of freezing and thawing. This method has been used to isolate, in high yield, 35 different recombinant proteins from *Escherichia coli* (Johnson and Hecht 1994). It would appear that freezing and thawing liberates the particles of recombinant protein, but not the proteins of the bacterial cytoplasm. This may be a very important development, for other recovery steps (e.g., cell rupture, recovery of inclusion bodies, and refolding of the protein) can be tedious, expensive, and require the use of chemicals best avoided if possible.

Solvent Extraction

The household example of solvent extraction is the removal of a grease spot from clothing with lighter fluid. A more interesting, and applicable, example is the extraction of DNA from an organism. The following *can* be done on an industrial scale and has been described by Alexander et al. (1985) and Wallace (1987a,b). Since DNA must be recovered if a gene is to be prepared for transfection, the separation method also is of interest at bench scale.

The TE buffer mentioned in the protocol is 10 m*M* Tris, 1 m*M* EDTA (pH 8.5). The alkaline pH prevents acid depurination of the DNA. The EDTA chelates Mg^{2+} and prevents $Mg(OH)_2$ from precipitating and entrapping the DNA. The EDTA also inhibits nucleases. Phenol of high purity, used in the procedure, is equilibrated with TE. [Yellow, pink or crystalline phenol must be redistilled prior to use (the contaminants cleave diester bonds).] Make the phenol 0.1 percent in 8-hydroxyquinoline to retard phenol oxidation and impart a yellow tinge to the phase. The phenol solvates the DNA. Chloroform denatures protein and increases the phase density. Isoamyl alcohol is added to the chloroform to prevent foaming. The ratio is phenol (TE-saturated) : chloroform : isoamyl alcohol (24 : 24 : 1, v/v/v). In ice-cold conditions, to one volume of cells add one volume of extraction solvent and mix gently. At pH 8.5, both DNA and RNA partition to the organic phase.

For tissue, chop the tissue into small pieces and add 15 g to a chilled blendor. Add 50 ml citrate saline (0.01 M *trisodium citrate with 9 mg NaCl/ml, pH 7.0) and homogenize 5 minutes at high speed. Centrifuge the homogenate at 6000–6500* $\times$ *g for 15 minutes. Remove the floating lipid with a spatula, pipette off the liquid, and resuspend the pellet in TE. Add an equal volume of phenol:chloroform:isoamyl alcohol and mix end-over-end for 15 minutes. Now centrifuge the system and gently pipette off the aqueous (top) phase. Per milliliter of the aqueous extract, add 0.1 ml of 3* M *sodium acetate (pH 5.5) and follow that with 2 volumes of ethanol. The DNA will form a white precipitate, the more rapidly if the preparation is centrifuged.*

Adsorption Chromatography

Adsorption chromatography on various silicas is the most common single method for the column chromatographic purification of organic drugs of low molecular weight. Good resolution often is obtained in a single step, but enantiomers are not resolved. The materials are alumina, silica and sometimes cellulose, with adsorption to the hydroxyl groups on the surface. To this end, the water content of the resin is important for variations can lead to irreproducible separations. The primary problem with unmodified adsorption resins is the time required for equilibration with the mobile phase. This also is the reason that gradient elution schemes are not preferred for production-scale adsorption chromatography. However, alumina and silica resins are inexpensive and are used widely (Wehrli 1987). The special advantage of cellulose is the potential for resolving enantiomers.

Equilibration and reconditioning time is briefer for the bonded phases, and they are more stable than the unaltered alumina or silica (Wehrli et al. 1978). Bonded phases (e.g., $-NH_2$, $-NO_2$, –CN, and diol) also are used for normal development, but are far more expensive than the unsubstituted materials so are used in small-scale preparations and for analyses. Reverse-phase separations with an acetonitrile gradient on, for example, C_{18}-substituted silica, are excellent. However, the resin is sufficiently expensive that applications are, in the main, reserved for analytical work.

The Sephadexes also exhibit adsorptive properties with aromatics of low molecular weight and are available with varying degrees of cross-linking and in an hydroxypropylated version (LH-20 and LH-60) suitable for use with polar solvents and in phase-partitioning. Although these are excellent resins, they are quite expensive and must be useful through many, many cycles to be cost-competitive with alumina or silica. The Sephadexes and related resins are in limited production use with nonbiologics. Loading limits are of the order of 0.1 to 1 mg of solute per gram of column packing. Most columns are less than 1 m in diameter, 0.5–2 m in length, and the particle diameters of column packings are in the 100 μm range.

Many authorities regard calcium phosphate (hydroxyapatite) crystals and gels as adsorbents rather than ion-exchangers. However, for reasons that will become apparent, this author classifies the hydroxyapatites as exchangers.

Ion-Exchange Chromatography

At this writing, ion-exchange chromatography is considered the single most powerful and cost-effective step in the purification of *biologics*. It is used less frequently in the purification of organics of less than 1,500 Da. The weak exchangers (diethylaminoethyl or DEAE, pK about 11; carboxymethyl or CM, pK 4.5) usually can be eluted at constant pH but with an increasing salt gradient.

Weak exchangers do not offer the resolving power of the strong exchangers (quaternary ammonium, pK greater than 13; sulfoethyl or sulfopropyl, pK about 13). The great advantage of the strong exchangers is that they maintain high capacity over a wide pH range. Strong exchangers often must be developed first with a salt gradient and then with a pH shift. That issue is avoided by choosing a pH not too distant from the pI of the protein of interest, so that development on a strong exchanger can be done under mild conditions. Unlike gel filters that usually have a low loading capacity, ion-exchangers may have a very high capacity for solute. A good example of an ion-exchange purification is presented in Lehman et al. (1993), wherein recombinant tick anticoagulant peptide (rTAP), a 7 kDa peptide with a pI of 4.9, is purified 33-fold from a fermentation broth in essentially two steps. Following equilibration in the starting buffer, loading and elution is on/from POROS strong cation exchanger (SCX). Capillary zonal electrophoresis indicates a single protein peak. Such levels of purification are adequate for allergens.

Teh and Froger (1994) developed a very interesting, and potentially powerful, method for the purification of clotting factor VIII by anion exchange chromatography on DEAE-Fractogel TSK 650M (Merck, Germany). They claimed recovery of 80.8 percent ± 12.9 (n = 4) from fresh frozen plasma, with 5 percent lost during the wash step of the column. This level of recovery of F-VIII by ion exchange chromatography alone is astounding. The method is *not* competitive with immunosorbent purification since the F-VIII eluate also contains clotting factors II, IX, X, fibronectin, and fibrinogen, hence would be unsuitable for the purification of recombinant product. However, where the plasma availability and costs are a consideration, the method is attractive for it can be done at a fraction of immunosorbent cost. Teh and Froger batch-stirred plasma with DEAE–Sephadex A-50 and loaded the *un*adsorbed plasma onto a column of the DEAE–Fractogel. The authors are of the opinion that the method is compatible with the solvent/detergent treatment of cryoprecipitate (Burnouf et al. 1991) or plasma, hence should be amenable of application to plasma from any source.

More commonly, a SAX or SCX is utilized in the purification to homogeneity of recombinant human proteins produced in eukaryotes such as Chinese hamster ovary (CHO) cells or prokaryotes such as *Escherichia coli*. One of the great advantages of ion exchange matrices is that they can be cleaned and many are autoclavable. An appropriate reagent for column cleaning is 2 percent Na lauroyl sarcosinate with 0.25 *M* NaCl, followed by isopropanol (25 percent) and then a small amount of 25 percent HCl to ensure removal of the detergent (Jacobs 1993). More commonly, polymer-based columns are cleaned with NaOH at concentrations up to 0.5 *M*.

Proteins can be removed from DNA on a strong cation exchanger (Kaczorowski and Szybalkski 1994). DNA is strongly electronegative, even at pH 2.0–3.0 when most proteins are strongly cationic. These authors used AG 50W-X2 (Bio-Rad, Hercules, CA). Where the fate of the proteins is immaterial, the ion exchange deproteinization of DNA should work well, at almost any scale.

There are some grey areas in which it is not clear whether an ion-exchange purification is on the basis of charge or a quasi-specific interaction. Dextran sulfate (DS) is a soluble SCX and, when it is coupled to cross-linked agarose, shows a sufficiently powerful affinity for von Willibrand's factor (vWF) that a single-step purification of vWF from human plasma cryoprecipitate appears possible (Harrison et al. 1989). Details were insufficient (e.g., no crossed immunoelectrophoresis or SDS–PAGE) to evaluate the purity of the vWF, but it is interesting that DS has proven particularly useful in the purification of vitamin K clotting factors, and Protein C which also is involved in clotting.

Viruses purified by salt elution from sulfated cellulose (and possibly pure enough for vaccine use) include human measles, adenovirus type 8, human parainfluenza type 3, human polio virus type 1, human echovirus type 8, and human RSV (O'Neil and Balkovic 1993). Human immunodeficiency virus (HIV) binds to dextran sulfate (Mitsuya et al. 1988) and to heparin (Baba et al. 1988). Herpes simplex virus binds to heparin sulfate (Lycke et al. (1991). It is highly likely that these are ion-exchange effects with SAX rather than biospecific interactions.

The matrices available commercially from BioRad, Pharmacia, Pierce Chemical and other biochromatography houses are well known to most chromatographers, but only Pharmacia pushes industrial sales. BioRad has, at this writing, downsized the industrial effort. However, some of the newcomers appear to be making the attempt. Among these are BioSepra (Marlborough, MA) and PerSeptive Biosystems (Cambridge, MA) with high flow, high capacity exchangers based on aggregated beadlets of polystyrene-divinylbenzene copolymers coated with a polyol. The new matrices may increase bed capacity and cut loading and elution time. Such materials may not be particularly amenable of steam autoclaving (which one can do with most of the polysaccharide matrices) but BioSepra's HyperD can be sterilized with peracetic acid, which is a very potent oxidizing agent used (as an aerosol) in the sterilization of the interiors of isolators and barrier systems.

The new product of interest is Macro-Prep Ceramic Hydroxyapatite (Bio-Rad Laboratories, Hercules, CA). Hydroxyapatite (HXT) traditionally has been a subtle yet widely useful separations tool for the researcher but rather unstable (HXT crystals fragment readily), hence of limited use for manufacturing. The new HXT matrix is produced from individual small crystals sintered into highly porous beads that are mechanically stable to high pressures and high flow rates. The beads are chemically stable to high pH and organic solvents, detergents and the like but not to acidic environments.

HXT is $Ca_{10}(PO_4)_6(OH)_2$, and is practically insoluble in water. From a pharmacologic viewpoint, a modest bleed from the column is of no particular concern. HXT is difficult to classify as a cation or anion exchanger on the basis of structure since Ca^{2+} and phosphate groups are interspersed on the surface of each crystal. However, at a pH of 6.8, various mouse ascites proteins (including albumin) elute quickly from HXT in 10 m*M* Na phosphate whereas a mouse monoclonal antibody (MAb) elutes in a phosphate gradient. On that basis, HXT would be considered a cation exchanger. The most obvious application is in the rapid purification of MAbs and polyclonal Abs.

Membrane proteins (e.g., antigens from a tumor line) sometimes are of interest. Membrane proteins, recovered in SDS (in this instance from erythrocytes), can be separated on HXT. The M_r of the eluting proteins tends to increase in a conductivity gradient, but fractions invariably contains a number of contaminating proteins as shown by SDS–PAGE. HXT probably should be used upstream of an organic ion exchanger for complex purifications of this type.

Size-Exclusion Chromatography

Size exclusion chromatography (SEC), also known as gel filtration, is used primarily in solvent-exchange for solutions of macromolecules (proteins, polysaccharides). The solvent exchange process on highly cross-linked gels, such as dextran or polyacrylamide resins, is relatively rapid and complete. An acid or alkali or salt can be removed from a macromolecule at a modest dilution (typically, a 25–50 percent increase in volume). Such systems are readily automated. The more porous gels (e.g., the high exclusion-limit Sephadexes, Sepharoses, BioGels) have use in the purification of macromolecules for research purposes but see very limited use in production-scale operations. One reason is that loading of the very porous gels must be limited to 1–5 percent of the column volume if good separations are required. The flow rates on the soft gels often are abysmally low, and the extent of purification is modest (typically 4–5 percent), hence the highly porous SEC resins are not cost-effective for proteins and polysaccharides. However, the solvent exchange resins (e.g., Sephadex G-25 Coarse) usually can be loaded to 25 percent of column volume and operated at relatively high linear flow rates.

Hydrophobic Interaction

The immobilization of alkyl or aromatic groups on stable matrices (silica, glass, cellulose) makes possible the separation of drugs, and even proteins, from molecular contaminants based on the propensity of the drug for an hydrophobic ligand. The ligands usually are alkyl (e.g., C_{18}) but may be aromatic (e.g., phenyl). Solute in an aqueous system sorbs to the ligand and solute sorption increases as the conductivity of the aqueous medium increases (i.e., as the aqueous phase becomes more polar). Development may be with an increasing or decreasing solvent gradient, and with solvents of increasing power. Such systems are in fairly widespread industrial use, usually with silica-based matrices, for the purification of synthetic drugs and secondary metabolites (SD–SM). Most pharmaceutical houses use reverse gradients of acetonitrile on C_{18} silica columns for analysis of SD–SMs *and* proteins, with excellent results. However, Scopes (1982) finds the resolving power of the technique to be inadequate for preparative purposes.

An interesting application that was not as effective as hoped was the removal of hepatitis B virus from human clotting factor IX concentrates on matrices substituted with alkylamino groups (Iwarson et al. 1987). The removal of virus onto octanohydrazide-Sepharose 4B, presumably by hydrophobic interaction of the alkyl ligand with the viral envelope, was successful for the material injected into two chimpanzees. However, unpublished followup by several blood houses suggested that heating of powders of clotting factors, or exposure to a solvent and a detergent, are more reliable for the destruction of HBV and HIV.

Chelate Chromatography

Metals, chelated to iminodiacetic acid on matrices such as agarose, are anion exchangers with a higher degree of specificity than conventional anion exchange ligands (diethylaminoethyl or DEAE; quaternary aminoethyl or QAE). Nickel, for example, shows a relatively high specificity for histidine. Diacetate-chelated Zn^{2+} has been found to have a high affinity for human clotting factor IX (Feldman et al. 1993; Roberts et al. 1994). The latter process is *approximately* as follows:

> *Plasma is 95 percent thawed and the cryoprecipitate removed from the plasma by centrifugation. One and one-half g diethylaminoethyl*

(DEAE) Sephadex A-50 per liter of cryosupernatant is equilibrated in 0.01 M *citrate buffer (pH 7.0) containing 0.2* M *NaCl (equilibration and starting buffer or SB). The cryosupernatant may be adsorbed to the A-50 in the batch mode but adsorption to the resin in a column is preferred, especially since DEAE resins are weak exchangers and operate more efficiently in the column mode. Elution of unbound proteins (including hepatitis B virus and surface antigen) is with the SB, followed by a small volume of the elution buffer (EB) of 2.0* M *NaCl in 0.015* M *citrate (pH 7.0). The EB contains essentially all of the clotting factors II, IX and X, and about 80 percent of VII. This mixture of clotting factors is called prothrombin complex concentrate (PCC). The PCC then is diafiltered, equilibrated in 0.01* M *citrate-phosphate buffer (pH 6.5) containing 0.5* M *NaCl and applied to the chelated Zn column equilibrated in the same buffer. The column is washed with the buffer adjusted to pH 4.4, then pH 7.0, and F-IX is eluted with 20 m*M *glycine (pH 7.0). The F-IX is diafiltered, concentrated, filled, frozen and dried.*

Other proteins that have been purified on chelated Zn^{2+} include alpha$_2$-macroglobulin and human fibroblast interferon.

In Vitro Biotransformations and Semisynthetics

Herein, biotransformations are taken to mean chemical alterations in an organic molecule (drug), tailored by means of an immobilized enzyme or cell, whether in an aqueous, organic, or mixed system. Dated but useful reviews are those of Abbott (1976), Chibata and Tosa (1977), Martin (1977), and Vandamme and Voets (1974). The best-known and best-documented works are those done on beta-lactam antibiotics (penicillins and cephalosporins), and on steroids. Most of the more current biotransformation work is held proprietary by various of the drug manufacturers. The recent review by Mazid (1993) provides a theoretical background but nothing in terms of applications presently in use.

The most practical guidelines are in Weetall and Pitcher (1986) and relate to scaleup of immobilized enzymes. Although the methods cited in Mazid and the other reviews remain in use—changing process is sufficiently expensive that it is seldom done—they are, to a large degree, technically passé. It remains practical and cost-effective to utilize immobilized cells to do complex chemical tailoring of semisynthetics (e.g., penicilline amidase), although prednisolone is produced with an immobilized steroid hydrogenase and immobilized ribonuclease is used in nucleotide production. One prefers an immobilized enzyme(s) to avoid side reactions that reduce yield. Intact organisms may add to the expense of product purification since, for example, phospholipids or peptides from the organisms need to be removed from the column effluent. Purified, immobilized enzymes are quite expensive due to the cost of enzyme purification and immobilization. Weetall and Pitcher (1986) pointed out that, as of 1986, only DEAE cellulose and glutaraldehyde had been designated "generally recognized as safe" (GRAS) for the immobilization of enzymes intended for use with foods. Many more immobilization systems have since been approved for use in the preparation of injectable drugs. However, the main impediment to the routine use of immobilized biologics in manufacturing is not regulatory but manufacturing cost.

One approach is to produce phage enzymes that are the counterparts of phage antibodies (see Chiswell and McCafferty 1992 on phage or coliclonal antibodies). What this involves is fusing the enzyme gene from any source (plant, animal, human) to gene 3 of the filamentous bacteriophage (fd) of *Escherichia coli*. Each iteration of the fd bacteriophage in the *E. coli* host would contain the enzyme of interest at the terminus of the phage. Alkaline phosphatase (McCafferty et al. 1991) and staphylococcal nuclease (Hoogenboom and Winter, personal communication cited in Chiswell and McCafferty 1992) have been prepared in this manner. The investment in generation of the enzyme is, for practical purposes, one-time since, to produce a fresh batch, one grows up the host and the phage in overnight culture. The methods of purification are those of purifying a phage (fd) of high molecular weight and aspect ratio, or those of the purification of the enzyme if an inhibitor that can be immobilized easily and cheaply is readily available.

There is a high probability that heat- or chemically sensitive lactones can be polymerized by means of bacterial lipases, according to Hiroshi Uyama and Shiro Kobayashi of Tohoku University, Sendai, Japan (Anonymous 1994). Although this method is not presently commercially viable, I suspect that it will become so.

Affinity Chromatography

An interesting paper, not reviewed in either of the affinity chapters in this text, concerns an unusual method of antibody purification. The complement system, together with specific antibodies, attacks bacteria and other protists that are exposed to the immune system. Complement protein C1q binds specifically to the Fc portions of immunoglobulin G (IgG) and immunoglobulin M (IgM). The Fc portions are remote from the antigen binding sites of the antibodies. Nethery et al. (1990) found that if C1q was purified by the method of Tenner et al. (1981), and the C1q was coupled to agarose by the CNBr method, IgM (and, less efficiently, IgG) can be purified on the immobilized C1q.

Because Protein A and Protein G are readily available in high purity, and work well in antibody purification, this application will remain a point of interest only.

Template Chromatography

A variety of messenger and ribosomal RNAs, DNAs, and DNA fragments can be purified by column methods (Schott 1984). There have been attempts at the production of vaccines based on the injection of microbial ribosomes, but these have not come to commercial fruition. Presumably, with the immobilized RNA from HIV, one might be able to examine unique drug interactions; however, much of the same study likely could be done in solution by spectroscopic methods. Template chromatography might be used in the purification of DNA or RNA probes for microbial identification or enumeration (Olson 1995), or in the purification of repressor proteins and the proteins that interact with promoters.

Chiral Separations

The attitude of the U.S. FDA in re unreactive sister species in an enantiomeric mixture presently is benign. This will change.

Historically, racemates are best resolved on a biopolymer [e.g., cellulose, or immobilized cyclodextrins (Quang and Khaledi 1993)], or immobilized serum albumin. Recently, Antus et al. (1992) have resolved racemic isoflavonones on a (+)-poly(triphenylmethyl methacrylate)-substituted matrix. The difficulty with resolution of racemates on a production scale is that the separations are subtle. The convention is that column loading must be very modest (1–2 percent of the column volume) to effect separations. Process columns making use of the *analytical* methodology would be prohibitively large. The approach of most manufacturers is to show that the racemate(s) to the drug is inactive or to direct a considerable effort to enantiomer-specific synthesis.

Alternatively, one might construct a dextrorotary enzyme intended to hydrolyze or otherwise alter the D-isomer of a racemic mixture. Celgene Corp. (Warren, NJ) reacts *p*-methanesulfonylbenzaldehyde with glycine to obtain racemic *threo-p*-methanesulfonylphenylserine. The D-enantiomer is degraded by a D-threonine aldolase, leaving the L-isomer in 98 percent enantiomeric excess. L-*threo*-phenylserines are precursors to some antibiotics and to methyldopa.

Preparative Ultracentrifugation for Plasmid DNA

Whenever DNA quality is critical, as in the preparation of DNA for transfection, ultracentrifugation is the preferred purification method

(Murray et al. 1993). Dorin and Bornecque (1995) purified plasmid DNA to homogeneity in a three-step CsCl gradient. Running time was 20 minutes as compared with 15–48 hours for isopycnic CsCl methodology. If one has an ultracentrifuge, the preparation is much simpler than the column method of Hines et al. (1992). The question for the biochemist preparing material for transfection is: can this method be scaled satisfactorily to prepare sufficient material for a large number of transfections? The answer is no, and the method of Hines et al., employing anion exchange chromatography, is preferred.

Continuous Flow Electrophoresis for DNA Fragments

DNA fragments generated by means of restriction enzymes can be purified in 10–100 μg amounts (yield of 25–40 percent of starting material) by preparative electrophoresis in the model 491 Prep Cell (Bio-Rad Laboratories, Hercules, CA). Schlax et al. (1995) provide adequate details of the buffers and running conditions. DNA fragments elute downstream of the RNA and 1 percent agarose diagnostic gels indicate excellent sample purity.

In the same issue of *BioTechniques*, Glasel (1995) shows that most of the values of the ratio of absorbance at 260 nm to that at 280 nm [*R*(260/280)], which have been used without question by biochemists for over 50 years, are in error. Sample values from Glasel's table indicate a ratio or 2 for pure nucleic acids, 1.98 for nucleic acids with 5 percent protein contamination, 1.87 for nucleic acids 50 percent contaminated with proteins, 1.32 for nucleic acids 10 percent contaminated with proteins, and 0.57 for pure proteins. The preparative biochemist will appreciate that this is not a trivial consideration.

Combined Methods for Intravenous Gamma Globulins and Other Products

The methods used by Uemura et al. (1994) at Alpha Therapeutics (Los Angeles, CA) and Green Cross (Osaka, Japan) to produce intravenous immunoglobulins (IVIGs) commence with the detergent/solvent system for virus inactivation developed by the New York Blood Center. This is a short series of steps for producing an injectable drug, commencing with precipitation and resolubilization.

For the production of what is called Venoglobulin S, Uemura et al. commence with Cohn fraction II or II + III (Cohn et al. 1946) to prepare an immunoglobulin-rich feed. The fraction II precipitate then is resuspended and put through an additional precipitation step with poly(ethylene glycol) 4000 (PEG) to 12 percent (w/v) at pH 8.3 to precipitate IgG aggregates and proteins of M_r significantly greater than 150 kDa. The reprecipitated immunoglobulins then are freed of PEG and trace protein contaminants on an ion exchange column. The IgGs are recovered, solvent exchange is done by UF (diafiltration), the IgGs are concentrated in the UF system, and the product is sterile filtered prior to filling and freeze-drying.

A variation on this (also in Uemura et al.) involves the addition of sorbitol (33 percent, pH 5.5) to the Cohn fraction II or II + III solution followed by 60°C pasteurization for 10 hours. [Human albumin, but without the sorbitol, is pasteurized and made virus-free under the same conditions.] In this variation, the pasteurization step is followed by 5 to 6 percent PEG (pH 5.5) to precipitate aggregated IgG and other contaminating proteins. The supernatant then is subjected to ion exchange chromatography to remove the PEG and trace protein contaminants (the PEG does not bind to the exchanger and is washed out during loading). The IgG peak is diafiltered, concentrated, filled, and freeze-dried.

The Alpha Therapeutics/Green Cross purification steps commence with Cohn fraction II + III which contains gamma globulins (IgG, IgA, IgM); clotting factors II, VII, IX, and X; and other alpha and beta globulins. The Cohn fraction is a starting material because human serum albumin is prepared from human plasma by the Cohn process. A modern facility, setting up *de novo* to produce gamma globulins and albumin from human plasma, likely would use fractionation steps similar to the ion-exchange methods of Falksveden and Lundblad (1980) and Curling (1980), respectively.

Why is the use of more than one method indicated consistently? The rationale is obvious to the experienced fractionator but may not be to the novice. When one precipitates contaminants of IgG [aggregates of IgG and some proteins of higher molecular weight than IgG (150 kDa)] by means of PEG, a small proportion of the contaminating proteins will escape precipitation and remain suspended. They must be removed—especially the aggregates, which can trigger undesired responses.

Removing the bulk of the contaminating proteins improves significantly the efficiency of the ion-exchange step(s) that follows because the isoelectric point of aggregated IgG differs only a bit from that of monomeric IgG. Elimination of the aggregates in the PEG step makes possible the loading of the ion-exchange matrix to near capacity, and increases the thoughput of the manufacturing process.

Zhang et al. (1992) developed a practical, economic, scalable three-step process for the purification to homogeneity of recombinant human interferon gamma (rhIFN-gamma) from inclusion bodies recovered from *Escherichia coli*.

> *Cells were homogenized in a Manton-Gaulin-type homogenizer, cooled and centrifuged at 1000 × g and 4°C for 30 minutes to*

*sediment the inclusion bodies (inclusion bodies sediment continuously, and more rapidly, in the Sharples™). The pellet was washed in 20 m*M *phosphate buffer (0.125* M *NaCl, pH 7.2), rehomogenized and resuspended in the same buffer containing 0.5* M *urea and 1 m*M *EDTA (pH 7.0). The washed pellet was resuspended and denatured in 7* M *guanidine hydrochloride in 50 m*M *boric acid (pH 7.2), rehomogenized, and centrifuged at 17,000 × g (4°C) for the removal of cell debris and insoluble protein. [The boric acid aggregates the cell wall debris so that it can be pelleted, together with insoluble proteins, while the rhIFN-gamma remains in solution, denatured.] The denatured extract was renatured by 70-fold dilution with the phosphate buffer/0.125* M *NaCl made 1 m*M *in EDTA and 5 percent (w/v) in sucrose. The renaturation buffer was then made to 0.2* M *in guanidine hydrochloride (pH 7.2). The renatured rhIFN then was centrifuged at 17,000 × g and 4°C for 30 minutes (clarification) and applied to S-Sepharose Fast Flow equilibrated in 20 m*M *phosphate buffer (pH 7.0). Two bed volumes eluted unbound proteins. Weakly bound impurities were eluted with 5 bed volumes of 20 m*M *phosphate buffer/0.3* M *NaCl (pH 7.0). The rhIFN-gamma was eluted with 3 bed volumes of 20 m*M *phosphate buffer/0.6* M *NaCl. Regeneration of the S-Sepharose Fast Flow was with 1* M *NaCl in the phosphate buffer. On scaleup, adsorption was in the batch mode. Nickel-chelating Sepharose Fast Flow was equilibrated with 0.2* M $NiSO_4$ *(pH 4.7) and washed with 2 bed volumes of distilled water. The column was washed with 50 m*M Na_2HPO_4*-0.5* M *NaCl (pH 4.0). The nickel column then was equilibrated with 20 m*M *phosphate buffer/0.3* M *NaCl (pH 7.5) and, following the addition of the crude rhIFN-gamma to the column, weakly bound proteins were eluted with 3 bed volumes of the same buffer. The IFN was eluted with 7 bed volumes of 20 m*M *phosphate buffer/9 m*M *imidazole/0.5* M *NaCl. This material was subjected to SEC on Superdex 75.*

In brief, the cells were disrupted by homogenization, the inclusion bodies spun down and washed. The inclusion bodies were denatured, renatured, and partially purified by IXC. The rhIFN-gamma peak was further purified on a Ni^{2+} column. There are more steps in this than one would desire, but much of the manipulation is required for the recovery, denaturation, and renaturation of the inclusion bodies.

The only single-step purification system for biologics that is successful is by means of immunosorbents. However, even with immunosorbents, buffer exchange and product concentration by ultrafiltration are required.

PURIFICATION STEPS AND YIELD

Even if the recovery of product is 95 percent at each of 6 steps in process, optimistically, the recovery is 74 percent. Therefore, reduction in process steps not only reduces costs but improves yield.

Where one uses column chromatography, the fouling of the bed with particles reduces effective column life and utility. Filtering all feed to the column adds an additional step and expense, and slows the manufacturing process. The equilibration of the resin before the feed is admitted to the column, and the washing of the column following the adsorption of the product, also tend to be slow steps if the feed percolates through a packed column. Pharmacia (Uppsala, Sweden and Piscataway, NJ) have developed a new column and process system (Streamline™; Figure 1.4). The new system *may* make possible a more rapid and efficient processing of cell culture media and other feedstocks, with the resins presently in use.

Conventional chromatographic beds are of fixed volume and the resin usually is compressed. The heart of the Pharmacia system is a retractable bed support that can be raised to accommodate moderate bed expansion as the resin is equilibrated in buffer, further expansion as the bed is equilibrated with the feedstock and subsequently washed to remove unadsorbed materials. Finally, the bed is compressed and eluted in a retrograde manner (Figure 1.5).

Such a system can operate efficiently only if the K_d is quite low, so that product adsorbs to the resin under starting conditions and does not dissociate and is not washed away in rinse steps. Consequently, this system should work best with strong exchangers or with the less expensive affinity resins. The EBA approach appears to combine the best features of fluidized bed operation with conventional column chromatography.

ISOLATORS

In the strict sense, isolators and barriers are not separations systems. They are systems for the exclusion of something (e.g., aerosolized microorganisms) or for the retention of something (e.g., a cytotoxic drug). An early view of isolators was presented in Oles et al. (1987). Isolators or barriers are unnecessary in the downstream purification of aseptically processed drugs until the purified drug is filled. However, the concept should be applied to the transport of product to and from columns and other purification systems. For example, product can be collected and transported in autoclave-sterilized 316L stainless steel

Figure 1.4. The Pharmacia Streamline™ expanded bed adsorption (EBA) system for macromolecule purification.

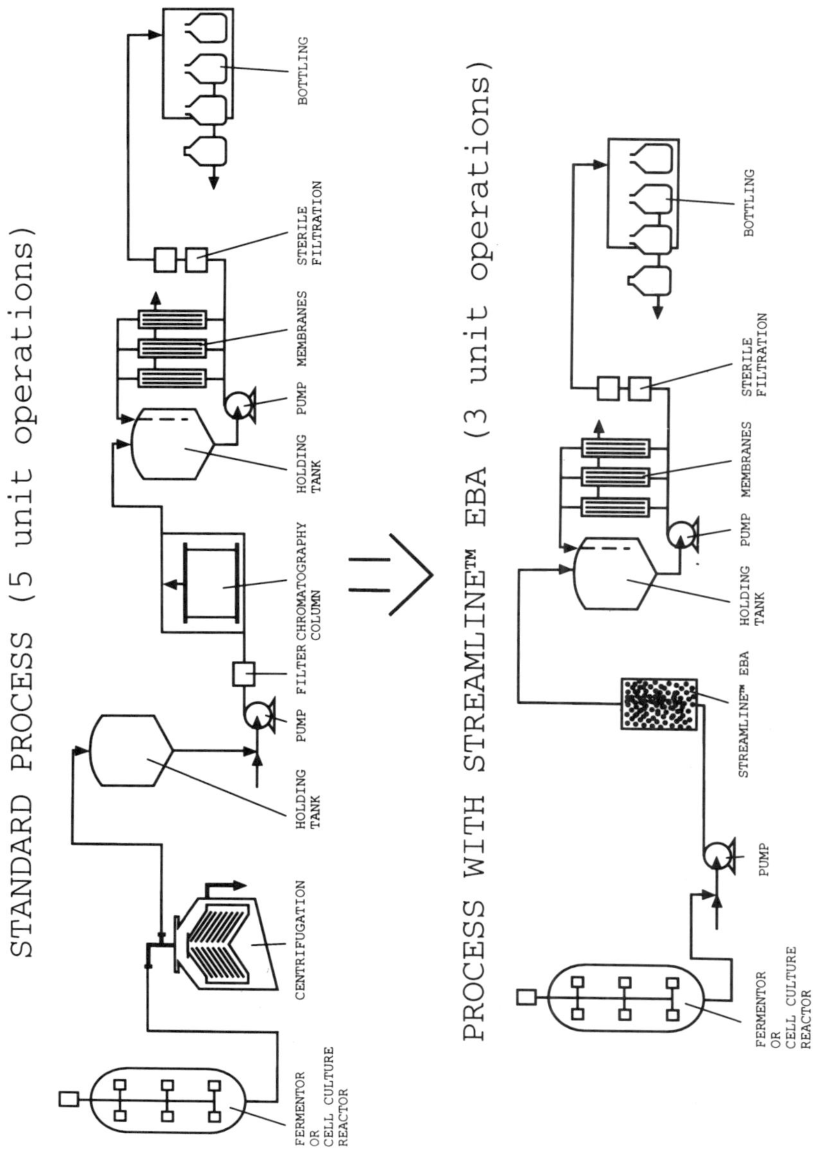

Figure 1.5. Operation of the Streamline™ EBA bed.

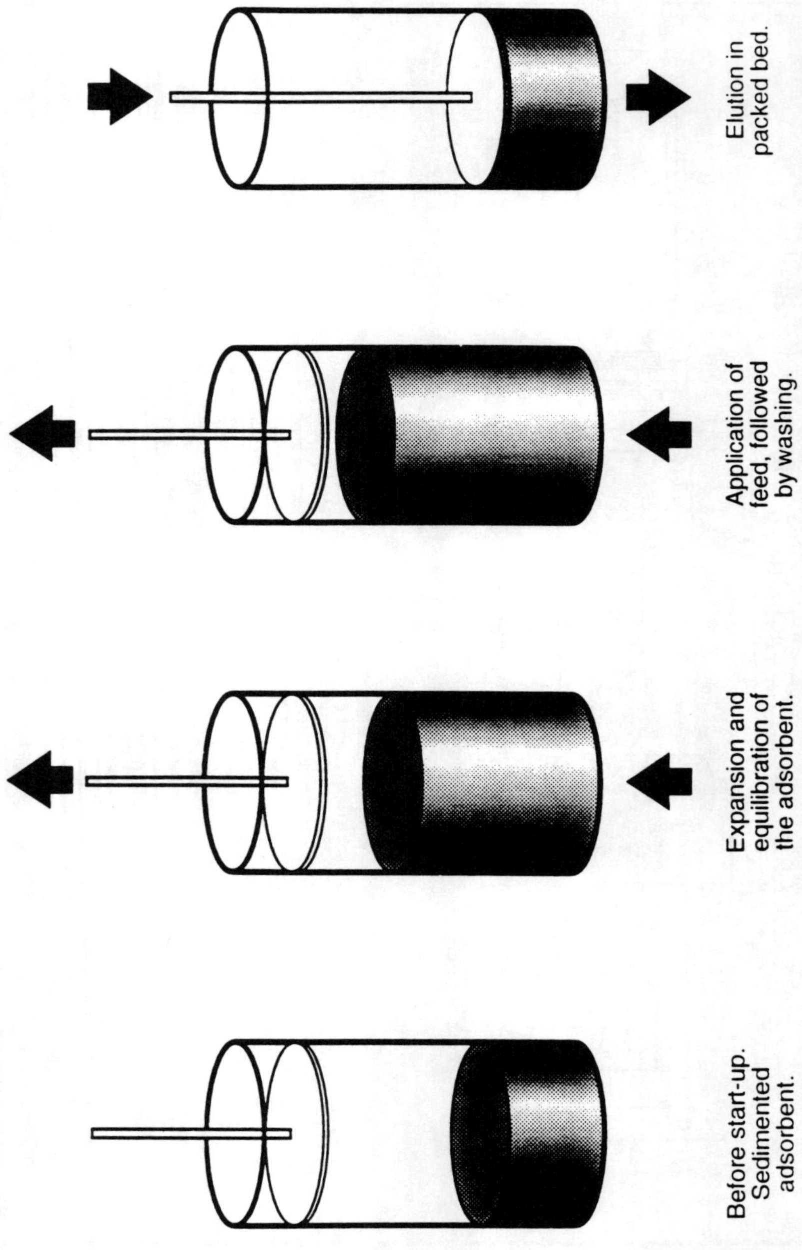

pressure vessels. Such systems protect the product and, in the case of cytotoxic drugs, also protect the worker.

SCALEUP FROM A PUBLISHED BENCH METHOD

Academics at the bench seldom are concerned with purity but seldom with scale, yield or cost. Translation of published bench purification methods to industrial scale can be satisfactory or unduly costly in terms of capital for the process and the product yield. The following is an analysis of the methods of Hayashi et al. (1994) from the perspective of scaleup (which was not the concern of those authors).

Bifidobacterium spp. are used in some health foods to produce panose and isomaltose from maltose by means of glucosyltransferase (EC 2.4.1.24). Panose is not fermentable by many oral bacteria, hence is an anticariogenic. Use of the enzyme, were it to be produced on an industrial scale, would be by the food industry which is highly cost-conscious. *Aureobasidium* sp. was used as a source of glucosyltransferase and lysed the cells with an endo-beta-1,3-glucanase.

> *Proteins were precipitated with ammonium sulfate to 80 percent saturation. The dialysate in 20 m*M *acetate, pH 4.5, was passed through S-Sepharose Fast Flow, reequilibrated in phosphate buffer (PB), pH 6.5, and applied to DEAE–Cellulofine A-800. The column was eluted with a NaCl gradient in PB. The active fraction was applied to Sephadex G-200 equlibrated in PB containing 0.1* M *NaCl.*

The purified enzyme was shown to be stable at pH 4–6 with optimum activity at 4.5–5.5. Optimum activity was at 65°C (the carbohydrate content was 32 percent) but decreased precipitously at higher temperatures. The M_r was estimated at 395 kDa. For a cost-effective scaleup using *conventional* packed bed columns, the cell debris must be removed. A continuous-flow centrifuge should suffice, with periodic bowl cleaning. A coarse, deadended filter would work but many of the matrices (e.g., fiberglass) adsorb proteins strongly. Any depth filter proposed in such a use requires advance testing of the tendency to adsorb product from the feedstock.

The ammonium sulfate step gave a 5.75-fold purification, which is consistent with the fold purifications that are possible with a single precipitation with ammonium or sodium sulfate. Because of the high M_r of the enzyme (almost 400 kDa), 20–25 percent saturation in place of 80 percent saturation might be sufficient to precipitate the enzyme yet leave most smaller proteins in solution. Ammonium sulfate precipitation might be retained if 10-fold or more purification could be done in this step, since the reagent is cheap. Tenfold purification might produce a utilizable product. If so, the diafiltered and concentrated enzyme, with impurities, might be immobilized for use.

The enzyme elutes in the void volume from a strong cation exchanger (S-Sepharose Fast Flow) at pH 4.5 and is retained on a DEAE column equilibrated at pH 6.5. The enzyme is positively charged at pH 4.5 but negatively charged at pH 6.5. If the enzyme has good activity at pH 7.5 and none of the substrates bind under those conditions, a QAE column might be used for immobilization of the enzyme. Alternatively, the enzyme could be covalently coupled to an activated polysaccharide matrix of the types available from Pharmacia (Uppsala, Sweden, and Piscataway, NJ), Pierce Chemical (Rockford, IL) or Bio-Rad (Hercules, CA).

For a food application, if the ammonium sulfate preparation was insufficiently pure, I would attempt a single-step purification at a pH of about 6.5 on an inexpensive, very porous strong anion exchanger (e.g., sulfopropyl or sulfoethyl [the authors used a cellulose matrix]). However, from the published data, this author would expect at best a 10-fold purification. This author would pack the resin in a tank with some headspace so that the bed could be resuspended during periodic backflush to remove cell debris not eliminated upstream. [See the Pharmacia EBA bed, Figures 1.4 and 1.5, which might make upstream centrifugation or filtration unnecessary.] If any ion-exchange step were to be used for this noninjectable, it would be one only, with no SEC since the purification factor for SEC is a trivial 1.1–1.2-fold in the published work. However, SEC can be extremely important for injectables (e.g., the elimination of aggregates of IgG in IVIG).

Ideally, the ion-exchange bed used to purify the enzyme may be used for immobilization of it. This overall approach reduces costs significantly.

AUTOMATION

Automation of purification systems is well-established practice. Off-the-shelf control systems exist (e.g., from Pharmacia). Fichtali et al. (1992) describe a straightforward automated ion-exchange system. Chapman et al. (1990) describe multistage systems.

REFERENCES

Abbott, B. J. 1976. Preparation of pharmaceutical compounds by immobilized enzymes and cells. *Adv. Appl. Microbiol.* 20:203–257.

Albertsson, P.-A. 1971. *Partition of cell particles and macromolecules*, 2d ed. London: Wiley-Interscience.

Alexander, R. R., J. M. Griffiths, and M. L. Wilkinson. 1985. *Basic biochemical methods*. New York: Wiley-Interscience.

Anonymous. 1994. Esterase enzymes catalyze polyester polymerizations. *Chemical and Engineering News*, 28 March: 41.

Antus, S., R. Bauer, A. Gottrsegen, H. Lotter, O. Seligmann, and H. Wagner. 1992. Enantiomeric separations of racemic isoflavanones and related compounds on (+)-poly(triphenylmethyl methacrylate)-coated silica. *J. Chromatog.* 603:133–137.

Baba, M., R. Pauwels, J. Balzarini, J. Arnout, J. Desmyter, and E. DeClercq. 1988. Mechanism of inhibitory effect of dextran sulfate and heparin on replication of human immunodeficiency virus *in vitro*. *Proc. Natl Acad. Sci. USA* 85:6132–6136.

Bonnerjea, J., J. Jackson, M. Moare, and P. Dunnill. 1988. *Enzyme Microb. Technol.* 10:357–343.

Boyum, A., T. Berg, and R. Blomhoff. 1983. Fractionation of mammalian cells. *Iodinated density gradient media—A practical approach*, edited by D. Rickwood. Oxford, UK: IRL Press.

Burnouf, T., M. Burnouf-Radosevich, J. J. Hart, and M. Goudemand. 1991. A highly purified factor FVIII:C concentrate prepared from cryoprecipitate by ion-exchange chromatography. *Vox Sang.* 60:8–15.

Burnouf-Radosevich, M., P. Appourchaux, J. J. Huart, and T. Burnouf. 1994. Nanofiltration, a new specific virus elimination method applied to high-purity factor IX and factor XI concentrates. *Vox Sang.* 67:132–138.

Chapman, G. E., P. Matjtschuk, J. E. More, and P. Pilling. 1990. Automated multi-stage chromatographic systems for production of biotherapeutics. *Separations for biotechnology*, edited by D. L. Pyle. London: Elsevier Applied Science.

Chibata, I., and T. Tosa. 1977. Transformations of organic compounds by immobilized microbial cells. *Adv. Appl. Microbiol.* 22:1–27.

Chiswell, D. J., and J. McCafferty. 1992. Phage antibodies: will new 'coliclonal' antibodies replace monoclonal antibodies? *Trends Biotechnol.* 10:80–84.

Cohn, E. J., L. E. Strong, W. L. Hughes, D. J. Mulford, J. N. Ashworth, M. Melin, and H. L. Taylor. 1946. Preparation and properties of serum and plasma proteins. IV. A system for the separat ion into fractions of the protein and lipoprotein components of biological tissues and fluids. *J. Am. Chem. Soc.* 72:459–475.

Curless, C. E., J. Pope, L. Loredo, and L. B. Tsai. 1994. Effect of preinduction specific growth rate on secretion of granulocyte macrophage colony stimulating factor by *Escherichia coli. Biotechnol. Prog.* 10:467–471.

Curling, J. M. 1980. Albumin purification by ion exchange chromatography. In *Methods of plasma protein purification,* edited by J. M. Curling. London: Academic Press.

Dorin, M., and C. A. Bornecque. 1995. Fast separations of plasmid DNA using discontinuous gradients in the preparative ultracentrifuge. *BioTechniques* 18:90–92.

Dwyer, J. L. 1983. Process separations of antibodies by HPLC process systems. *Bio/Technology* 2:957–961.

Falksveden, L.-G., and G. Lundblad. 1980. Ion exchange and polyethylene glycol precipitation of immunoglobulin G. In *Methods of plasma protein fractionation,* edited by J. M. Curling. London: Academic Press.

Feldman, P. A., L. Harris, D. R. Evans, and H. E. Evans. 1993. Preparation of a high purity factor IX concentrate using metal chelate affinity chromatography. *Colloques Inserm* 227:63–68.

Fichtali, J., E. A. Charter, K. V. Lo, and S. Nakai. 1992. Separation of egg yolk immunoglobulins using an automated liquid chromatography system. *Biotechnol. Bioeng.* 40:1388–1394.

Fiore, J. W., W. P. Olson, and S. L. Holst. 1980. Depth filtration. In *Methods of plasma protein fractionation,* edited by J. M. Curling. London: Academic Press.

Fisker, S., K. Kudahl, and O. Sonne. 1990. Isolation of rat peritoneal mononuclear and polymorphonuclear leucocytes on discontinuous gradients of Nycodenz. *J. Immunol. Methods* 133: 31–38.

Garvan, J. M., and B. W. Gunner. 1963. Intravenous fluids: "A solution containing such particles must not be used." *Med. J. Australia*: 140–145.

Garvan, J. M., and B. W. Gunner. 1964. The harmful effects of particles in intravenous fluids. *Med. J. Australia*:1–8.

Glasel, J. A. 1995. Validity of nucleic acid purities monitored by 260nm/280nm absorbance ratios. *BioTechniques* 18:61–63.

Groves, M. J. 1993. *Particulate matter: Sources and resources for healthcare manufacturers.* Buffalo Grove, IL: Interpharm Press.

Harrison, P., R. H. Saundry, and G. F. Savidge. 1989. The purification of the VIII/vWf complex using dextran sulphate Sepharose chromatography. In *Biotechnologie des proteins du plasma,* edited by J. F. Stoltz, and C. Rivat. Colloque INSERM Vol. 175.

Hayashi, S., T. Hayashi, Y. Takasaki, and K. Imada. 1994. Purification and properties of glucosyltransferase from *Aureobasidium. J. Industr. Microbiol.* 13:5–9.

Hearse, D. J., C. Erol, L. A. Robinson, M. P. Maxwell, and M. V. Braimbridge. 1985. Particle-induced coronary vasoconstriction during cardioplegic infusion. *J. Thoracic Cardiovasc. Surg.* 89:428–438.

Hines, R. N., K. C. O'Connor, G. Vella, and W. Warren. 1992. Large-scale purification of plasmid DNA by anion-exchange high-performance liquid chromatography. *BioTechniques* 12:430–434.

Iwarson, S., M. Einarsson, L. Smallwood, P. Snoy, and R. J. Gerety. 1987. Chromatographic removal of hepatitis B virus from a factor IX concentrate. *Transfusion* 27:171–173.

Jacobs, W. 1993. Chromatography: Separation and purification/ BIOQ&A. *Bio/Technology* 11:1061.

Johnson, B. H., and M. H. Hecht. 1994. Freeze-thaw cycles to isolate recombinant proteins from *Escherichia coli. Bio/Technology* 12:1357–1362.

Kaczorowski, T., and W. Szybalski. 1994. Rapid removal of proteins from DNA sequencing mixtures using a strong cation exchanger. *BiTechniques* 17:34.

Koolwijk, P., E. Rozemuller, R. K. Stad, W. B. M. de Lau, and B. J. E. G. Bast. 1988. Enrichment and selection of hybrid hybridomas by Percoll density gradient centrifugation and fluorescent-activated cell sorting. *Hybridoma* 7:217–225.

Lehman, E. D., J. G. Joyce, D. K. Freymeyer III, F. J. Bailey, W. K. Herber, and W. J. Miller. 1993. A novel process for the large-scale purification of recombinant tick anticoagulant peptide using perfusion chromatography. *Bio/Technology* 11:207–212.

Lycke, E., M. Johansson, B. Svennerholm, and U. Lindahl. 1991. Binding of herpes simplex virus to cellular heparan sulfate, an initial step in the adsorption process. *J. Gen. Virol.* 72:1131–1137.

Martin, C. K. A. 1977. Microbial cleavage of sterol side chains. *Adv. Appl. Microbiol.* 22:29–58.

Mazid, M. A. 1993. Biocatalysis and immobilized enzyme/cell bioreactors. *Bio/Technology* 11:690–695.

McCafferty, J., R. H. Jackson, and D. J. Chiswell. 1991. *Protein Eng.* 4:955–961.

Meltzer, T. H. 1987. *Filtration in the pharmaceutical industry.* New York: Marcel Dekker.

Meltzer, T. H. (1991) Pharmaceutical water: generation, storage, distribution and quality testing. In *Sterile pharmaceutical manufacturing: Applications for the 1990s,* edited by M. J. Groves, W. P. Olson, and M. H. Anisfeld. Buffalo Grove, IL: Interpharm Press.

Mitsuya, H., D. J. Looney, S. Kuno, R. Ueno, F. Wong-Stahl, and S. Broder. 1988. Dextran sulfate suppression of viruses in the HIV family: inhibition of virion binding to CD4+ cells. *Science* 240:646–649.

Murray, M., G. Molloy, and E. Orr. 1993. Purification of cloning quality lambda *gt*11 DNA. *BioTechniques* 15:606–607.

Nethery, A., R. L. Raison, and S. B. Easterbrook-Smith. 1990. Single-step purification of immunoglobulin M on C1q Sepharose. *J. Immunol. Methods* 126:57–60.

Oles, P., B. St. Martin, J. Teillon, D. Meyer, and C. Picard. 1987. Isolator technology for manufacturing and quality control. In *Aseptic pharmaceutical manufacturing: Technology for the 1990s,* edited by W. P. Olson, and M. J. Groves. Buffalo Grove, IL: Interpharm Press.

Olson, W. P. 1987. Sterilization of small-volume parenterals and therapeutic proteins by filtration. In *Aseptic pharmaceutical manufacturing: Technology for the 1990s,* edited by W. P. Olson, and M. J. Groves. Buffalo Grove, IL: Interpharm Press.

Olson, W. P. 1995. Gas chromatography, substrate utilization, and systems for microbial identification: some observations. In *Microbial identification and quantitation,* edited by W. P. Olson. Buffalo Grove, IL: Interpharm Press.

O'Neil, P. F., and E. S. Balkovic. 1993. Virus harvesting and affinity-based liquid chromatography. *Bio/Technology* 11: 173–178.

Quang, C., and M. G. Khaledi. 1993. Improved chiral separation of basic compounds in capillary electrophoresis using beta-cyclodextrin and tetraalylammonium reagents. *Anal. Chem.* 65:3354–3358.

Roberts, P. L., C. P. Walker, and P. A. Feldman. 1994. Removal and inactivation of enveloped and non-enveloped viruses during the purification of a high-purity factor IX by metal chelate affinity chromatography. *Vox Sang.* 67(suppl 1):69–71.

Schlax, Jr., P. E., M. W. Capp, and M. T. Record, Jr. 1995. Preparative-scale purification of DNA restriction fragments by continuous-flow gel electrophoresis. *BioTechniques* 18:94–100.

Schott, H. 1984. *Affinity chromatography: Template chromatography of nucleic acids and proteins.* New York: Marcel Dekker.

Scopes, R. K. 1982. *Protein purification, principles and practice.* New York: Springer-Verlag.

Teh, L.-C., and M. Froger. 1994. Direct capture of plasma factor VIII:C by ion exchange chromatography. *Vox Sang.* 67:8–13.

Tenner, A. J., P. H. Lesavre, and N. R. Cooper. 1981. Purification and radiolabelling of human C1q. *J. Immunol.* 127:648–653.

Terstappen, G. C., R. A. Ramelmeier, and M.-R. Kula. 1993. Protein partitioning in detergent-based two-phase systems. *J. Biotechnol.* 28:263–275.

Titchener-Hooker, N. J., M. Hoare, and P. Dunnill. 1990. New approaches to the more efficient purification of proteins and enzymes. *Ann. N.Y. Acad. Sci.* 589:157–171.

Uemura, Y., Y. H. J. Yang, C. M. Heldebrant, K. Takechi, and K. Yokoyama. 1994. Inactivation and elimination of viruses during preparation of human intravenous immunoglobulin. *Vox Sang.* 67:246–254.

Vandamme, E. J., and J. P. Voets. 1974. Microbial penicillin acylases. *Adv. Appl. Microbiol.* 17:311–369.

Wallace, D. M. 1987a. Large- and small-scale phenol extractions. *Methods Enzymol.* 152:33–41.

Wallace, D. M. 1987b. Precipitation of nucleic acids. *Methods Enzymol.* 152:41–48.

Weetall, H. H., and W. H. Pitcher, Jr. 1986. Scaling up an immobilized enzyme system. *Science* 232:1396–1403.

Wehrli, A. 1987. Preparative liquid chromatography in the pharmaceutical industry. In *Preparative liquid chromatography,* edited by B. A. Bidlingmeyer. *J. Chromatog. Library* 38. Amsterdam: Elsevier.

Wehrli, A., J. C. Hildenbrand, R. Keller, R. W. Stampfil, and J. Frei. 1978. Influences of organic bases on the stability and separation properties of reversed-phase chemically bonded silica gels. *J. Chromatog.* 149:199–210.

Wheelwright, S. M. 1989. The design of downstream proceses for large-scale protein purification. *J. Biotechnol.* 11:89–102.

Wolf, F. J. 1969. *Separation methods in organic chemistry and biochemistry.* New York: Academic Press.

Zhang, Z., K.-T. Tong, M. Belew, T. Pettersson, and J.-C. Janson. 1992. Production, purification and characterization of recombinant human interferon gamma. *J. Chromatography* 604:143–155.

2

Membrane Pervaporation

Timothy deVilliers Naylor

The pharmaceutical industry is highly competitive and produces high value products in relatively small quantities, often at low concentrations. The intermediates necessary for producing these products are often unstable and require great care in their handling. Traditional techniques used to concentrate these intermediates include distillation, evaporation, crystallization, and centrifugation. Recently, the membrane process known as pervaporation has been recognized as a potentially valuable new way of concentrating pharmaceutical intermediates and products (Rautenbach et al. 1992). For instance, the process has been demonstrated to be highly cost effective, in terms of capital and energy, for removing water from aqueous solutions where distillation is difficult. Examples include distillation applications where the vapor-liquid equilibrium curve is narrow, requiring high reflux ratios, or particularly where there is a pinch point indicating azeotrope formation. In such cases energy savings can be as much as 60–65 percent (Pearce 1989). Pervaporation avoids the use of entrainment chemicals that may be toxic or environmentally damaging. Pervaporation also is effective at removing water from complex mixtures since operational temperatures often are low, this can be beneficial with heat-sensitive components. The modular nature of pervaporation makes scaleup/scaledown very easy. In particular, pervaporation is very effective at debottlenecking existing plants, by reducing the loading of a column.

PRINCIPLES OF PERVAPORATION

Pervaporation is a membrane separation process for miscible liquids. It differs from other membrane separation processes, such as ultrafiltration, in that it is capable of breaking azeotropic mixtures by selectively permeating only one component (usually water) through the membrane. In addition, the permeate is removed in the vapor phase and not the liquid phase as in ultrafiltration.

A schematic diagram of a pervaporation membrane is shown in Figure 2.1 in which a heated liquid feedstock circulates over the membrane surface. By suitable choice of the membrane polymer, the selective sorption (i.e., adsorption *onto* the membrane surface followed by absorption *into* the membrane) of the feed component(s) occurs. The use of appropriate polymers leads to very significant perturbation of

Figure 2.1. Principle of pervaporation.

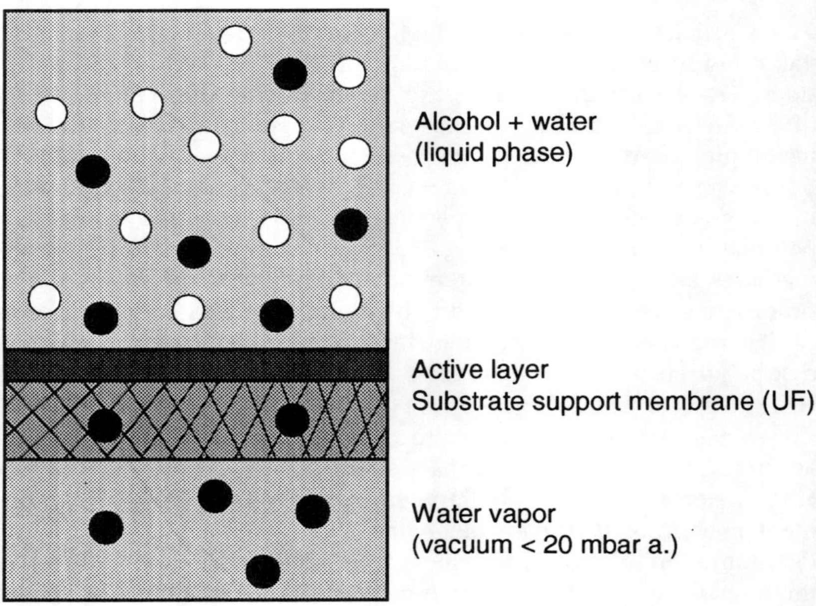

The preferentially permeated water is evaporated from the underside of the membrane.

the liquid-vapor equilibrium between feed and permeate. Figure 2.2 illustrates the level of perturbation possible. Application of a vacuum to the underside of the membrane leads creates a driving force for the permeation of the sorbed feed component (usually water) through the membrane. The magnitude of the driving force is determined by the difference in the partial pressure of the permeating component in the feed and downstream (permeate) sides of the membrane (Huang and Rhim 1991; Aptel et al. 1976). Since the permeate is removed as vapor, sensible heat must be supplied to replace the latent heat of vaporization that is lost on permeation. The permeate may be condensed or, if pure enough, removed through the vacuum pump. Figure 2.3 illustrates a typical dehydration process scheme.

The factors that affect pervaporation membrane performance are as follows:

- Feed flow rate over the membrane surface
- Feed composition and temperature
- Membrane type (both chemical composition and thickness)
- Downstream pressure

Figure 2.2. Isopropanol/water phase diagram showing range of permeate water contents achievable with pervaporation membranes.

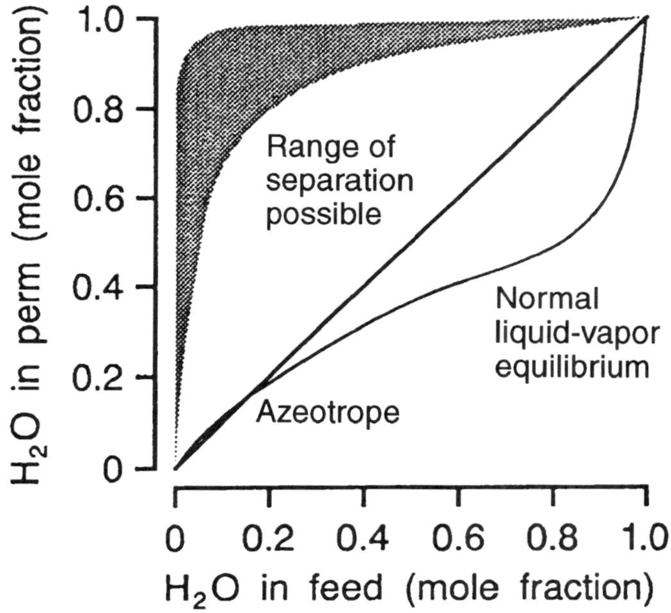

Figure 2.3. Schematic diagram of the pervaporation process.

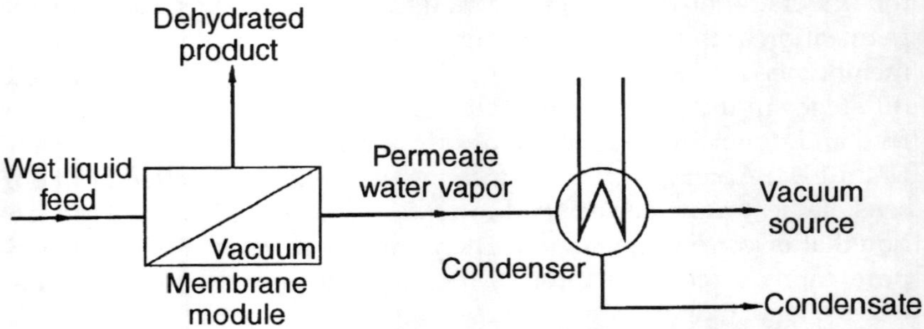

These in turn are affected by the module design.

It is important to understand these so as to use pervaporation effectively within a process stream, and to choose equipment correctly for the desired application.

FACTORS AFFECTING MEMBRANE PERFORMANCE

Feed Flow Rate

Feed flow rate can have a significant effect on membrane performance due to its influence on the concentration polarization boundary layer that will exist on the upstream side of the membrane (Colman et al. 1988; Naylor et al. 1989; Colman and Mitchell 1991; Colman and Naylor 1991). The boundary layer arises from the build-up of rejected species adjacent to the membrane surface and decreases the concentration of the permeating species, lowering the driving force for permeation hence lowering the permeate flux. The severity of the influence of concentration polarization on permeate flux depends up the following factors:

- The nature of the membrane—the influence being less for low flux membranes than for high flux membranes.
- The viscosity of the feed—the more viscous the feed the greater the influence of concentration polarization.
- The type of module being used can have a very significant effect on the influence of concentration polarization—flat sheet modules can have much poorer liquid flow patterns than spiral wound or tubular modules and so will suffer more from concentration polarization.

Figure 2.4 illustrates the effect of both feed flow rate and feed water concentration on the water flux for a tubular polyacrylate pervaporation membrane with an isopropanol/water feed at 70°C (Colman and Naylor 1991).

Feed Temperature and Composition

All the transport processes in pervaporation are influenced by temperature (Blume et al. 1991; Franken et al. 1990; Kalantzi et al. 1989). While *sorption* of the permeate tends to be reduced by an increase in temperature, *diffusion* of the permeate molecules through the membrane is an activated process and has an Arrhenius-type dependency on temperature. The removal of permeate molecules from the downstream face of the membrane occurs by evaporation and increases with temperature. Overall, the net effect of a temperature increase of 10°C tends to result in the permeate flux increasing by a factor of 1.5.

It has been stated that feed composition has no effect on permeate flux (Rautenbach et al. 1992), since the driving force for permeation is determined by the partial pressure difference across the membrane. However, other work has shown that in the case of a feed containing a high (15 percent wt.) concentration of a hydrophilic salt, water permeate fluxes were significantly reduced (Naylor et al. 1989). Generally, for any application that requires the pervaporation of a feed that contains

Figure 2.4. Variation of permeate water flux with feed water concentration and feed flow rate for tubular polyacrylate pervaporation membranes (Colman and Naylor 1991).

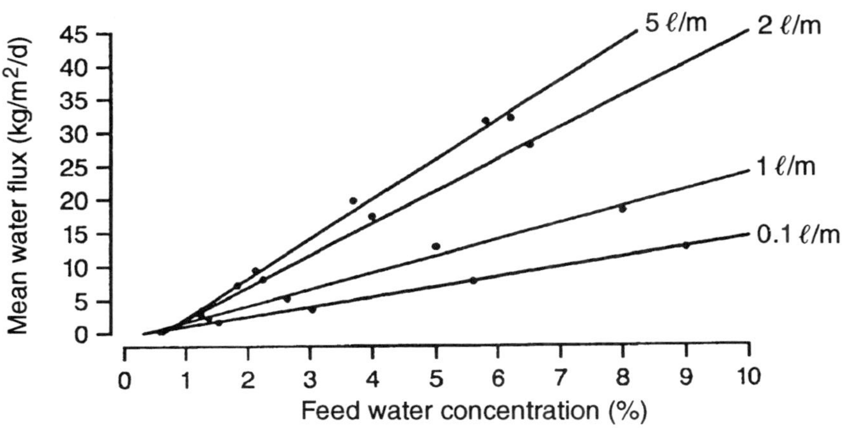

a dissolved solute, preliminary experiments should be carried out to determine effects that the solute might have on performance due to its hydrophilic or hydrophobic nature, as well as concentration polarization effects due to the viscosity of the feed.

Downstream Pressure

A typical pervaporation system is required to run at as low a downstream pressure as possible (typically 10 mbar absolute) in order to achieve the maximum driving force for permeation. In general, permeate flux decreases as downstream pressure increases. Additionally, the negative influence of downstream pressure increases as the feed water concentration increases (Arbouchar and Bruschke 1992). The actual variation in permeate flux with downstream pressure tends to vary with the type of membrane used. The results shown in Figure 2.5 for a tubular polyacrylate membrane are illustrative (Colman and Naylor 1991). The design of the permeate removal channels must ensure that pressure losses are minimized.

Depending on the nature of the pervaporation membrane used, the composition of the permeate may be adversely affected by an increase in the downstream pressure. For instance, with the above polyacrylate membrane, no change in permeate composition was found. In contrast, for a less selective polyvinyl alcohol membrane with a

Figure 2.5. Variation of permeate water flux with downstream pressure for a range of feed water concentrations for tubular polyacrylate pervaporation membranes (Colman and Naylor 1991).

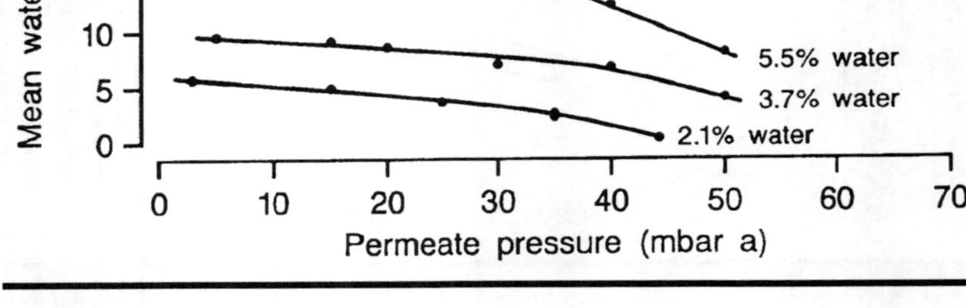

85 percent wt. feed, the permeate isopropanol (IPA) concentration increased from 0.3 percent to 2.5 percent wt. for a pressure increase from 0 mm Hg to 60 mm Hg. At higher IPA concentrations (98 percent wt.) the permeate IPA concentration increased to 6 percent wt. over the same pressure range (Shah 1992).

Membrane Type

Hydrophilic

Hydrophilic membranes are used to selectively permeate water. The four main types of polymers used are shown together in Table 2.1. Many other membrane materials have been described in the literature. The most recent and potentially promising one is a zeolite membrane (Kita et al. 1995) (The Smart Chemical Company, London).

Hydrophobic

A range of polymer membranes have been studied for the removal of organics from water. The most common ones available commercially are polydimethyl siloxane (Deutsche Carbon, Neunkirchen, Germany), filled polydimethyl siloxane (Deutsche Carbon), proprietary silicone (MTR, Menlo Park, CA), polyetherblockamide (GKSS, Geesthacht, Germany) and polybutadiene (Bayer AG, Leverkusen, Germany). These have been used for the removal of a range of organic compounds and solvents from water, such as alcohols, acetone, tetrahydrofuran (THF), lactones, aniline, and phenol. The degree of selectivity achieved with the membranes is not as high as that obtained by the hydrophilic

Table. 2.1. Classification of Pervaporation Membranes

Polymer	Water Flux	Selectivity	Solvent Range
Polyvinyl alcohol	intermediate	intermediate	wide
Plasma	high	high	wide
Poly-acrylate	high	high	not acidic feed
Polyimide	intermediate	high	not chlorinated solvents
Zeolite	very high	very high	wide

membranes for the removal of water. Consequently, the range of applications to date has been limited. The development of membranes with higher selectivity will undoubtedly lead to more opportunities.

Module Design

Two designs of pervaporation modules are readily available, namely plate and frame (Deutsche Carbon) and spiral wound (Texaco, Beacon, NJ). Other types such as hollow fiber (Ube Industries, Chiba, Japan) and tubular (Kalsep, Surrey, UK), have been or are produced.

Each manufacturer provides a detailed description of each type of module and its particular merits. This section simply highlights the major factors in considering a particular design for use in a pharmaceutical application.

Plate and Frame

Plate and frame systems consist of flat sheets of membrane sealed onto porous permeate plates by O-rings that are compressed onto the membrane surface. Typical packing densities are 80 m^2/m^3 of permeate vessel. While this design is good on the permeate removal side (having a low pressure drop), it has a poor feed flow design. Little turbulence occurs over the membrane surface. Consequently, concentration polarization can occur and may reduce membrane performance. The other main drawback with this system is the potential for leaks to occur around the very many O-ring seals that are used. However, it is relatively easy to obtain such seals in a wide range of materials so that feed/seal compatibility problems can be of minor concern.

Spiral Wound Modules

Spiral wound modules consist of a flat sheet of membrane that is folded so that the two free ends are glued onto a porous tubular permeate collection pipe, the membrane then is wound around the central pipe. Usually two or four such leaves are attached and wound around the core. The wound module is placed in a tubular pressure vessel and feed liquid flows axially down the module across the envelope. The portion of the feed that permeates the membrane envelope spirals towards the center and exits via the collection tube. Spacers are interleaved between the membrane envelopes and the membrane sheets so as to reduce the pressure drop on the feed and permeate sides, and to promote turbulence on the feed side. A schematic diagram of a spiral wound module is shown in Figure 2.6. In gas separation systems

Figure 2.6. Spiral wound module element.

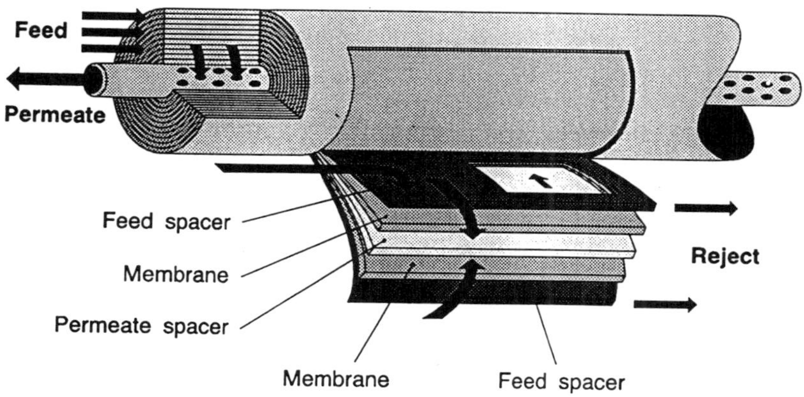

spiral wound modules have a significantly larger surface-area-to-volume ratio than flat sheet modules; however, because of the imperative need to avoid permeate pressure drops in the pervaporation modules, the packing density is limited to 100–120 m^2/m^3. While this confers some advantage over plate and frame systems in terms of a more compact module, the module only accounts for approximately 20 percent of the overall volume of the complete system. The other factor to consider in pharmaceutical applications of spiral wound systems is: (a) the nature and stability of the glue used with the process solvent(s), and with any intermediates present, and (b) the possibility of leaching of contaminants from the glue.

It has been claimed that spiral wound modules are unable to operate at temperatures as elevated as those used with plate and frame systems, due to glue and spacer limitations (Arbouchar and Bruschke 1992). The relatively long and constricted feed path length can lead to significant feed side pumping costs.

Hollow Fiber Systems

Hollow fiber systems for gas separation applications have very high membrane area/volume ratios. The need to limit pressure (and temperature) drop down the fiber results in the use of short fibers. Again the need to avoid significant permeate pressure drop results in spacing the fibers out much more than in gas systems. The result is a module that has similar or worse packing density to that of the plate and frame system. Hollow fiber modules, like spiral modules, require the use of glues to pot the fibers.

Tubular Membranes

Tubular membranes based on coated ceramic monolith elements offer excellent thermal and chemical resistance properties as well as good feed and permeate flow geometries. The main drawback to this type of system is the high cost of the ceramic substrate.

ENERGY REQUIREMENTS

The dehydration of aqueous solvents by membrane pervaporation offers significant potential for both capital and operational savings when compared with conventional technology such as distillation. *Pervaporation is unique among commercial membrane processes in that a phase change occurs during permeation.* This section briefly describes the energy requirements of pervaporation.

Pervaporation

The outline process flow diagram in Figure 2.3 indicates that there are four energy consuming operations.

Liquid Side Pumping

Pumping power requirements are normally low. Pressure is provided on the liquid side only to ensure the feed remains liquid at elevated temperature and to overcome any pressure drop due to the feed flow path. Liquid side pressure does not provide the driving force for pervaporation.

Feed and Interstage Heating

Permeating molecules are removed from the downstream surface of the membrane as a vapor. The latent heat for the phase change is obtained from the sensible heat of the feed. It is advantageous for the pervaporation process to operate at as high a temperature as possible due to the temperature dependency of the permeate flux. Therefore, feed normally is heated to the maximum temperature possible for the membrane. Interstage heaters are used between the module stages to compensate for heat losses. The heat requirement in pervaporation is significant, but often can be obtained at low cost by using low grade or waste/recovered heat since operating temperatures are lower than for distillation.

Condensation

The permeate vapor normally is condensed for recovery/recycle or for disposal. The energy requirement for this step is equivalent to the heating requirement for feed and interstage heating. Condensation is normally carried out at temperatures between 0°C and 10°C to achieve operating permeate pressures between 10 mbar Absolute and 20 mbar Absolute. Because condensing medium temperatures are lower than the tower cooling water used for distillation column condensers, a refrigeration energy input is required. This can be minimized by using the low temperature cooling services that often are available on pharmaceutical and fine chemical sites.

Permeate Pumping

The vacuum pump in the pervaporation process serves to remove noncondensable gases from the system resulting from leakages. The condenser creates the vacuum, The permeate pressure is determined by the condenser temperature. Vacuum pump power requirements are, therefore, very low. In some circumstances, (e.g., for small-scale applications), the condenser may be omitted if the membrane is highly selective. In such a case, vacuum pump power requirements can be significant, but the overall process is simplified.

Distillation

Figure 2.7 shows a distillation scheme for drying a 50 percent wt. alcohol stream. In distillation energy enters the system at the base of the column and is withdrawn from the condenser at the top. The reboiler operates hotter than the pervaporation system, thereby incurring a cost penalty, but the condenser can use tower cooling water and is, therefore, cheaper to run than the pervaporation condenser. A continuous distillation column needs reflux to separate the components efficiently. The closer the boiling points of the components to be separated and the greater the required purity of the head product, the greater the amount of head product that must be returned as reflux.

Azeotrope formation is a special case where the advantage of pervaporation becomes very significant. Approaching the azeotropic composition, the separation factor between different components in distillation becomes very low, as shown in Figure 2.10. High reflux

Figure 2.7. Distillation process scheme.

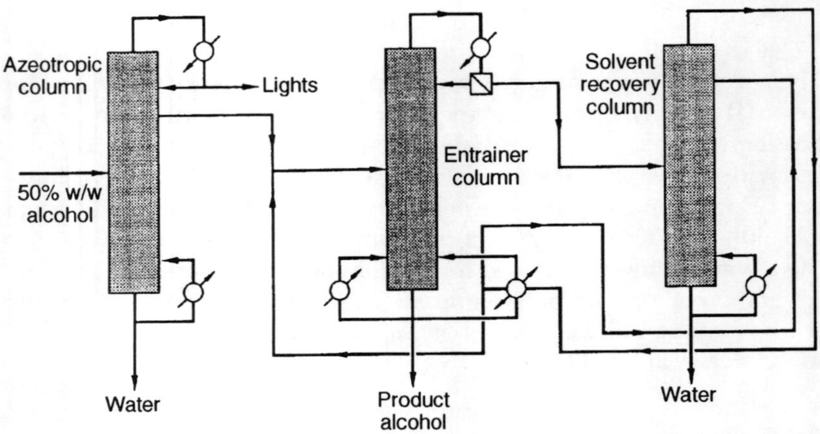

ratios are required to reach the azeotropic composition; then a separate azeotrope-breaking column is required, which incorporates an entrainer. The entrainer itself must be purified for recycling in a separate solvent recovery column. Energy consumption in reaching the azeotrope, breaking the azeotrope, and recovering the entrainer in the latter stages of a distillation process are much greater than in the first stages.

APPLICATION OPPORTUNITIES

The main opportunities for the use of membrane pervaporation are as follows:

- Solvent recovery—campaign operation
- Process integration—improved efficiency/productivity
- Solvent production—replacement of distillation and debottlenecking

In the fine chemical/pharmaceutical industry opportunities will often fall into the process integration category.

Feedstocks

The primary feedstocks for pervaporation are alcohols (ethanol and isopropanol), THF, pyridine and dimethyl formamide (DMF). Apart from DMF, all form azeotropes; DMF has a particularly narrow vapor-liquid equilibrium curve over a very wide composition range. Many other feedstocks also can be suitable for pervaporation, including other alcohols, ketones, esters, amines, acetic acid, acetonitrile, toluene, ethers, and dimethyl sulfoxide.

Case Studies

Process Integration

This application required the dewatering of an IPA/reaction intermediate mixture in which the IPA was used as a carrier solvent. In the existing process water was created in the reaction stage and was removed from the reactor as an IPA azeotrope. The azeotrope was recovered by campaign distillation. The new process (Figure 2.8) currently is running alongside the former process. Membrane pervaporation removed the requirement for campaign recovery on the IPA azeotrope by dewatering the reaction mixture in a single pervaporation stage by reducing the water content of the batch from 7 percent wt. to 2 percent wt. The permeate from the plant has a high water content (99 percent wt. water) and is disposed of directly to the effluent plant. Batch times for the pervaporation process are significantly lower than for the original process. Together with the benefit of lower labor costs as a result of improved automation, a <18-month payback time for the pervaporation plant was achieved.

Figure 2.8. Process integration scheme.

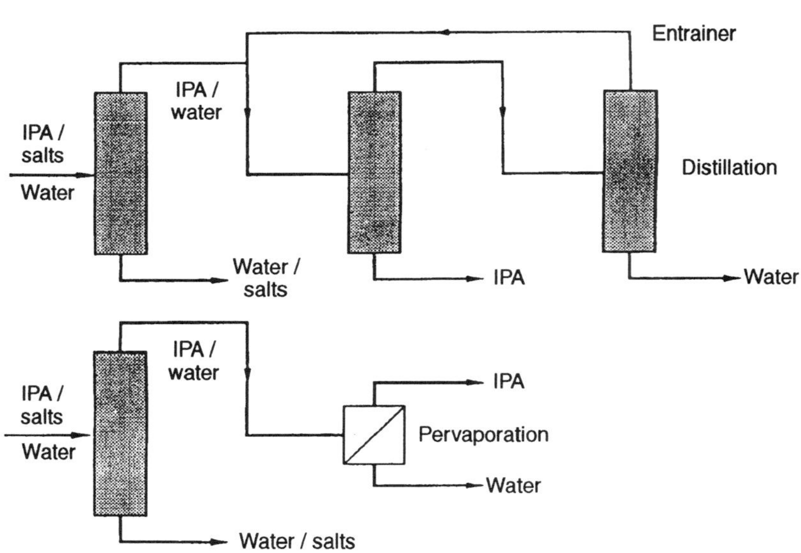

IPA Debottlenecking

Plant production was bottlenecked by the azeotrope column at the trays above the feed point. By enriching the composition of the feed to the azeotrope column from 87.5 percent wt. IPA to 89.8 percent wt. IPA, the production of pure IPA from the base of the column could be increased from 215 tpd (tonnes per day) to 270 tpd, without increasing the liquid loading of the trays above the feed point. The process flow diagram incorporating pervaporation is shown in Figure 2.9 with the before and after mass balance in Figure 2.10.

Productivity Improvements in the Pharmaceutical Industry

IPA was used as a reactant in an esterification reaction. Water created in the process needed to be removed more cost-effectively than conventional azeotropic distillation. The dewatered IPA could be recycled to the process, provided the water content was reduced to 1–2 percent wt. Process flow diagrams for the existing and new process schemes are shown in Figure 2.11. Pervaporation times of approximately 10 hr were significantly below the 16 hr batch times for azeotropic distillation.

For a capital outlay of $300,000 the payback resulting from the extra capacity of the plant and the payback time for this project was less than 2 years (Pearce 1989).

Figure 2.9. Debottlenecking isopropanol/water dehydration.

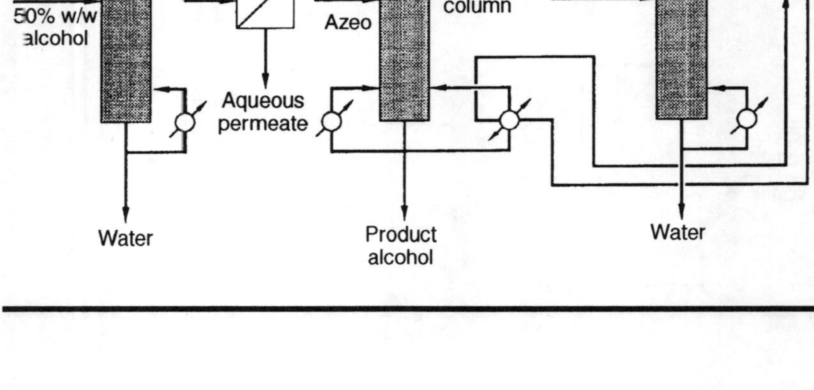

Figure 2.10. Mass balance for debottlenecking scheme.

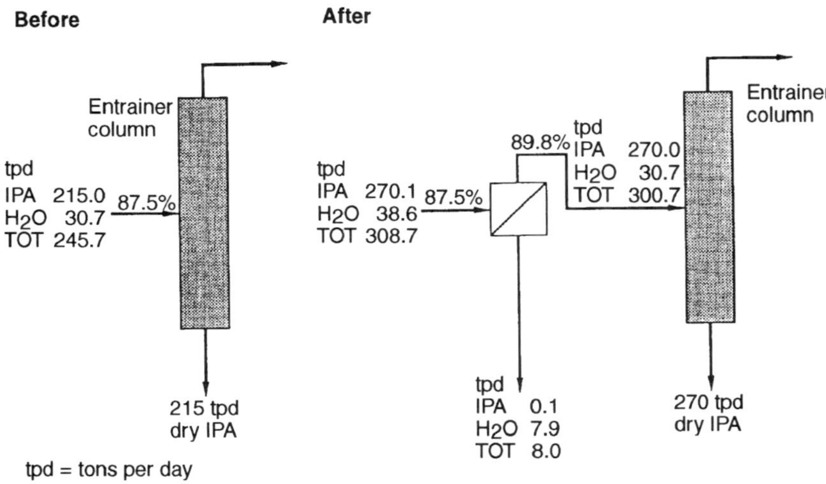

Figure 2.11. Productivity improvement via membrane pervaporation.

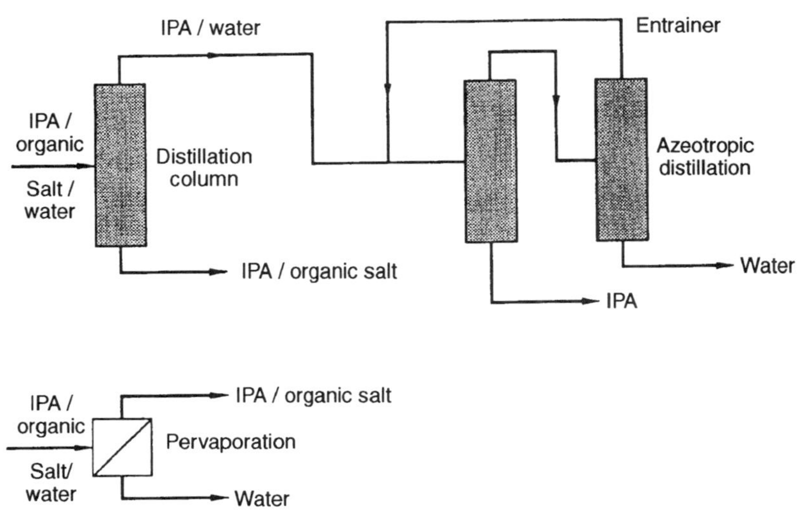

CONCLUSIONS

Pervaporation is an established commercial separation process that can be highly cost-effective in the pharmaceutical industry for a range of applications. Dewatering is the best established with applications, ranging from simple azeotrope breaking to displacement of equilibrium reactions. The removal of organic components from aqueous solutions represents a wide-open field of opportunity, provided sufficiently selective membranes can be produced.

REFERENCES

Aptel, P., N. Challard, J. Cuny, and J. Neel. 1976. Application of the pervaporation process to separate azeotropic mixtures. *J. Membr. Sci.* 1:271.

Arbouchar, R., and H. Bruschke. 1992. Long-term experience with industrial pervaporation plants. *Proceedings of the 6th International Conference on Pervaporation Processes in the Chemical Industry,* 27–30 September, Ottawa, edited by R. Bakish. Englewood, NJ: Bakish Materials Corp.

Blume, I., A. Bos, P. J. F. Schwering, M. H. V. Mulder, and C. A. Smolders, 1991. Transport phenomena of vapour and liquid permeants in elastomeric membranes. *Proceedings of the 5th International Conference on Pervaporation Processes in the Chemical Industry,* 11–15 March, Heidelburg, edited by R. Bakish. Englewood, NJ: Bakish Materials Corp.

Colman, D. A., and W. S. Mitchell. 1991. Mass transfer enhancement by pulsatile flow in baffled channels: The effect of Schmidt number variation and scaling. *Industrial Chemical Engineering* Symp. Series No. 118:119–134.

Colman, D. A., and T. deV. Naylor. 1991. The influence of operating variables on flux and module design in a high performance pervaporation system. *Proceedings of the 5th International Conference on Pervaporation Processes in the Chemical Industry,* 11–15 March, Heidelburg, edited by R. Bakish. Englewood, NJ: Bakish Materials Corp.

Colman, D. A., T. deV. Naylor, G. K. Pearce, R. D. Whitby. 1988. *Proceedings of the International Conference on Membrane Separation Processes,* Brighton, U.K., edited by A. Green, B.H.R.A. Cranfield, Bedford, UK: The Fluid Engineering Centre.

Franken, A. C. M., M. H. V. Mulder, and C. A. Smolders. 1990. Pervaporation process using thermal gradient as the driving force. *J. Membr. Sci.* 53:127–141.

Huang, R. J. M., and J. W. Rhim. 1991. Separation characteristics of pervaporation membrane separation processes. In *Pervaporation membrane separation processes,* edited by R. J. M. Huang. Amsterdam: Elsevier.

Kalantzi, K., M. Marin, and H. Gilbert. 1989. Pervaporation Module with an Internal Heating Source. *Proceedings of the Fourth International Conference on Pervaporation Processes in the Chemical Industry,* 3–7 December, Fort Lauderdale, edited by R. Bakish. Englewood, NJ: Bakish Materials Corp.

Kita, H., K. Horii, K. Tanaka, and K.-I. Okamoto. 1995. Pervaporation of water/organic liquid mixtures using a zeolite NaA membrane. *Proceedings of the Seventh International Conference on Pervaporation Processes in the Chemical Industry,* 26 February–2 March, Reno, edited by R. Bakish. Englewood, NJ: Bakish Materials Corp.

Naylor, T. deV., F. Zelaya, and G. J. Bratton. 1989. The BP-Kalsep Pervaporation System-Membrane Performance and Properties. *Proceedings of the Fourth International Conference on Pervaporation Processes in the Chemical Industry,* 3–7 December, Fort Lauderdale, edited by R. Bakish. Englewood, NJ: Bakish Materials Corp.

Rautenbach, R., S. Klatt, and J. Vier. 1992. State of the Art of Pervaporation 10 Years of Industrial PV. *Proceedings of the Sixth International Conference on Pervaporation Processes in the Chemical Industry,* 27–30 September, Ottawa edited by R. Bakish. Englewood, NJ: Bakish Materials Corp.

Shah, V. 1992. Influence of Process Parameters on Texsep Pervaporation membranes. *Proceedings of the Sixth International Conference on Pervaporation Processes in the Chemical Industry,* 27–30 September, Ottawa edited by R. Bakish. Englewood, NJ: Bakish Materials Corp.

3

Tangential Flow Filtration

Stephen L. Michaels
Chris Antoniou
Vinay Goel
Paula Keating
Ralf Kuriyel
Alan S. Michaels
Steven R. Pearl
Gaston de los Reyes
Eric Rudolph
Marty Siwak

The objectives of this chapter are as follows:

1. To discuss the process of tangential flow filtration (TFF) and the physicochemical principles governing its performance.
2. To describe the filtration media, equipment, and systems currently available for carrying out TFF.
3. To outline TFF's capabilities and limitations for important solid/liquid separations in the pharmaceutical industry.

THE HISTORY OF TANGENTIAL FLOW FILTRATION

Separation of solids from liquids by filtration is among the oldest chemical engineering unit operations. Until the discovery of reverse osmosis (RO) for water desalination (the earliest practical molecular-filtration process) in the late 1950s, filtration was regarded and exclusively employed as a dead-end pressure-driven solid/liquid separation process. In this process the suspended particles were retained and accumulated on the surface of the filter medium while the particle-free liquid was collected as the filtrate.

With the advent of RO, and soon thereafter ultrafiltration (UF), as practical pressure-driven membrane separation processes to remove or concentrate a solute present in solution, the concept of continuously fractionating a solution via a semipermeable barrier into a solute-free "permeate" and solute-enriched "retentate" without accumulation of solute on the barrier was accepted as a viable solid/liquid separation process. This process is generally accomplished by passing the pressurized feed liquid over the membrane surface while solute-depleted liquid is continuously removed through the membrane as permeate.

The performance penalties associated with retained-solute accumulation on the upstream membrane surface during RO and UF were evident very early in the development of these processes. Much experimental and analytical effort has been (and continues to be) expended to find equipment design and liquid management strategies that minimize this accumulation and its detrimental consequences. Without these strategies, neither RO nor UF would have their current stature as useful industrial separation processes.

For pressure-driven molecular separations involving liquids, the phenomena (principally liquid convection and diffusion) responsible for solute accumulation at the upstream membrane surface—so-called "solute polarization"—are reasonably well understood and predictable from the basic principles of liquid mechanics and mass transport. Present-day RO and UF module designs have evolved from rational engineering analysis of these phenomena.

Application of the principles of liquid management to the conduct of solid/liquid separations involving suspensions of particulate solids (as differentiated from dissolved solutes) did not receive attention as a practical alternative to conventional accumulative filtration until well after the introduction of RO and UF into industrial practice. Filtration of this type has come to be known as "cross-flow" or "tangential flow" filtration.

This delay in application has multiple reasons. Many important filtration operations involve the removal of small amounts of suspended particles from liquids (e.g., clarification and cold sterilization), while others involve the removal of relatively large, rigid particles from suspension, forming highly permeable filter cakes of high-solids density. For such systems TFF offers few benefits.

On the other hand, suspensions of exceedingly fine particles (1.0 μm and less) or of elastically compliant or gelatinous particles are poor candidates for conventional filtration because they form highly solvated filter cakes of exceedingly low hydraulic permeability. Suspensions of this kind (typified by fermentation broths, cell cultures, and animal/vegetable tissue homogenates) are widely encountered in the food and pharmaceutical industries. It is not surprising, therefore, that TFF applications in these industries have achieved substantial importance over the past several years. Today, tangential flow microfiltration (TFM) has substantially displaced conventional filtration for the dewatering of fermentation broths and concentration and recovery of biomass from cell suspensions and is competitive with centrifugation for many solid/liquid separations.

Tangential flow filtration is also challenging conventional microfiltration (MF) in such applications as clarification and cold sterilization of biological media, parenteral liquids, liquid foodstuffs, and beverages. More recently, TFF has proven uniquely applicable to the development of continuous extractive fermentations for production of antibiotics (and other secondary metabolites) and the development of mammalian cell-culture reactors for the production of vaccines and other biopharmaceuticals.

PRINCIPLES AND THEORY

Interplay of Hydrodynamics and Mass Transport in Tangential Flow Filtration

When a liquid containing suspended particles or a dissolved solute flows under pressure over a membrane that retains the particles or solute but allows permeation of the solvent, liquid convection parallel and normal to the membrane surface transports the suspended particles or solute molecules both parallel and toward the membrane surface. Because the membrane is stationary with respect to the flowing liquid, in the absence of slip at the membrane surface, a gradient in liquid velocity will be established that causes liquid close to the surface to move slower than liquid further from the surface.

Because particles or solute molecules reaching (and rejected by) the membrane surface are being transported down the feed channel at a lower velocity than those above it, the local concentration of solutes in solution adjacent to the membrane will rise relative to that in the bulk solution moving above it. This phenomenon will establish a gradient in solute or particle concentration near the membrane surface.

If the particles are free to diffuse in solution (or if liquid turbulence creates eddies near the surface that cause liquid flow normal to and away from the membrane surface), such diffusional transport will tend to relieve the solute concentration buildup due to convection and rejection. In time a steady state will be reached in the feed channel where, in a thin layer near the membrane surface, convection of solute toward the membrane will be balanced by the diffusive traffic of solute in the opposite direction. In this zone a gradient in solute (particle) concentration, increasing in the direction of the membrane surface, will be established (Figure 3.1). This constitutes "polarization" of the membrane; the layer of solution within which the majority of this concentration gradient exists is regarded as the mass-transport boundary layer or polarization layer. The ratio of the solute concentration at the membrane surface to that in the bulk solution above it is termed the polarization modulus.

Polarization invariably has adverse effects on the filtration process, as will be discussed. Solute convection to the membrane surface is governed by the transmembrane solvent flux, while the back transport rate is determined by both solute and solvent physicochemical parameters (liquid viscosity and density, solute diffusivity) and liquid channel hydrodynamics (liquid velocity and channel dimensions); the polarization modulus is determined by the ratio of these two quantities.

For the case of a membrane-impermeable solute, the relationship is

$$(C_w/C_b) = \exp(J/k)$$

where J is the transmembrane solvent flux, k is the local mass transfer coefficient for the solute, and (C_w/C_b) is the polarization modulus.

The mass transfer coefficient, k, can be defined by the relationship

$$k = D/\delta$$

Figure 3.1. Polarization during ultrafiltration.

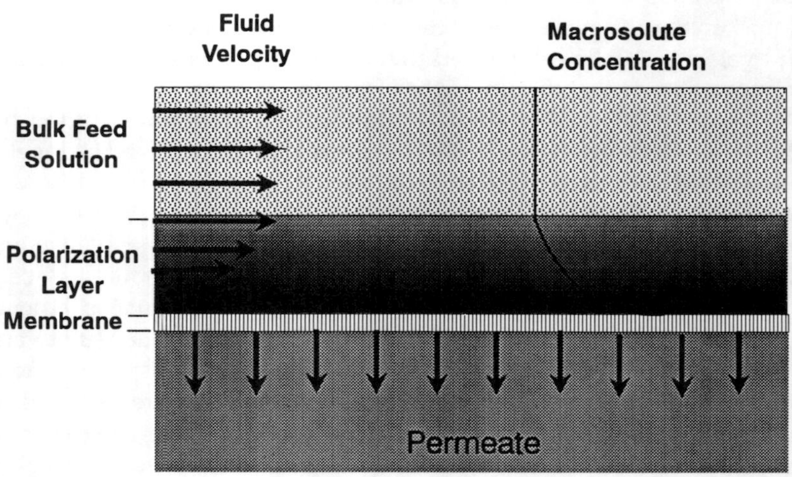

where D is the solute diffusivity in solution and δ is the effective thickness of the polarization layer. The dimensionless parameter that ultimately governs the magnitude of the polarization modulus is the Sherwood Number,

$$Sh = k_c/D = d_c/\delta$$

where d_c is the hydraulic diameter of the feed channel. (For thin rectangular channels of depth t, $d_c = 2t$.) The Sherwood Number is a unique function of the Reynolds Number and the Schmidt Number for various flow regimes and channel geometries. (See, for example, Cheryan 1986 for details on these relationships.) It is shown that polarization is strongly dependent upon solute molecular size and liquid viscosity (which determine solute diffusivity), membrane permeability and transmembrane pressure (TMP) drop (which determine permeation flux), and liquid velocity and channel dimensions (which determine polarization layer thickness). Tangential flow filtration performance will thus be enhanced by measures that minimize liquid channel dimensions and maximize feed-channel liquid velocity or otherwise increase the back transport of solute from the membrane surface into the bulk of the flowing liquid.

Strategies for achieving these results by membrane module design are described later in this chapter; all involve techniques for localizing liquid kinetic energy close to the membrane surface so that convective transport of solute back into the bulk flowing liquid can enhance mass transfer. This process can be achieved by controlling feed-channel configuration or spacer geometry, by modifying the surface texture of the membrane, or by employing means other than liquid circulation to create turbulence in the flowing liquid. Such performance improvements come at a price, either as increased module/hardware cost or increased power requirements. For high-value pharmaceutical products, however, the improvements in product yield or purity that result from these strategies usually compensate for the higher operating costs.

Polarization: Its Effects and Significance in Membrane Filtration

Since polarization raises the solute concentration at the membrane surface above that of the overlying bulk solution, the permeation flux for a given hydrostatic pressure difference across the membrane will be depressed. This is due to the elevation in osmotic pressure of the (retained) solute relative to that of the bulk solution. If the membrane is not totally impermeable to the solute, its elevation in concentration at the upstream surface of the membrane will raise the solute flux across the membrane, thus the concentration of solute in the permeate. Therefore, the apparent rejection of solute (based upon the bulk upstream solution concentration) will be depressed.

As the feed-side pressure is raised (under constant feed-side flow conditions), the permeation flux will rise less than proportionately with pressure and the solute rejection will decrease as shown in Figure 3.2. If the rise in solute osmotic pressure at the membrane surface were to be equivalent to the increase in the transmembrane hydrostatic pressure difference, there would be no increase in permeation flux with increasing pressure.

For incompletely retentive membranes, as the feed-side pressure is increased without limit, the solute concentration at the membrane surface will increase to a value at which the apparent rejection drops to zero (the solute concentration in the permeate equals that in the bulk feed solution). On the other hand, if the membrane is totally impermeable to the solute, the membrane surface concentration (and the polarization modulus) will increase rapidly with increasing TMP until the solute's limit of solubility is reached. At this point precipitation or crystallization of the solute will occur without any further elevation of its concentration in solution. In principle, under these conditions the

Figure 3.2. Effects of polarization on flux and rejection in tangential flow filtration.

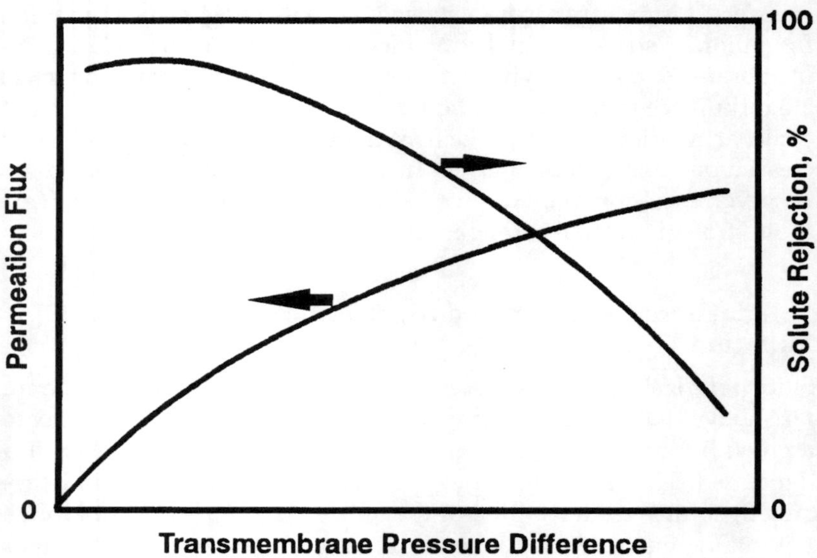

osmotic pressure of the solute becomes constant while the hydrostatic pressure in solution continues to increase; the permeation flux should increase linearly with pressure. These situations are schematically represented in Figure 3.3, where solute rejection and permeation rate have been plotted vs. TMP pressure for solute-permeable and solute-impermeable membranes.

In RO, where solute diffusivities are large and membrane hydraulic permeabilities are low, the classical model of polarization is adequate to explain the experimental observations. (In most applications of RO, solute polarization is minor in determining process performance and economics.) However, in UF of macrosolutes or MF of dispersions, where solute diffusivities are low or nonexistent and membrane permeabilities high, classical treatments of polarization are inadequate to describe the observed phenomena. In these applications the effects of polarization on TFF performance are so great that they

Figure 3.3. Flux and rejection vs. transmembrane pressure difference for totally and partially rejecting membranes.

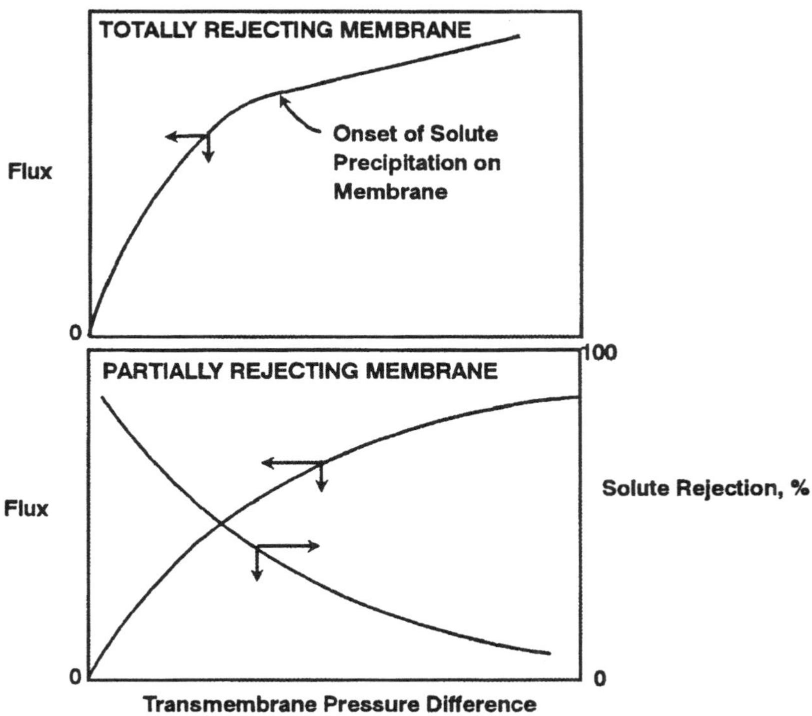

often limit the feasibility of this process for important solid/liquid separations.

Tangential flow ultrafiltration (TFU) of solutions of proteins and other macrosolutes is invariably characterized by a permeation flux-vs.-TMP dependency, illustrated in Figure 3.4. The flux/pressure curve is concave to the x-axis; the flux becomes asymptotic to the x-axis at higher pressures. The limiting flux is usually far less than that predicted from the intrinsic membrane solvent permeability; it is strongly dependent upon the solute concentration in the feed solution and increases monotonically with increasing liquid velocity in the feed channel (or decreasing channel dimension).

Blatt et al. (1970) first proposed that, with macrosolutes, increasing polarization with increasing transmembrane flux leads to a sufficiently high concentration of solute at the membrane surface. This yields a cohesive and adherent but hydraulically permeable gel layer of constant solute concentration, above which lies a conventional polarization layer of declining solute concentration (Figure 3.5). Because convection and diffusion govern solute mass transport within the classical polarization layer, as that model predicts, any attempt to increase flux by raising pressure will be accompanied by a corresponding increase in gel-layer thickness sufficient to reduce the flux to the same level as

Figure 3.4. Flux vs. pressure relationships in tangential flow ultrafiltration: Effect of solute concentration at constant feed channel flow rate.

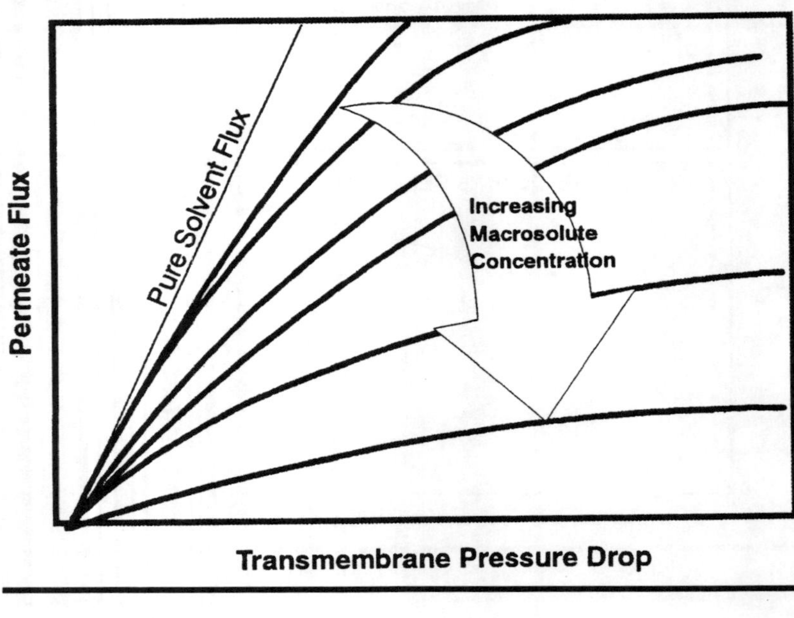

at the lower pressure. If the hydraulic permeability of the gel layer is low and its thickness small with respect to the flow channel dimensions, this behavior is satisfactorily represented by the relation

$$J_{\text{lim}} = k \ln (C_g / C_b)$$

where J_{lim} is the limiting flux, C_g and C_b are the solute concentrations in the gel-layer and bulk solution, respectively, and k is the mass transfer coefficient in the feed channel.

The log-linear correlation between limiting flux and feed solute concentration (which is obeyed for a larger number of macrosolutes, membranes, and membrane module configurations, as illustrated in Figure 3.6) and the fit between measured values of k and those predicted from mass transport theory, have given considerable credence to this gel-layer-polarization model. Also, according to the preceding equation, the limiting flux should go to zero when the bulk solute concentration reaches the gel-layer concentration; extrapolation of limiting flux/concentration correlations to zero flux yields, for many macrosolutes, yields gel-layer concentrations that are quite reasonable.

With the development of more sophisticated treatment of the thermodynamic, diffusional, and transport properties of concentrated solutions of macrosolutes and their application to the polarization boundary layer, the limiting flux phenomenon can be predicted on the basis of a refined osmotic pressure model, without invoking the existence of a gel layer. Indeed, Wijmans et al. (1985) were able to show from irreversible thermodynamic analysis of transport in a classical

Figure 3.5. Gel-layer polarization model.

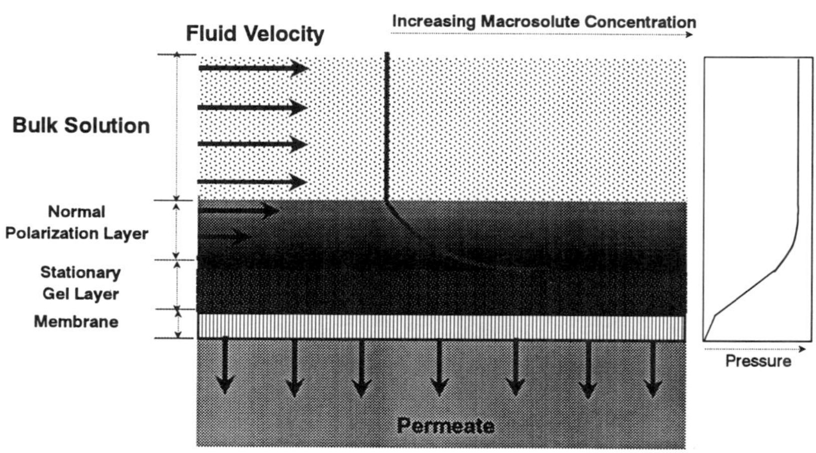

polarization layer that the elevation in solute osmotic pressure within the layer was the exact equivalent of a decline in hydrostatic pressure of solvent within the layer. This conclusion renders differentiation between the osmotic and gel-layer models of UF flux limitation difficult.

The following observations do, however, support the gel-layer model:

- Detection of the presence of a true gelatinous layer of macrosolute on a polarized membrane surface
- Failure of the polarization layer to spontaneously "relax," or back diffuse into the bulk solution when permeation is interrupted

Both models are probably applicable, the osmotic model appropriate for relatively low-molecular-weight macrosolutes and/or low TMPs, and the gel-layer model appropriate for large molecules and/or high TMPs.

With TFM of suspensions of particles greater than approximately 0.2 μm in size, the flux/pressure/flow-rate relationships qualitatively resemble those observed with RO and UF; the predictive models based on diffusion and convection, however, are neither applicable nor relevant. Such particles have negligible diffusivities in solution; thus a

Figure 3.6. Flux vs. macrosolute concentration during tangential flow filtration with gel-layer polarization.

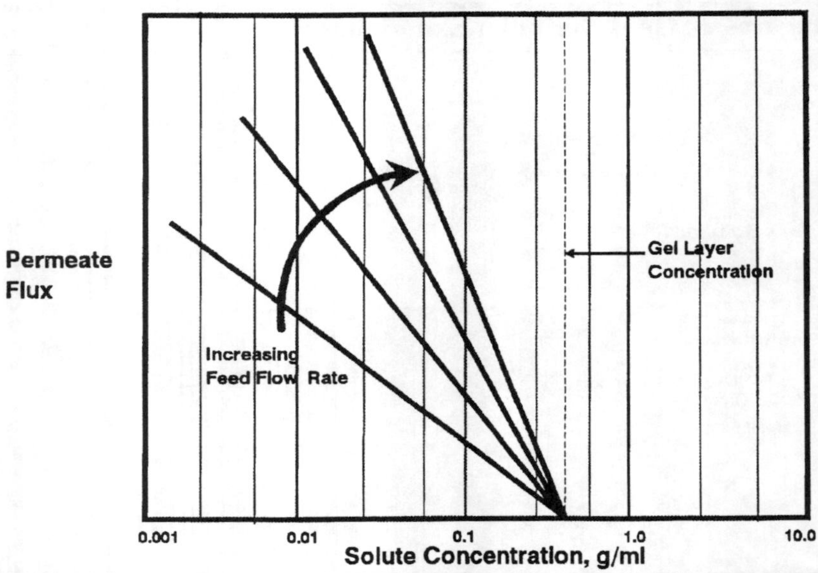

diffuse boundary layer of rising particle concentration near the membrane surface cannot exist. Particles retained at the membrane surface may be transported convectively parallel to the membrane surface by the flowing liquid and (in turbulent flow) normal to the membrane surface by eddy diffusion.

Because membrane permeation drives particles toward the membrane surface, however, these particles will accumulate on the membrane as a hydraulically permeable filter cake through which the suspending liquid must pass before permeating the membrane. The pressure drop through the filter cake has two important effects: (1) it reduces the available hydrostatic pressure gradient driving liquid through the membrane (thereby reducing membrane flux) and (2) it imposes a normal stress on the filter cake (greatest at the membrane surface, declining to zero at the cake/liquid boundary) that causes it to resist tangential or normal displacement.

Because the cake behaves as a viscoelastic solid, a steady-state condition will be reached at which the cake thickness will remain constant. At this condition the rate of particle deposition by convective flow of liquid through the cake and membrane equals the rate of tangential flow of particles in the upper layer of the cake under the shear force exerted by the flowing liquid. This is shown schematically in Figure 3.7. Therefore, the steady-state cake thickness will increase with increasing feed-channel pressure and decrease with increasing flow velocity in the feed channel. If the channel dimensions are relatively

Figure 3.7. Tangential flow filtration of particulate suspensions: solids distribution on the membrane surface.

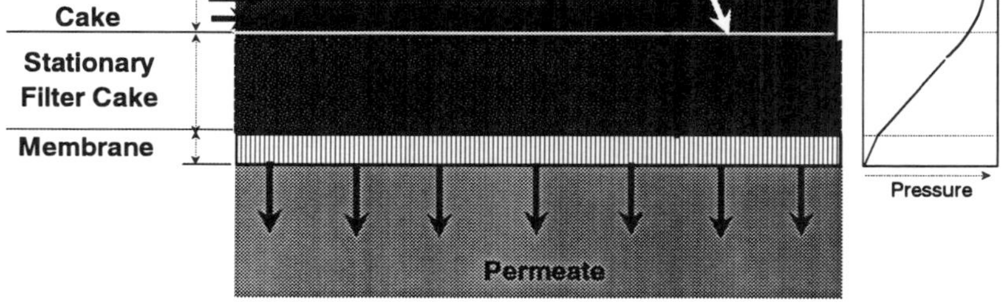

small, the TMP drops large, and the liquid velocities in the feed channel low, the cake thickness can increase to occupy most of the channel cross-section. In that event feed flow through the channel will virtually cease; the channel will be completely filled with solid filter cake.

The dynamics of TFU and TFM are not determined simply by the rheological and diffusional properties of the solution or dispersion being processed. They are also critically dependent upon the hydrodynamic and viscoelastic properties of the particle concentrate deposited on the membrane surface. These properties, in turn, are dependent upon the size, shape, and size distribution of the particles present; the elastic compliance or plasticity of those particles; and the surface chemical variables that affect both particle/particle and particle/membrane-surface interactions. Therefore, solutions of two different proteins of virtually identical molecular weight, when processed under virtually identical hydrodynamic conditions with the same membrane, can display widely disparate flux/pressure/flow-rate dependency.

Similar disparities can be found between solutions of the same protein that have different pH levels or salinities or between two different membranes of approximately equal permeabilities and retentivities, when they are used to process the same feed solution or suspension. While the interplay of variables that influence membrane filtration behavior is usually too complex to allow quantitative prediction of system performance, a qualitative understanding of how these variables affect the separation process is nonetheless helpful in system and membrane selection and in optimizing operating conditions.

Physicochemical Factors Influencing Polarization

As the primary consequence of polarization is the concentration of solute at and near the membrane surface, the effects of such concentration on the state of aggregation of the solute are important in determining the structure and properties of the polarization layer. If the concentration of solute at the membrane surface exceeds the critical nucleation concentration for crystallization or precipitation, a variably dense, solid film can form on the membrane surface that can significantly obstruct liquid transport. In the case of ampholytic macromolecules, such as proteins, the propensity to aggregate and precipitate is dependent not only on their concentration in solution but also upon the pH and ionic strength of the suspending medium.

Proteins generally display minimum water solubility at or near their isoelectric points (pIs), and their solubility is usually depressed by increased ionic strength or solution salinity. Under these conditions, the formation of dense, coherent, and low-hydraulic-permeability polarization layers is favored, leading to major flux loss and poor clearance of lower-molecular-weight solutes. In many cases precipitation under these conditions is also accompanied by irreversible changes in molecular conformation (denaturation), leading to further layer consolidation, formation of insoluble adherent films, and loss in biological activity of the protein.

In general, uncharged, highly hydrophilic macromolecules (e.g., dextrans, polyethylene glycols [PEGs], polyacrylamide) have little propensity to precipitate out of aqueous solution and tend to form low-density, high-permeability polarization layers that have little effect on transmembrane flux and solute retention. In contrast, amphipathic proteins that tend to denature easily are far more likely to form dense, obstructive, polarization layers.

Similar considerations apply to the filtration of particulate suspensions, but with different consequences. Conditions that favor flocculation of such solids frequently lead to the formation of open-structured, porous layers, resulting in high permeation fluxes and low dissolved solute retention; those favoring stable dispersion tend to yield dense, low-permeability layers, resulting in low permeation fluxes and elevated dissolved-solute retention. Therefore, deliberate addition of flocculating agents to fine particle suspensions prior to UF or MF is frequently beneficial to flux performance and separation efficiency.

Suspensions comprising rigid colloidal particles (e.g., clays, pigments, other mineral solids) tend to form fairly rigid cakes of reasonably high hydraulic permeability, thus displaying high permeation fluxes in TFM. In contrast, suspensions comprising readily deformable or elastically compliant particles (e.g., hydrogel fragments, tissue homogenates, lysed microbial cells) will usually yield gelatinous, compressible layers of low hydraulic permeability, whose porosity and permeability decline very rapidly with compression. Tangential flow filtration of such suspensions is invariably associated with low filtration fluxes, filtration rates that frequently decline with increasing TMP, and a tendency to retain solutes that would otherwise freely permeate the membrane. Such "secondary membrane" formation on polarization is a major source of difficulty in recovering macromolecular products from fermentation broths and tissue culture media by TFM.

The Role of Membrane Surface Texture in Polarization

The preceding discussion of polarization dealt with the phenomenon solely in terms of the properties and characteristics of the layer of concentrate deposited on the membrane surface during filtration. So far, no role has been attributed to the membrane (other than its intrinsic retentivity to solute and permeability to solvent) in influencing the polarization process. The surface geometry of the membrane, however, can have a significant effect upon polarization.

For example, most asymmetric UF and MF membranes have barrier layers whose porosity is substantially less than that of the porous sublayer. Because of the extreme thinness of this barrier layer, however, low porosity will still yield a membrane of very high intrinsic hydraulic permeability. Electron microscopic evidence indicates that some asymmetric membranes have porosities as low as 1 percent of the barrier layer surface area, while others exhibit surface porosities of 10 percent or higher.

For low-hydraulic-permeability polarization layers, the layer thickness required to depress the permeation flux to a very small fraction of that measured for pure solvent is only a few micrometers; thus, polarization of a membrane with low surface porosity will have a far greater effect on flux than that of a high-surface-porosity membrane. For a given average permeation rate the local permeation velocity through the pores of a low-porosity barrier will be much higher than through a high-porosity barrier; this leads to a locally thicker polarization layer on the former, relative to the latter, membrane (see Figure 3.7). This is one explanation for the frequent observation that, of two UF or MF membranes with the same retentivity and nearly equivalent solvent permeability, one will suffer a far greater flux loss on polarization with a particular macrosolute or colloid than will the other.

Another important factor in polarization is the topography of membrane surface. If the membrane surface is smooth, flow in the feed channel will result in the full development of laminar flow near the surface, thus full polarization layer development. A rough or undulating surface (of the proper scale and periodicity), however, will induce flow separation and eddying (and thus enhanced back transport and reduced polarization) at lower liquid velocities than hydrodynamic theory would predict. Induced surface texture of this sort—with accompanying improvements in filtration flux—has been achieved with certain plate-and-frame module designs in which the membrane support plates contain periodic ridges that cause the membrane to conform to these irregularities. A recently developed ceramic/metallic composite MF membrane incorporating a wire mesh (Cowieson and Gallagher 1991) possesses similar surface topography; the unusually high permeation flux observed with this membrane on filtration of concentrated suspensions appears attributable to this flow-separation phenomenon.

Membrane/Solute Interactions and Membrane Fouling

The preceding discussion might convey the impression that the membrane is a passive element in the filtration process and that its composition, microstructure, and surface chemistry are of no importance to process performance. This is hardly the case; the membrane/solution interface is in many instances a dominant factor, determining whether a given separation can be accomplished by UF or MF or whether the filtration process will be economically viable.

Most UF and MF membranes are produced from low-water-absorptivity polymeric materials, which are relatively hydrophobic. The surfaces of such membranes (including the surfaces of the pores within the membrane microstructure as well as the external membrane surfaces) tend to be strongly sorptive for amphipathic molecules such as surfactants and for oleophilic substances such as water-immiscible oils. Many water-soluble macrosolutes (in particular, proteins and lipopolysaccharides) also possess amphipathic properties and tend to adsorb strongly (and irreversibly) onto such polymer surfaces. In addition, if the membrane surface contains ionic functions, it will also tend to adsorb and bind polyelectrolytes (or colloidal particles) of opposite sign of charge by electrostatic forces. Because proteins are polyampholytes, they tend to adsorb to charged surfaces irrespective of their sign of charge, even at pH levels far removed from their pIs.

If the adsorbing solute is too large molecularly to enter the membrane pores, adsorption will necessarily be confined to the upstream surface of the membrane. The extent of adsorption will depend on the concentration of adsorbate in the contacting solution. At high levels of adsorption, the layer so formed will both obstruct permeation of solvent and impede or block the passage of smaller solute molecules that—but for the presence of the adsorbed layer—would freely permeate the membrane. The net result of the adsorption process will be a decrease in hydraulic permeability of the membrane and a shift in the sieving curve of the membrane toward retention of smaller molecules.

The same results will be observed if the membrane is partially or completely permeable to the adsorbing solute, in which event adsorption will take place on the walls of the membrane pores as well as on the geometric membrane surfaces. In this case permeability loss and increased solute rejection is attributable to a reduction in the effective pore diameter. In any event the consequences of macrosolute adsorption by the membrane will be the following:

- Reduction in membrane permeability to solvent (process flux)
- Tightening of the membrane-rejection spectrum

Adsorption can, therefore, render the concentration of a particular solute by UF (or a particular suspension by MF) impractical or the separation of two macrosolutes of differing molecular size with a membrane of appropriate rejection profile impossible. Because the adsorptive affinity of a given membrane for a given macrosolute is a unique property of the solute and membrane, it is virtually impossible to predict initially whether a particular membrane will be suitable for use in concentrating or purifying a particular solute.

Most cases of macrosolute adsorption and particle-membrane adhesion involve hydrophobic or ionic interactions; electrostatically neutral, hydrophilic membranes are generally among the least adsorptive, so-called nonfouling structures. Among the more attractive of these is regenerated cellulose. Because highly hydrophilic polymers such as cellulose have limited utility as UF and MF membranes due to their low mechanical strength and chemical durability in aqueous environments, much effort has been expended over the past few years in developing chemical techniques for rendering the surfaces of refractory, hydrophobic polymeric membranes (e.g., polysulfone, polyvinylidene difluoride, polyacrylonitrile) hydrophilic and electrostatically neutral. These membranes have the dual advantage of low macrosolute (protein) sorptivity coupled with superior thermal, mechanical, and chemical durability; they are rapidly becoming the workhorses of TFU/TFM in the food, pharmaceutical, and aqueous-waste-treatment fields.

Adsorption is perhaps the most important factor in membrane "fouling"; it is not, however, the sole cause of the problem. All too frequently, when a dilute protein solution is submitted to concentration by TFU with a retentive membrane, the permeate flux undergoes a slow, continuous decline over days or weeks to unacceptably low values; efforts to restore the membrane to its original flux performance by even vigorous cleaning procedures are of limited utility.

Examination of the surface of such a fouled membrane reveals a dense, coherent, strongly adherent film of the protein. It appears that the processes of adsorption and concentration polarization initiate a cascade of protein denaturation on the membrane surface that renders the protein water insoluble, strongly adherent to the membrane, and resistant to chemical or enzymatic proteolysis. This time-dependent irreversible fouling can usually be ameliorated by using a nonsorptive membrane and performing the filtration at low TMPs and high feed-recirculation rates. The consequent reduction in initial permeation flux is more than compensated by a marked reduction in long-term flux decline.

The fouling problem is exacerbated when sterilizing or clarifying protein solutions by TFM using membranes that should be freely permeable to such macrosolutes. After a relatively short period of operation at high permeation fluxes, yielding clarified sterile filtrate without loss of protein, a rapid decay of permeation flux can result, with the production of a permeate markedly depleted in protein. Similar results are often observed with efforts to recover, by TFM, a protein present in the supernatant solution of a fermentation broth or cell-culture suspension. Examination of the fouled membrane reveals that the membrane pores are virtually blocked by precipitated protein—a supposedly impossible consequence with a membrane freely permeable to that protein. These effects appear again to be the consequence of a polarization → precipitation → denaturation "cascade," in this case occurring within the membrane pores (Zydney 1992). A microparticle in the feed solution enters a pore and lodges in a downstream pore constriction. The microparticle then restricts passage of the soluble protein, which then polarizes within the pore, precipitates, and slowly denatures. The precipitated protein can then retain dissolved protein entering from adjacent pores, repeating the process. Ultimately, a large fraction of the total pore volume becomes obstructed by precipitated, denatured protein. Here again, the cascade can often be averted by carrying out the filtration with a nonsorptive membrane at low TMPs and performing periodic backwashes of the membrane with permeate to displace accumulated particulate solid from the membrane.

Product Recovery by Diafiltration: Effects of Polarization and Fouling

In many practical situations, UF or MF is employed either (1) to recover, from a suspension or macrosolute solution, a product of lower molecular weight than most of the components present (e.g., recovery of an antibiotic from a fermentation broth or of a recombinant oligopeptide from a mammalian cell-culture medium) or (2) to remove from such a suspension or solution low-molecular-weight solution components (e.g., desalting a protein solution or removal of culture medium components from a mammalian cell suspension). The standard method for this is diafiltration, where the retentate is continuously recycled and diluted with fresh wash solution to replace that removed as permeate. Rapid and efficient product recovery by diafiltration requires virtually unimpeded passage of the desired product into the permeate.

Gel-layer polarization or membrane fouling can thwart such a separation by either reducing permeation flux to impracticably low levels or, more seriously, by obstructing passage of the product into the permeate. For diafiltration applications of these types, carrying out the filtration with minimum possible polarization at low TMPs and fluxes is the most practical solution (van Reis et al. 1992). Here, maximizing liquid shear at the upstream membrane surface (to maximize mass transfer) while minimizing TMP drop has particular merit. Vortex flow filtration (Müller and Kroner 1992) is one answer to this requirement. Recent evidence also suggests that pulsed electric fields applied to the polarization layer can effectively depolarize a membrane, with marked enhancement in steady-state permeation flux and clearance of membrane-permeable solutes (Siegel et al. 1992). This method may prove to be another solution to this important practical problem.

A temptation exists to equate diafiltration with dialysis; although in reality the processes differ significantly only on a microscopic scale, the results achieved using the two processes are likely to be substantially different. In dialysis every membrane-permeable solute is amenable to removal or recovery in the dialysate without altering the concentration or otherwise affecting the membrane-impermeable components. In diafiltration polarization at the membrane surface causes marked local increase in the concentration of retained solutes, which can cause irreversible aggregation and precipitation of those components. The polarization layer of retained solutes or colloids can impede or prevent passage of solutes that would otherwise clear the membrane. The higher the TMP during diafiltration, the greater the likelihood of these consequences. These factors should be kept in mind when considering diafiltration for such separations and when selecting equipment and operating conditions for performing the process.

Polarization and Fouling in Perspective

In summary, retained-solute or retained-particle polarization during TFU and TFM is an inescapable consequence of these pressure-driven membrane separation processes and has generally adverse effects (e.g., flux depression, poorer solute retention) upon process performance. With macrosolutes or colloids of limited stability, polarization can lead to irreversible aggregation, precipitation, and/or denaturation, with significant losses in product yield. Polarization can, however, be mitigated (but never eliminated) by rational membrane module design, proper liquid-flow management, control of TMP drop, and appropriate membrane selection.

Membrane fouling phenomena have much greater impact on UF/MF performance. Macrosolutes or colloids undergo irreversible adsorption on membrane surfaces, depressing membrane permeability and altering sieving properties, then gradually precipitate and accumulate upon the membrane surface to form dense, adherent, secondary transport barriers of very poor solvent transmissivity. Fouling is a particularly serious problem with relatively hydrophobic membranes when exposed to amphipathic or oleophilic macrosolutes or colloids or with ionically charged membranes when exposed to macrosolutes and colloids that carry charges of opposite sign.

Membranes with highly hydrophilic uncharged surfaces are least susceptible to fouling by these adsorptive mechanisms. However, microporous membranes, even if highly hydrophilic, are prone to irreversible fouling by macrosolutes that are normally too small to be retained, unless the solution to be filtered contains even trace amounts of submicroscopic particles that can obstruct pores and cause macrosolute polarization and precipitation within the membrane. As a rule, operation of such filtration processes at high liquid shear rates and low TMP drops will minimize fouling of this nature.

TANGENTIAL FLOW FILTRATION EQUIPMENT

Membrane Types and Materials

Historical Background

A broad variety of natural membranes have been performing intricate and highly selective transport functions for centuries. Synthetic membranes, however, are very recent—less than 60 years old—with the vast majority of commercial developments having occurred since 1970.

The first laboratory versions of what are currently known as MF membranes were developed in Germany during World War II. Engineering scaleup, however, occurred later in the United States; these membranes were developed and commercialized in the 1960s. Asymmetric UF membranes came next, in the 1970s; interfacially polymerized composite RO and nanofiltration (NF) membranes are the most recent, having been commercialized in the 1980s.

The proliferation of membrane types, materials, and manufacturers has been so rapid that it would be virtually impossible to list all the membrane manufacturers and their offerings. This section will highlight only the types of membranes and membrane materials most widely used.

Definitions

The origins of the terms *microfiltration*, *ultrafiltration*, *nanofiltration*, and *reverse osmosis* are probably more historic and application-oriented rather than a systematic classification of membranes based on pore size. Consequently, there are no rigid definitions for these membranes, just qualitative descriptions broadly accepted by the membrane community.

Microfiltration membranes are designed to remove suspended solutes, generally ranging in size from 0.05 μm to about 5 μm, from liquids. The most widely used MF membranes are homogeneous, but there are some skinned-asymmetric or anisotropic membranes on the market.

Ultrafiltration membranes are invariably asymmetric and cover the pore size range from about 0.001 μm to about 0.05 μm, although there can be some overlap with NF on the small pore size end and with MF on the large pore size end.

Nanofiltration membranes are also highly asymmetric and bridge the gap between UF and RO. Nanofiltration membranes originated as an offshoot of RO membranes and, therefore, do not separate species on the basis of size alone.

Reverse-osmosis membranes are also highly asymmetric and were designed originally for rejecting dissolved ions from water. They do not necessarily separate species on the basis of size; some recent variants of RO membranes are designed for the separation of organic molecules rather than ions.

Membrane Materials and Formation Processes

Materials used to make commercial membranes can be broadly classified into either organic polymeric or inorganic materials. Although the overwhelming majority of current commercial membranes are organic polymeric, the use of inorganic membranes is increasing as this technology advances.

Organic polymeric membranes are formed by one of the following processes:

- Phase inversion (vapor induced, liquid induced, or temperature induced)
- Sintering and stretching
- Track etching
- Thin-film composites (interfacially polymerized or thin coatings)

Inorganic membranes are generally made from ceramics using sintering and/or sol-gel techniques.

Membrane Geometries

Membranes are manufactured in one of the following three geometric forms:

- Flat sheet
- Hollow fiber
- Tubular

Flat-sheet membranes are the most versatile and can be fabricated into a variety of devices, some of which will be described in the next section.

Hollow fibers have the advantage of high packing density (membrane area/unit volume), hence their devices are quite compact. However, hollow fibers suffer from high pressure drop in the lumen and often have lower burst pressures than flat-sheet membranes because hollow fibers are unsupported.

Tubular membranes have very low packing densities, higher costs per unit membrane area, and higher pumping requirements. They are used for a small number of specialized applications in which the feed streams are either very viscous or have heavy suspended matter loads.

List of Commercial Membranes

A selected list of popular commercial membranes is presented in Table 3.1. For more exhaustive lists of MF, UF, and RO membranes, please refer to a recently published handbook (Ho and Sirkar 1992).

Device Types

The term *filtration device* refers to the fabricated product that contains the membrane and allows it to function in a separation. For membranes to be useful, they must be sealed and contained in a device. Internal sealing is required for the separation to be reliable; external sealing ensures containment. The device also provides support for the membrane and, thereby, protection from damage caused by hydraulic and thermal stresses.

A wide variety of TFF devices are available and can be categorized by configuration. There are five basic configurations—stirred cell, flat plate (plate and frame), spiral, tubular, and hollow fiber. Devices can be further characterized as conventional TFF devices or enhanced TFF devices, based on whether depolarization is managed primarily by moving the feed liquid over the membrane or by other means.

Devices have relative advantages or disadvantages primarily due to configuration combined with materials of construction. For example, certain flat-plate devices may be solvent resistant as a result of their materials of construction. The three most important factors to consider when selecting a TFF device are reliability, performance, and cost.

The TFF manufacturer must consider several important factors during development. Discussion of these factors, provided below, offers some insight for the selection of the device by the TFF user. Table 3.2 summarizes each device type and its general characteristics.

Table 3.1. Selected Commercial Membranes

Type	Material	Geometry	Features
Microfiltration	Mixed cellulose esters	Flat sheet Hollow fiber	High permeability Poor chemical resistance Very brittle
	Modified polyvinylidene fluoride	Flat sheet	Medium permeability Good chemical resistance Extremely low protein binding
	Nylon 66	Flat sheet	Medium permeability Very good chemical resistance Very high protein binding
	Modified polysulfone	Flat sheet	High permeability Good acid-alkali compatibility Low protein binding
	Modified polyethersulfone	Flat sheet Hollow fiber	High permeability Good acid-alkali compatibility Low protein binding
	Polytetrafluoroethylene	Flat sheet Tubular	Excellent chemical resistance High protein binding Very hydrophobic
	Polypropylene	Flat sheet Hollow fiber	Very good chemical resistance Poor oxidative stability High protein binding Hydrophobic
	Polyethylene	Flat sheet Hollow fiber	Similar to polypropylene except lower temperature and better oxidative stability

Continued on next page

Continued from previous page

Type	Material	Geometry	Features
Micro-filtration (cont.)	Nylon 6	Flat sheet Hollow fiber	Similar to Nylon 66
	Cellulose acetate	Flat sheet	Poor chemical resistance Low protein binding
	Polycarbonate	Flat sheet	Track-etched membranes Low permeability
	Alpha alumina Zirconia Silicon carbide	Tubular	Ceramic membranes Excellent resistance to organic chemicals Good acid-alkali compatibility Extremely high temperature resistance
Ultra-filtration	Cellulose acetate	Flat sheet	Low permeability Hydrophilic Poor chemical resistance Low protein binding
	Regenerated cellulose	Flat sheet	Low permeability Hydrophilic Good chemical resistance Very low protein binding
	Polysulfone	Flat Sheet Hollow fiber Tubular	High permeability Excellent alkali compatibility Poor organic solvent resistance High protein binding
	Polyether-sulfone Modified polyether-sulfone	Flat sheet Hollow fiber Tubular	Similar to polysulfone with slightly better organic solvent resistance

Continued on next page

Continued from previous page

Type	Material	Geometry	Features
Ultra-filtration (cont.)	Polyvinylidene fluoride Modified polyvinylidene fluoride	Flat sheet Tubular	Moderate permeability Good chemical resistance
	Gamma-alumina	Tubular	Low permeability Excellent resistance to organic solvents Poor acid-alkali compatibility Very high temperature resistance
Nano-filtration	Polyamide (interfacially polymerized)	Flat sheet	High permeability Good chemical resistance Poor oxidative stability Limited alkali compatibility
Reverse osmosis (asymmetric)	Cellulose acetate Cellulosic blend Cellulose triacetate	Flat sheet	Medium permeability Moderate rejection Moderate chlorine tolerance Poor organic solvent resistance Poor acid-alkali compatibility
	Aromatic polyamide	Hollow fiber	Low permeability Poor chlorine tolerance Moderate to high rejections Good pH stability Moderate organic solvent resistance
Reverse osmosis (thin film composites)	Polyamide (interfacially polymerized)	Flat sheet	High permeability Poor chlorine tolerance High rejections Good pH stability Poor organic solvent resistance (limited by substrate)

Table 3.2. General Characteristics of Different Tangential Flow Devices

Device Type	Device Characteristics						
	Reliability	Flux Performance	Cost	Compactness	Ease of Use	Scalability	Containment
Stirred cells	H	L	L	L	H	P	G
Flat plate: molded	M/H	M	M	M	M/H	G	A
Flat sheet (cassettes)	H	H	M	H	H	E	A
Spiral wound	M	M	L	M	M	G	E
Tubular	M	M	H	L	M	G	G
Ceramic	M	M	H	L	M	G	E
Hollow fiber	L	L	L	H	H	G	E
Enhanced TFF	L	H	H	L	L	P	A

KEY
H = high; L = low; M = moderate; P = poor; G = good; E = excellent; A = adequate

Stirred Cells

Stirred cells (Figure 3.8) typically consist of a cylindrical pressurized vessel with a porous support on the bottom. The vessel contains the liquid to be filtered and the porous support accepts a membrane disk that is sealed with O-rings. A magnetically coupled stirring bar depolarizes the surface of the membrane by stirring the liquid above the membrane. Stirred cells have been available for many years. Although this configuration is most often used for UF, high-pressure RO stirred cells are also available.

The performance of a stirred cell does not scale up predictably to large devices due to the variable and often poor depolarization

Figure 3.8. Stirred cells. (Courtesy of Amicon, Inc., Danvers, MA)

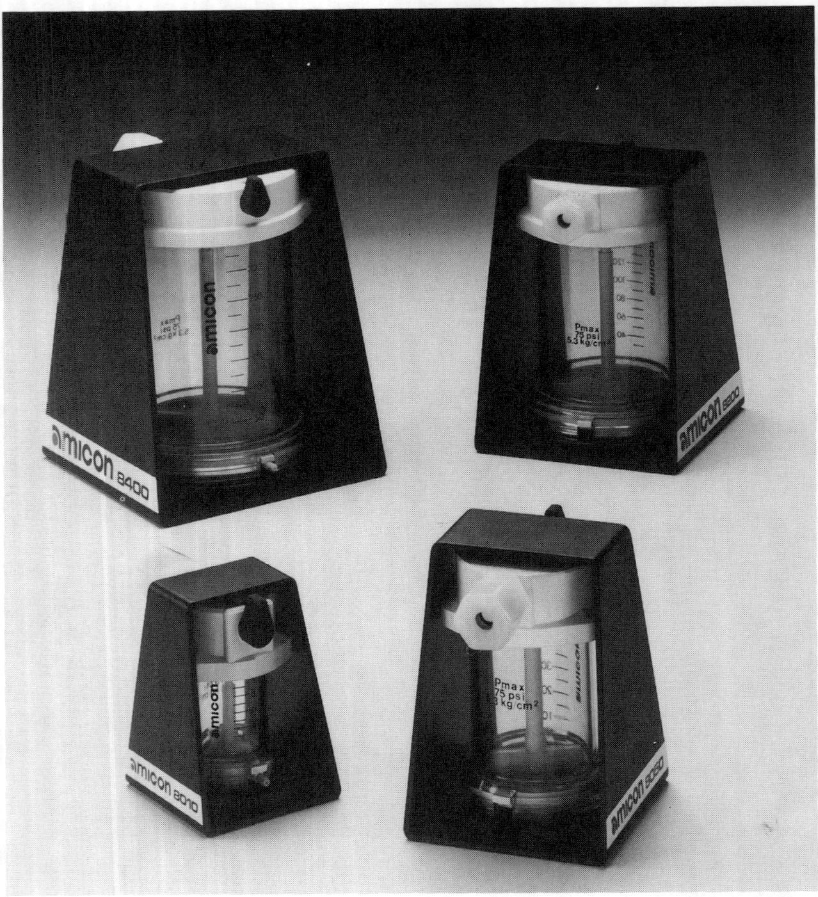

provided by the stirring bar. However, stirred cells provide a convenient and cost-effective way to concentrate and diafilter small volumes of samples in the laboratory. Stirred cells are limited in volume and provide fluxes lower than those of other TFF devices.

Flat-Plate and Flat-Sheet Devices

There are two different styles of flat devices (Figure 3.9). So-called flat-plate devices (Figure 3.9a) are characterized by the use of a molded support plate that may be rectangular or circular. The membrane must be sealed to the plates reliably. The plates are stacked with or without screens in the feed channels. Screened channels provide enhanced

Figure 3.9a. Flat plate TFF device.

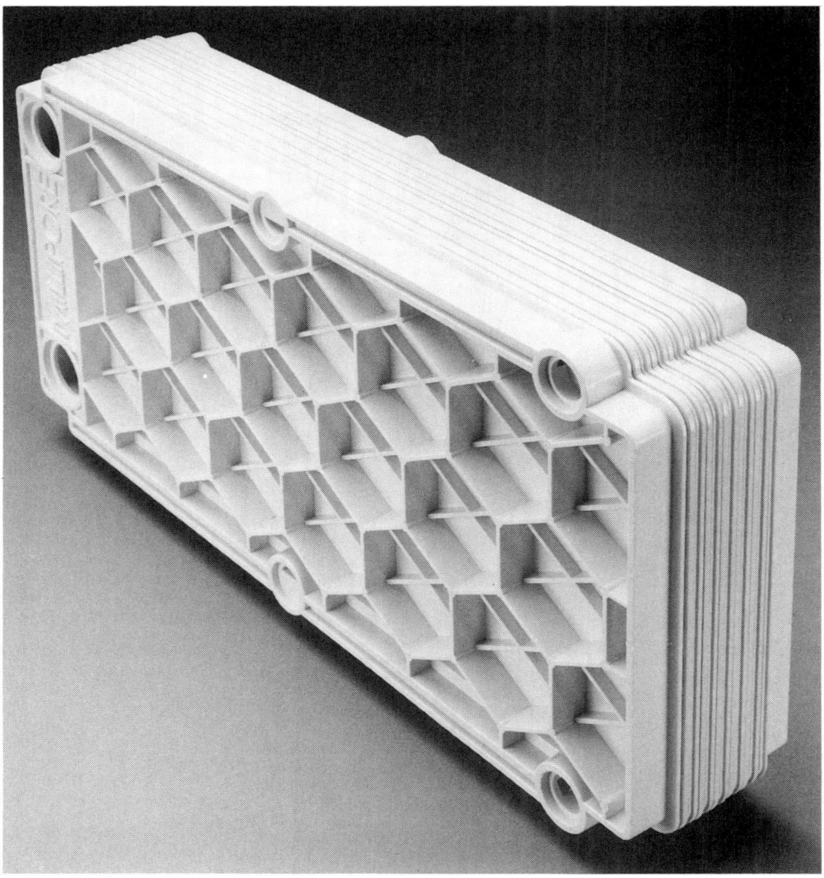

depolarization and higher fluxes at lower cross-flow rates, suitable when filtering clear solutions. In contrast, channels without screens (so-called "open channels") are less prone to plugging if suspended solids are present in the feed stream.

Seals between the plates can be elastomeric, requiring compression, or thermoplastic (thermally, vibrationally, or ultrasonically welded). Manifolding for the feed/retentate channels is provided by ports in the plates. Permeate may be individually removed from each plate or may be manifolded for central collection.

A typical flat-sheet device (Figure 3.9b) is a cassette in which membranes and screens are stacked with an appropriate glue pattern for internal and external sealing. Cassette devices lack a rigid plate with which to support the membrane; instead, flexible screens separate each sheet of membrane in the device from the adjacent sheet.

Figure 3.9b. Flat sheet TFF device.

Manifold holes are located on the screens and on the membranes. These holes collect retentate and permeate and distribute feed.

The flat-plate or flat-sheet configuration is most often used for microporous and UF membranes, although a RO plate-and-frame device suitable for high-pressure operation is also available. Unlike the flat-sheet device, the molded flat-plate design is often less compact and may have many parts with elastomeric seals. Assembly and reliable sealing can be complex and difficult with some designs. However, some flat-plate designs with molded plates (e.g., Prostak™ modules, Millipore Corp., Bedford, MA) have thermoplastic seals, enabling the modules to be installed as integrated blocks and significantly limiting the number of seals and parts to be installed.

In contrast, cassette-type flat-sheet devices can be very compact, but are prone to sealing inconsistencies and performance variation induced by the design of their holders and how much the devices are compressed by the holders from run to run.

Other considerations for flat-plate devices include channel deflection when pressurized and channel dimensional consistency. Deflection needs to be minimized by proper holder design. Channel height control—defined by the plate-to-plate interaction and distance—needs to be carefully considered in the design of the filter and holder. Tightness of the screen against the membranes in the channels of screened-channel devices is affected by compression and other design aspects and can significantly affect performance and process consistency.

Spiral-Wound Cartridges

A spiral-wound cartridge (Figure 3.10) consists of multiple layers of folded membrane, feed screen, and permeate screen wound around a center permeate collection tube. Glue is applied to three edges of the folded membrane, permitting permeate flow into the center tube via the permeate screen. This configuration is most often used for UF and RO membranes. Containment is reliable due to the use of a sealed cylindrical housing (pipe) with end caps. Spiral-wound cartridges are compact and typically cost much less per membrane area than flat-plate, flat-sheet, and tubular configurations. Performance depends on screen selection, amount of bypass flow (through the annular space outside the cartridge and inside the housing), flow design, and membrane area control (glue application), as well as operating conditions. Spiral-wound cartridges typically contain thermosetting adhesives and are more difficult to manufacture reliably than other devices due to their mechanical complexity. Some large-area spiral-wound cartridges also have nonsanitary areas that are not swept by the feed or permeate flow, making them difficult to apply to pharmaceutical applications.

Tubular Elements

Tubular membrane elements (Figure 3.11) are available with inside diameters varying from 0.3 cm to 2.5 cm. Tubes are manufactured by casting a membrane onto the inside or outside diameter of porous support tubes.

In a broad sense there are two types of tubular elements. The first is polymeric, in which the porous support tube is either an epoxy-coated knit material or a nonwoven material. Tubular elements consist

Figure 3.10. Spiral-wound cartridge.

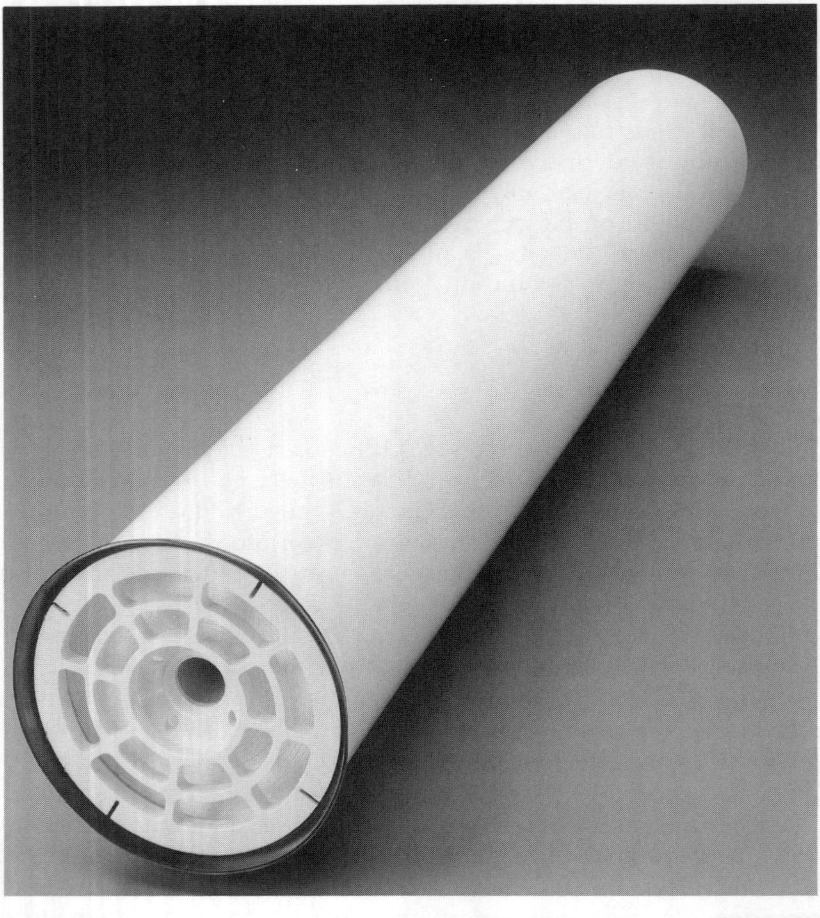

Figure 3.11. Tubular ceramic TFF device.

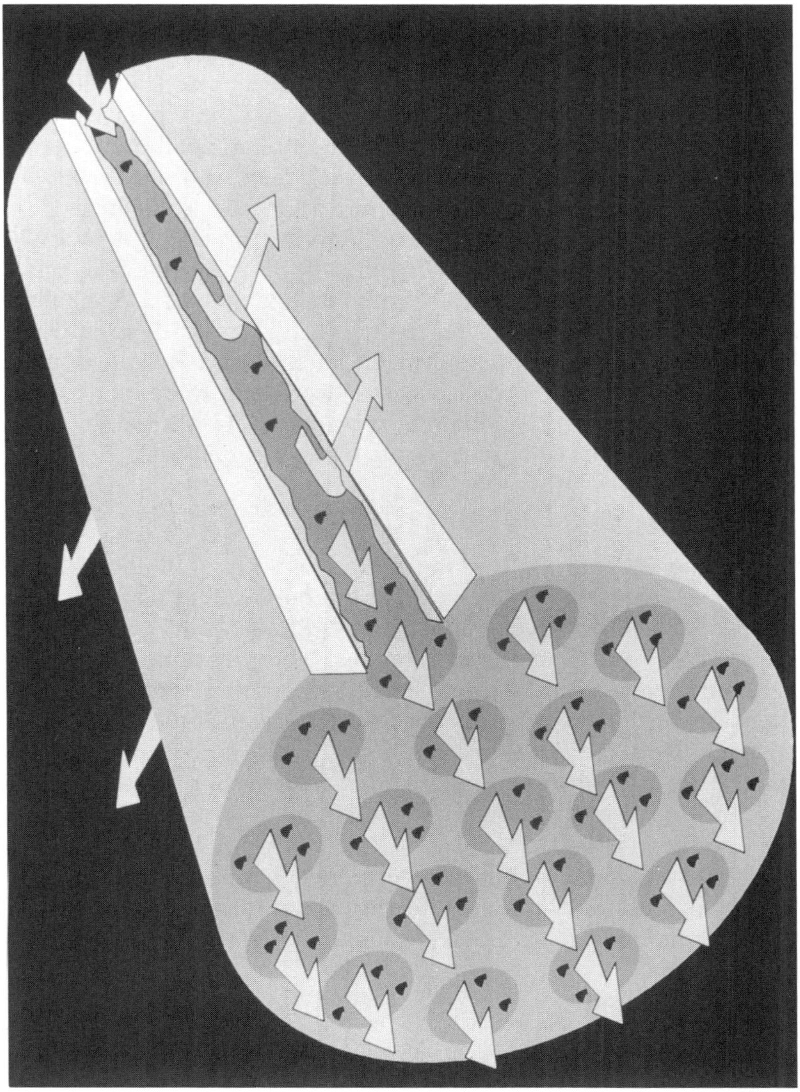

of single or multiple tubes sealed in a cylindrical housing. Seals are either elastomeric or adhesive (potting). Tubes are available with UF and RO membranes. This configuration has very limited resistance to reverse pressure (permeate pressure greater than feed pressure) due to the lack of membrane support on the feed side. With reverse pressure the tube will collapse or the membrane will delaminate.

The second type is a ceramic tubular membrane, in which a ceramic monolith has multiple tubular lumens running axially along its length. The ceramic membrane is slip cast onto the internal lumen surface of the monolithic support. Lumens range in size from 2 mm to 6 mm in diameter. Ceramic devices can withstand high permeate back pressure (including backpulsing) due to the strength of the sintered ceramic material. Both UF and microporous membranes are available. Membrane materials are generally alumina, titania, or zirconia.

Tubular devices are excellent for filtering process streams with suspended solids. Cleaning can be very aggressive, including physically scouring the surface of the membrane (e.g., with sponge balls in the case of 0.5–1.0-inch-diameter tubes). Cost is generally high due to individual casting of the membrane on the tube and lack of compactness due to large open feed channels. Fluxes tend to be low because tubes are not as efficiently depolarized by normal tangential flow rates as are screened-channel or narrow, open channel devices.

Hollow Fiber

Hollow fibers (Figure 3.12) are narrow-diameter tubes that are unsupported (e.g., membrane fibers). Very fine hollow fibers are typically used for RO, while larger-diameter fibers (0.3–2 mm) are typically used for UF or microporous membranes. Fibers can have membrane on the internal surface, external surface, or both, or can be isotropic with equal porosity throughout the wall thickness. Some two-sided or isotropic fibers provide superior fouling resistance and strength. Traditionally, hollow fibers have had low operating pressure limits (e.g., 175 kPa); recently, however, hollow fibers with higher pressure units have been introduced.

Fibers are typically bundled and potted with an adhesive into a plastic shell with the appropriate feed/retentate and permeate fittings. The potting/fiber interface is a high-stress area and is susceptible to fiber breakage.

This configuration is very compact and low cost. Fluxes are generally lower than those of screened-channel devices because hollow fibers operate in a laminar-flow regime. The shell side can have poor, variable flow distribution leading to cleaning difficulties. Some fibers can withstand back pressure and can thus be backflushed with permeate as a cleaning technique.

Enhanced Tangential Flow Devices

A variety of enhanced TFF devices (Figure 3.13) is now becoming more common. These devices use mechanically induced forces instead of

Figure 3.12. Hollow fiber module.

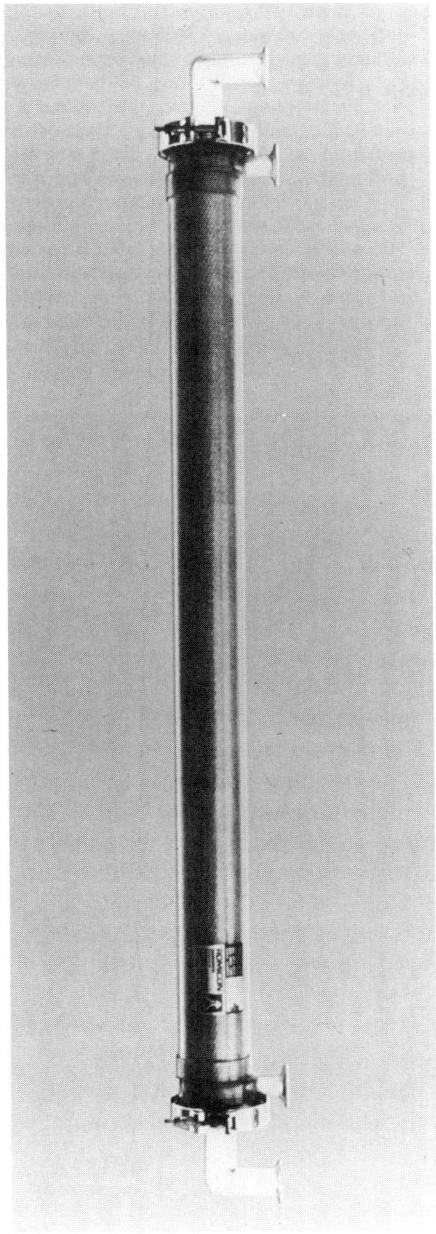

Figure 3.13. Schematic of enhanced TFF device.

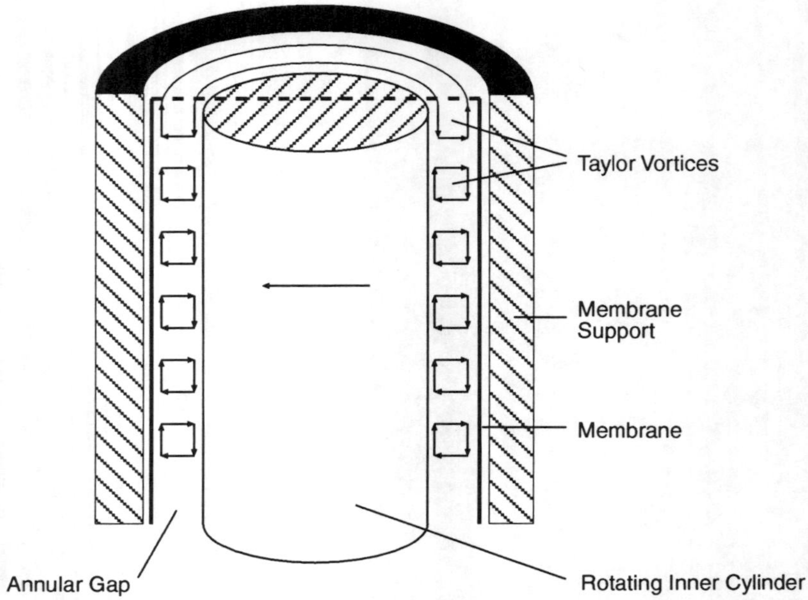

tangential flow to depolarize the surface of the membrane. One version uses a spinning cylinder adjacent to a concentric fixed cylinder. The membrane is mounted on one of the cylinders. The spinning cylinder creates local Taylor vortices above the membrane surface that effectively depolarize at low tangential flow rates.

A second type uses spinning stacked disks with fixed disks in between the spinning disks, similar to a stack of stirred cells in parallel. As with stirred cells, the spinning disk creates turbulence at the membrane surface.

A third version uses a stack of disks, but uses a torsional vibrating motion of the stack to generate shear and turbulence at the membrane surface.

All enhanced TFF devices have the primary advantage of significantly reduced pumping (low tangential flow rates) requirements, since the turbulence for depolarization is created by a secondary means. Additionally, the promotion of turbulence is decoupled from the operating transmembrane pressure, unlike normal TFF devices. Well-proven large-scale systems using these devices do not yet exist, and scaleup of these devices is more complex than scaleup of normal TFF devices because their depolarization behavior is controlled by a greater number of variables. Also, many enhanced TFF devices do not adequately support the membrane at low TMPs, allowing the membrane to lift and tear or delaminate when operated below 35 kPa.

Selection Criteria for Tangential Flow Devices

The three most important factors to be considered in selecting a TFF device/system are reliability, performance, and cost. Reliability is generally a result of good design and good manufacturing practices, coupled with an understanding of the application. Reliability can be improved for all devices; adhesives that are not qualified and matched to the needs of the application can cause problems, as can stressing the membranes beyond the design capability.

Performance depends on more than just the flux from the device. First, performance means that the device can achieve the needed separation by retaining or passing the desired species. Second, flux must be normalized to account for the pumping required to generate it. A good technique for assessing the efficiency of a device is to analyze the flux vs. the conversion (permeate flow/feed flow). Figure 3.14 shows this type of analysis for two different devices. Device 1 has higher performance. The flux is permeate-flow normalized for membrane area and the conversion is permeate-flow normalized for feed flow. This type of analysis takes into account the membrane area and flow (pumping) requirements for the particular device. These two factors are of primary importance in balancing the economics of a system. In other words, high fluxes (low membrane area) can be obtained by high feed flows (large pumps and piping). On the other hand, the pumping can be minimized by using large amounts of membrane area (low fluxes). There is an economic optimum that balances these two factors.

The overall operating cost of a TFF system always includes the cost of the filter device. However, operating costs for some applications may be dominated by capital costs, energy costs, or maintenance costs, rather than by the membrane replacement cost alone. Spiral-wound cartridges and hollow fibers are the lowest-cost devices, as the membrane is made in a continuous mode and the device can incorporate a large amount of membrane area in a one-step winding or bundling/potting operation. Tubular devices are higher cost, since the membrane is produced on a piece basis and large membrane areas require multiple tubes. Flat-plate devices are intermediate in cost—they use a continuously produced membrane, but require multiple manufacturing steps to obtain large-area modules.

In selecting a device, it is important to consider other important design factors encountered during development. These factors provide insight into the limitations of device type.

Figure 3.14. Plotting flux as a function of conversion to compare efficiencies of TFF devices.

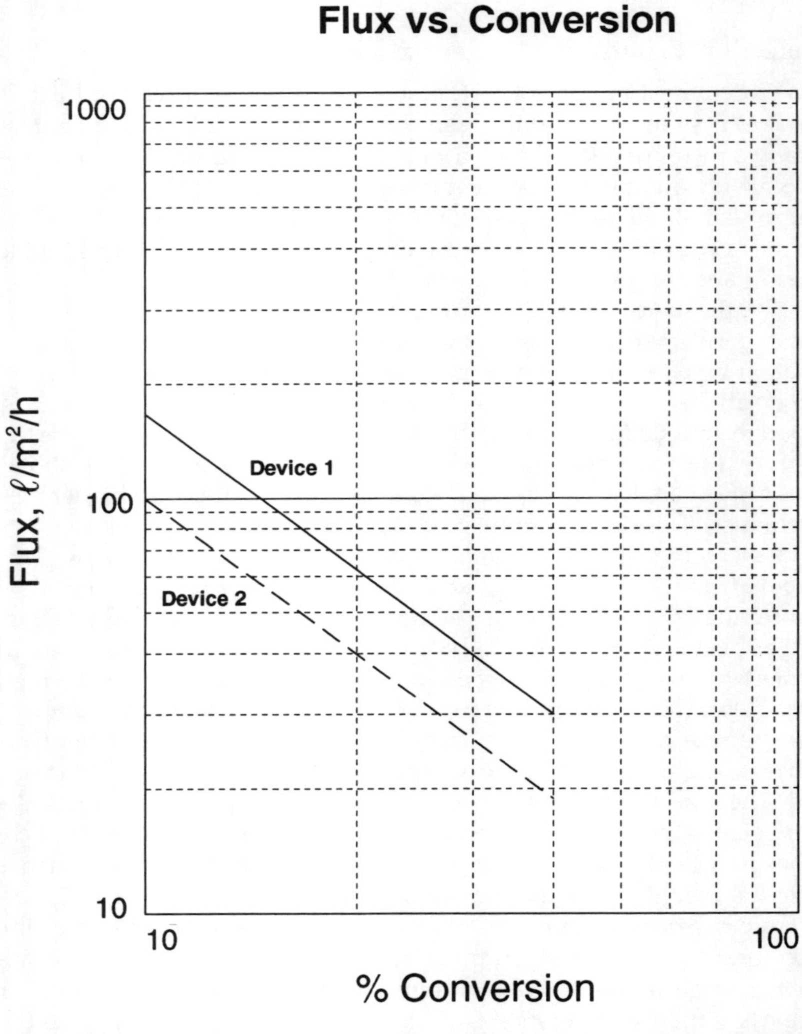

Complexity

In general, increased complexity leads to poorer reliability. For example, enhanced TFF devices have significant mechanical complexity, which complicates the design of large-scale systems and could lead to premature failure of components. On the other hand, a device that consists of hollow fibers potted into a housing has relatively few parts and is quite simple.

Chemical Compatibility

Devices that have a single material of construction are preferable. The selected material should be resistant to attack from process streams and cleaning chemicals. The membrane should limit the type of applications that can be served, not any other device components. Adhesives (polyurethane, epoxy) generally limit chemical compatibility and must be validated for compatibility with a given application.

Durability

The device must survive the various stresses to which it will be exposed in its applications. These stresses include normal operating conditions, cycling of temperatures and pressures, chemical cleaning, permeate back pressure, high cross-flow, deflection of holders, and vibration. Appropriate testing should be done to validate the use of the device for the application.

Cleanability

Tangential flow devices are cleaned on a regular basis to remove fouling material from the membrane and to restore performance. The design or configuration of the TFF device can enhance cleanability. Stagnant areas should be avoided in the design. Open tubular devices can be scoured mechanically, if necessary. Ceramic devices can be backflushed or backpulsed to improve cleaning effectiveness. Some devices are more chemically resistant and can, therefore, be cleaned more aggressively.

Feed-Channel Design

In many cases flux is controlled by the feed-channel design. For open channels the height of the channel is important. For screened channels the characteristics of the screen are important; the compression of the screen in the channel can have a significant effect. The density and thickness of the screen must be optimized for the particular application.

Feed-channel length is an important determinant of operating condition control. A longer path results in different operating conditions from feed end to retentate end, yet requires a lower tangential flow rate for a given total membrane area, which can reduce the cost of the system significantly.

The user must assess all of these factors before selecting the device for a particular application.

Components in Tangential Flow Filtration Systems

A TFF system requires a minimum number of components to operate and control it effectively. The set and type of components, however, depend on the process mode selected, which in turn depends on the application.

The application classes of interest are matched with recommended process modes in Table 3.3. This table provides a guide to which process modes are typically chosen for each application class. The fit presented in the matrix will become clearer once each process mode is described.

This section discusses each of the five TFF process modes listed in Table 3.3. Preparation of TFF systems for operation (e.g., priming and pump startup) or shutdown (e.g., draining, cleaning, and sanitizing) are not discussed here. Therefore, all diagrams are "process schematics" that specifically omit drain lines and any additional lines or components needed to fill and clean the system.

Consult your TFF supplier for more detailed advice on the system design that will best suit your application.

Type I, or Simple, Batch Process

Figure 3.15 shows flow schematics for a simple batch process. This is the simplest, most versatile TFF process, for many years preferred for all small-scale TFF processes (less than 10,000 liters per day). It can be run in either concentration or wash mode as shown in Figures 3.15(a) and 3.15(b), respectively.

The type I batch process is well suited for protein concentration and buffer exchange due to its simplicity and low holdup volume. A typical protein concentration and buffer exchange process has three steps:

1. Initial concentration (the batch is concentrated to the protein concentration best suited for washing[1])
2. Wash (while holding the concentration and, therefore, the batch volume, constant)

---

[1]The flux of a TFF module decreases monotonically with increasing protein concentration until the "zero flux" concentration (c^*) is reached (at which the flux approaches 0). An analysis of the situation reveals the existence of an optimum concentration for washing: If the solution is too dilute, large volumes of wash solution are required to accomplish the job, leading to a large filtration area; on the other hand, if the solution is too concentrated, the flux is very low, again leading to a large filtration area. The optimum wash concentration is $\approx {}^1/_3\, c^*$.

Table 3.3. Application vs. Process Mode Matrix Showing the Best Fit

Application	Process Mode				
	Type I Batch	**Type II Batch**	**Type III Batch**	**Feed & Bleed**	**Multi-stage**
Protein clarification	OK to use	—	Most reliable control	—	—
Mammalian cell harvest	OK to use	—	Most reliable control	—	—
Bacterial cell harvest	Best for <10 kiloliters	Best for >10 kiloliters	—	—	—
Protein concentration and diafiltration	Most common	—	—	Ok but not typical	Low residence time

3. Final concentration (to reach the desired concentration)

This process is also the best choice for small-scale bacterial cell harvest, which comprises at most two steps:

1. Concentration (of cell mass)
2. Wash (to remove contaminants in the supernatant)

In general, this process is a good choice when any one of the following conditions are met:

- The filtration conversion[2] is relatively high (10–25 percent).
- The scale is relatively small (≤10,000-liter batches).
- Accurate control of the permeation rate is not required (when accurate control is required, the type II batch process is the best choice).

[2]The "conversion" (*E*) is defined as the fraction of the feed stream that is filtered in one pass. Specifically, E = (permeate flow rate/circulation flow rate) $\times$ 100 (to express it in percent).

Figure 3.15. Flow schematics: Type I batch TFF processes.

(a) In concentration mode—permeate is withdrawn continuously until the desired concentration is reached.

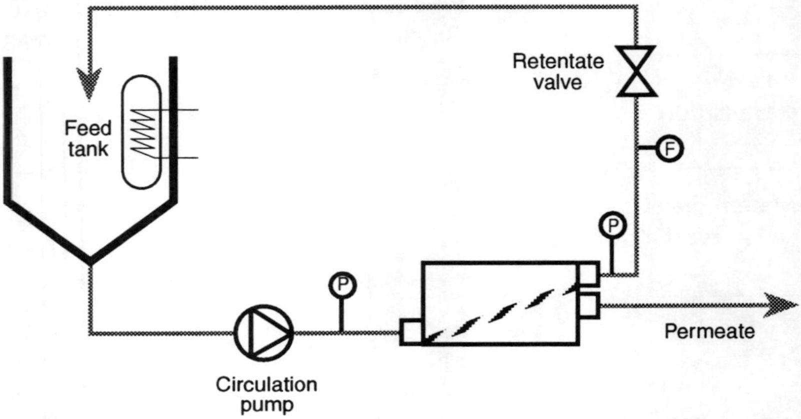

(b) In wash mode—wash solution is added simultaneously as permeate is withdrawn, usually at the same rate.

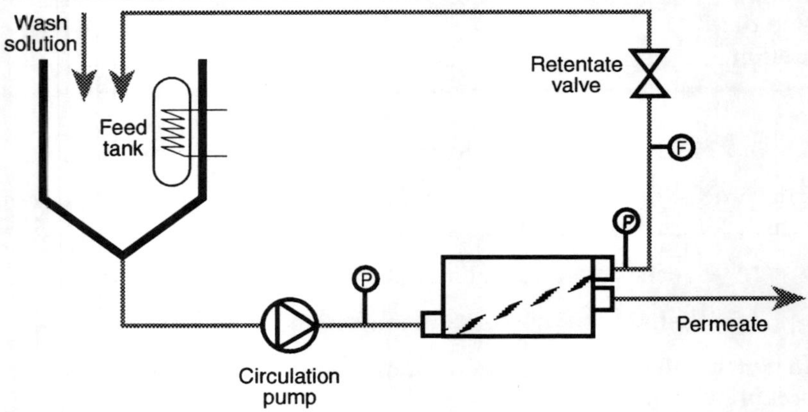

A TFF system designed to run a type I batch process requires at least four components: the feed tank, the circulation pump, the TFF module, and the retentate valve. It is good practice to also install a heat exchanger (for cooling) with a temperature sensor and control[3], feed and retentate pressure sensors, and a circulation flow meter.

[3]Even though it is shown here in the feed tank, a preferred location for the heat exchanger in on the retentate line downstream of the retentate valve.

Figure 3.15 is a diagram for a system designed to run manually; the user sets the circulation flow rate and the retentate pressure by appropriate adjustments of the circulation pump and the retentate valve; the user operates the system manually to switch from one step to another (the user is the controller). However, it is becoming increasingly popular to reduce operator intervention. At one extreme, a fully automatic system can sequence through all process steps according to a predefined set of parameters (initial concentration factor, wash volume, final concentration factor). Automatic processes require more sensors in addition to the controller itself, however.

Table 3.4 describes the types of key components most often selected for pharmaceutical processing in TFF systems.

Type II Batch Process: Decoupled Circulation Loop

Figures 3.16(a) and 3.16(b) show flow schematics for the type II batch process, in which the circulation loop is decoupled from the feed tank. According to this design, only a fraction of the retentate goes back to the feed tank; the remainder is sent directly to the circulation pump.

Table 3.4. Recommended Components for Tangential Flow Systems

Component	Recommended Type(s)	Comments
Circulation pump	Rotary lobe	Most popular due to "low" shear
	Centrifugal	Good choice for many proteins
Retentate valve	Diaphragm	Sanitary Easy to adjust
Pressure sensors	Isolated (mechanical) or electronic)	Sanitary Easy to adjust
Flowmeter	Turbine Magnetic	Good choice for proteins Best for cell harvesting
Tank	Smooth, conical bottom	316L stainless steel or polyolefin
Heat exchanger	Shell and tube	

Figure 3.16. Flow schematics: Type II batch TFF processes.

(a) In concentration mode—of multiple flow-ratio-control schemes possible, this is the most reliable.

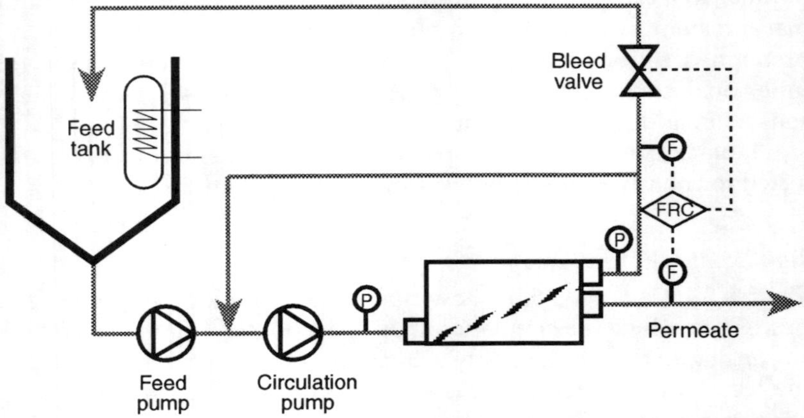

(b) In wash mode.

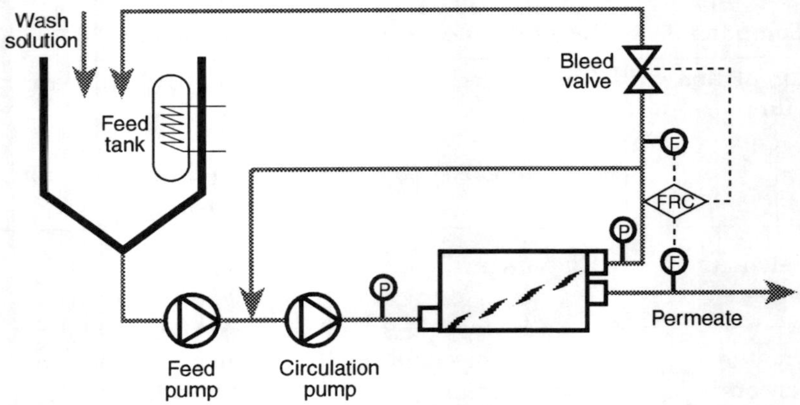

There are two advantages of this process over the simple batch process.

First, the flows into and out of the tank are an order of magnitude smaller than the circulation flow, reducing the formation of vortices in the tank. In general, vortex formation is undesirable and sometimes completely unacceptable (foams are a leading cause of protein denaturation). Additionally, this arrangement eliminates the potential for cavitation, which can lead to product denaturation and severe mechanical damage to the circulation pump.

Second, the energy loss across the retentate valve is significantly reduced; the circulation stream is maintained at the high operating pressure, rather than being reduced to atmospheric pressure as it is returned to the feed tank in the simple batch process. This represents a 25 percent to 75 percent energy savings compared to the type I batch process (energy costs are typically the third highest operating cost factor for TFF systems).

These benefits come at the expense of increased complexity, since the process now has to manage four streams instead of three streams. Additional components required are as follows:

- Pump (the feed pump)
- Flowmeter
- Automatic[4] flow-ratio controller
- Automatic bleed valve (instead of a manual retentate valve)

Additionally, the flux will be somewhat lower, due to the increased concentration that results in the circulation loop relative to that in the feed tank.

The type I batch process is preferred when the circulation flow is large, which occurs when the filtration capacity is large and/or when the conversion is low. Low conversions are typical when using open-channel modules in either tubular or rectangular configuration, which is the channel geometry required when filtering streams with suspended solids. Of the application classes shown in Table 3.3, only bacterial cell harvesting possibly meets the two requirements of large size and suspended solids; the remaining applications are typically much smaller than the 10,000-liter benchmark.

In principle, the flow-ratio control can be based on any two of the following three streams: permeate, feed, or retentate. In practice, however, controlling the bleed-to-feed ratio is least desirable, since small errors in the ratio lead to large concentration changes in the circulation loop. Of the other two combinations, the permeate-to-bleed-ratio control shown in Figures 3.16(a) and 3.16(b) is most often recommended. Typical flow ratios are as follows:

- Permeate rate of approximately 1 percent of circulation rate
- Feed rate of approximately 10 percent of circulation rate

[4]Even though it is possible to control the flow ratio manually, this is a poor practice. A small change in the viscosity or hydraulic resistance across the retentate valve can lead to plugging so quickly that an operator may not have time to react.

- Bleed rate of approximately 9 percent of circulation rate
- Permeate-to-bleed ratio of approximately 1:9

As in type I batch processes other than for the flow-ratio controller, the type II batch system can be manual or automatic depending on the requirements of the operating environment.

Type III Batch Process

Some applications require fine control of the flux to prevent damage to the product. An example is the harvesting of mammalian cells; if the flux is too high, the cells will lyse. Early in the 1980s, Millipore Corporation (Bedford, MA) introduced the permeate-flow-control technique for the processing of mammalian cells,[5] which is defined here as the type III batch process.

As shown in Figures 3.17(a) and (b), the process looks much like the simple batch process, except that a positive-displacement pump has been added to control the flow rate of the permeate stream and the retentate valve has been eliminated. The permeate pump provides positive control of the flux, which could also be accomplished with a flow-control valve. One particular benefit of the permeate pump is that it eliminates the retentate valve; rather than applying back pressure, the permeate pump generates a small vacuum to obtain the desired permeation rate. Given that the permeate pressure is no longer fixed, but rather fluctuates to meet the desired permeation rate, a pressure sensor is installed on this stream. This is possible only because the subject applications require very low operating pressures, on the order of a few psi at the most.

Peristaltic pumps are typically used in the harvesting of mammalian cells; this is one of the few applications in which peristaltic pumps are practical production-scale pumps due to the small batch sizes (10–1,000 liters). At larger-scale production, low-speed rotary lobe pumps are most often selected to prevent shear-induced cell damage.

Another application for the type III process is the clarification of protein solutions. In this case flux control is not used to prevent damage to the proteins; it maintains protein permeation through the filter. Proteins, in spite of their size, can become retained by a microfilter. This happens because the suspended solids being retained become a rejective layer to the proteins if their concentration at the membrane surface becomes high enough. Flux control prevents or reduces this phenomenon.

---

[5]This technique was originally developed for processing blood cells; at that time Millipore was working at developing a membrane plasmapheresis device as well as a wash filter for red blood cells.

Figure 3.17. Flow schematics: Type III Batch TFF processes.

(a) In concentration mode.

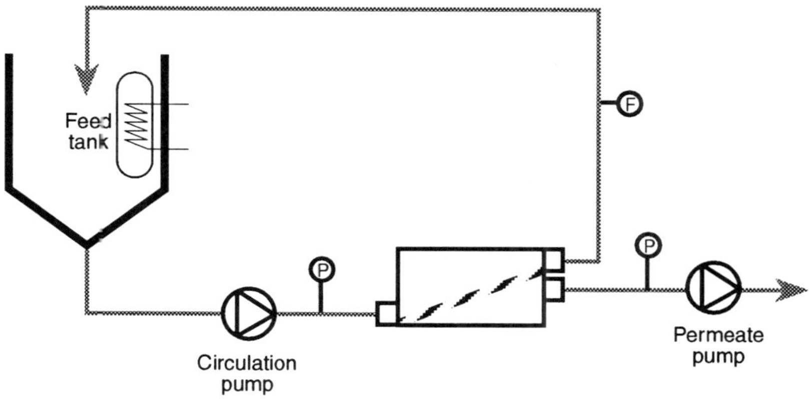

(b) In wash mode.

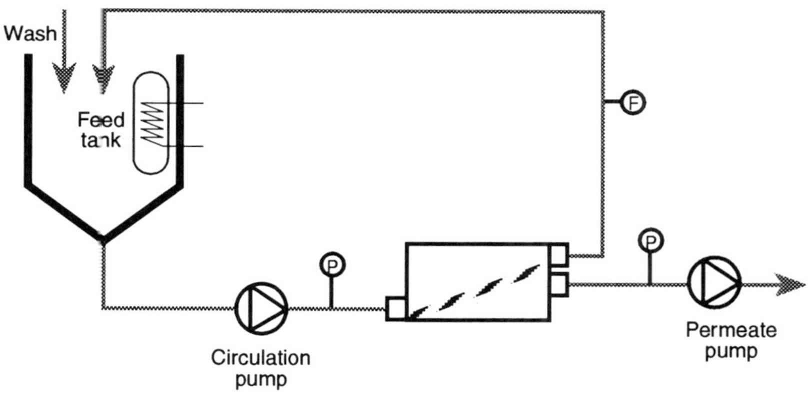

The Continuous Feed-and-Bleed Process

Figure 3.18 shows a flow schematic for the feed-and-bleed process, which resembles the type II batch process except that it is continuous rather than batch. Since the bleed stream is not returned to the feed tank, the concentration of the feed stream is constant, and a steady-state concentration quickly develops in the circulation loop.

As shown in Table 3.3, this process is not very common due to several inherent disadvantages. First, the flux is low because the TFF module is operated at the final (maximum) product concentration.

Figure 3.18. Flow Schematic: Feed and bleed TFF process.

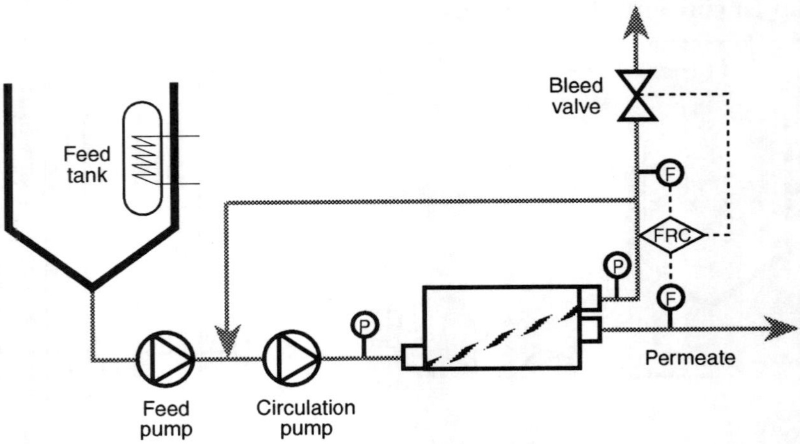

Second, washing is very inefficient, as shown in Table 3.5; much larger quantities of wash solution are required to achieve the same impurity removal compared to the simple batch process.

The only benefit of this process is that the residence time is low, which is important if the product degrades upon exposure to processing conditions. Table 3.3 shows this as a possible process for the harvesting of bacterial cells; it is of interest only if the flux is not overly dependent on concentration, which is the case if the final

Table 3.5. Impurity Removal (%) as a Function of Number of Wash Volumes (with no Additional Concentration): Comparison of Simple Batch vs. Continuous Feed and Bleed Process

Wash Volumes	Process Mode	
	Simple Batch	Feed and Bleed
1	63.2	50
2	86.5	67
3	95.0	75
4	98.2	80
5	99.3	83

concentration is relatively low (less than or equal to 10 percent solids by volume).

The Continuous Multistage Process

If low residence times are required, the best choice is the multistage process. To overcome the inherent limitations of the feed-and-bleed process, a cascade of feed-and-bleed units can be used, as shown in Figure 3.19; the greater the number of stages, the closer the efficiency approaches that of the simple batch process (the most efficient).

Each stage contains almost all the elements of the feed-and-bleed process. The retentate from each stage becomes the feed to the next stage, while the permeate is collected in a common manifold in a cross-flow configuration. If washing is required, the additional wash stream is added simultaneously, again in cross-flow configuration.

The advantages of this process over the single-stage feed-and-bleed process is that it can approach the efficiency of the batch process while running continuously, thereby providing a low residence time (typically 15–30 minutes). These benefits are obtained at the expense of much more cost and complexity:

1. Cost is associated with each stage, regardless of its size. Therefore, the more stages the higher the cost, even while maintaining the same capacity.

2. An additional flow-ratio controller is required to carry out the wash step.

3. Heat exchangers must be installed on every stage to remove the energy generated by the circulation pump.

Therefore, for the multistage process to be a good choice, there must be a need for a low residence time. Also, the size of the system should be very large; for a sufficiently large size even a batch system would have to be split into multiple "skids." Standard pumps come in limited sizes, thereby losing one of the inherent economic advantages over the multistage process.

Finally, there is an interesting variant, known as the "countercurrent wash multistage" process, in wash mode. According to this variant, the wash stream is fed only to the last stage while the permeate streams are added back countercurrently to the retentate stream. The benefit of this variant is to reduce the volume of wash solution required at the expense of significantly higher membrane area. Countercurrent washing is only worth considering if the wash solution is very expensive or if the product of interest is in the permeate

Figure 3.19. Flow schematic: Multistage feed and bleed TFF process. Stages 1–3 concentrate the feed stream while stages 4–5 provide the wash.

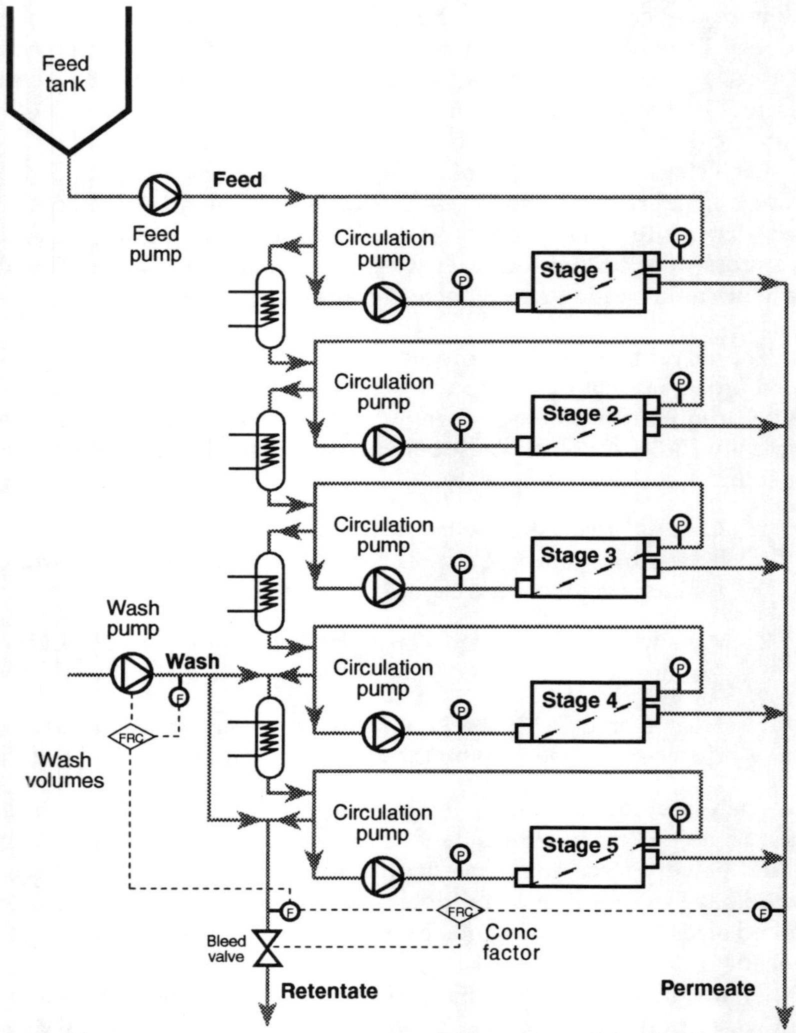

and its dilution is highly undesirable, neither of which are typical situations.

Table 3.6 lists the major manufacturers of TFF membranes, devices, and systems around the world.

Table 3.6. Manufacturers of Tangential Flow Filtration Equipment

Company	Location	Membrane Type				Filter Module Geometries Supplied				
		MF	UF	NF	RO	Flat Sheet	Hollow Fiber	Spiral-Wound	Tubular	Other
A/G Technology	Needham, MA	P	P				X			
Alfa-Laval	Tumba, Sweden	P, C	P, C	p	p	X	X	X	X	
APV	Crawley, W. Sussex, U.K.	P, C	P, C	P	P	X		X	X	
Asahi	Tokyo, Japan		P	P			X			
Asahi Glass	Tokyo, Japan		C	C	C				X	
Bio-Engineering	Wald, Switzerland	P	P			X				
Bio-Flo	Glasgow, Scotland	P, C	P				X			X

Continued on next page

Continued from previous page

Company	Location	Membrane Type				Filter Module Geometries Supplied				
		MF	UF	NF	RO	Flat Sheet	Hollow Fiber	Spiral-Wound	Tubular	Other
Bio-Molecular Dynamics	Beaverton, OR			P			X			
Bio-Recovery	Northvale, NJ	P	P			X				
BioSepra	Marlboro, MA		P				X			
Carbone Lorraine	Paris, France	C, M	C	C	C				X	
Celfa	Seewen-Schwyz, Switzerland		P			X		X		
Ceramem	Waltham, MA	C	C							monolith
Cuno/Bioken	Norwood, MA		P					X		
Dow/DDS/ Film Tec	Midland, MI		P	P	P	X		X		

Continued on next page

Continued from previous page

Company	Location	Membrane Type				Filter Module Geometries Supplied				
		MF	UF	NF	RO	Flat Sheet	Hollow Fiber	Spiral-Wound	Tubular	Other
duPont	Wilmington, DE	C/M			P		X		X	
Envirogenics Systems	El Monte, CA		P					X		
Filtration Engineering	Minneapolis, MN		P	P	P	X		X		
Filtron Products	Northborough, MA	P	P			X	X			s, c
Grace/Amicon	Danvers, MA	P	P			X	X	X		s, c
Hoechst-Celanese	Charlotte, NC	P	P			X	X	X	X	
Ionics	Watertown, MA				P			X		
Koch-Romicon	Wilmington, MA		P				X	X	X	

Continued on next page

Continued from previous page

Company	Location	Membrane Type				Filter Module Geometries Supplied				
		MF	UF	NF	RO	Flat Sheet	Hollow Fiber	Spiral-Wound	Tubular	Other
Kodak (Fastek)	Rochester, NY	P						X		
Kuraray	Tokyo, Japan		P				X			
Membrane Products Kiryat Weizmann	Rehovot, Israel		P	P	P			X	X	
Membrane System Specialists	Wisconsin Rapids, WI		P					X		
Membrex	Fairfield, NJ	P	P							v
Memron	Amherst, MA	P					X		X	

Continued on next page

Continued from previous page

Company	Location	Membrane Type				Filter Module Geometries Supplied				
		MF	UF	NF	RO	Flat Sheet	Hollow Fiber	Spiral-Wound	Tubular	Other
Memtex (Memcore)	S. Windsor, New South Wales, Australia	P	P				X			
Microdyn	Wuppertal, Germany	P					X		X	
Microgon	Laguna Hills, CA	P	P				X			
Millipore	Bedford, MA	P, C	P, C	P	P	X		X	X	s, c
Minntech	Minneapolis, MN		P				X			
Mitsubishi Rayon	Tokyo, Japan	P					X			

Continued on next page

Continued from previous page

Company	Location	Membrane Type				Filter Module Geometries Supplied				
		MF	UF	NF	RO	Flat Sheet	Hollow Fiber	Spiral-Wound	Tubular	Other
Mott	Farmington, CT	M							X	
New Brunswick Scientific	Edison, NJ	P	P							v
New Logic	Emeryville, CA	P	P			a				
NGK	Tokyo, Japan	C	C						X	
Niro Filtration	Hudson, WI	P, C	P, C	P	P			X	X	
Nitto Denko	Osaka, Japan		P				X			
Osmonics	Minnetonka, MN		P	P	P	X		X		
Perstorp Biolytica	Stockholm, Sweden		P				X			

Continued on next page

Continued from previous page

Company	Location	Membrane Type				Filter Module Geometries Supplied				
		MF	UF	NF	RO	Flat Sheet	Hollow Fiber	Spiral-Wound	Tubular	Other
Progress Technologies	St. Petersburg, FL				P			X		
Rhone-Poulenc-Rorer	Miribel, France	P, C	P, C			X			X	
Rochem (DT)	Torrance, CA		P		P	X				
Sartorius	Göttingen, Germany	P	P			X				
SRT	Morrisville, NC	P	P			X		X		
Sumitomo Bakelite	Tokyo, Japan		P				X			
Toray	Tokyo, Japan	P	P				X			
Toyo Advantec	Tokyo, Japan	P					X			

Continued on next page

Continued from previous page

Company	Location	Membrane Type				Filter Module Geometries Supplied				
		MF	UF	NF	RO	Flat Sheet	Hollow Fiber	Spiral-Wound	Tubular	Other
UOP Fluid Systems	San Diego, CA				P				X	
U.S. Filter	Rancho Mirage, CA	C	P, C	P	P				X	X
X-Flow	Enschede, Netherlands	P	P						X	

KEY

P = Polymeric membranes
C = Ceramic membranes
M = Sintered metal membranes
s = Stirred cells
c = Centrifuge-driven sample preparation devices;
v = Taylor vortex-flow (rotating membrane assembly) device;
a = Device is a stack of circular flat plates that vibrates around its axis.

APPLICATION CLASSES IN PHARMACEUTICAL SEPARATIONS

Clarification of Antibiotic Fermentation Broths

Separation Objective

Antibiotics are secondary metabolites produced by such organisms as *Penicillium chrysogenum* (Penicillin G or V) and *Cephalosporium acremonium* (Cephalosporin C). The recovery and purification of antibiotics from fermentation broths is a complex process (Vandamme 1984) involving several separation steps and is outlined in general terms in Table 3.7. In terms of separation process method, antibiotics can be classified as two types:

1. **Type I:** The antibiotic molecule is in the aqueous phase of the fermentation, whether expressed by the biomass or expressed by pH change. The product is water soluble. Isolation is through ion exchange and or liquid/liquid extraction in rotary extractors.

Table 3.7. Purification Methods for Antibiotic Recovery

Broth clarification	Tangential flow filtration Filtration with diatomaceous earth (filter aid) Centrifugation
Primary isolation	Ion-exchange adsorption Solvent extraction Precipitation
Purification	Liquid-liquid extraction Fractional precipitation Chromatography
Final product isolation	Crystallization Spray drying Lyophilization

2. **Type II:** The antibiotic molecule is contained in the biomass of the fermentation. It is not water soluble and is isolated through direct solvent extraction of the biomass. This may be a multistep process depending on the miscibility of the required solvents.

The quantity and composition of the fermenter broths can vary widely. The range of typical values are as follows:

- Volume of broth: 5–200 m^3
- Solids concentration: 40–100 kg/m^3
- Antibiotic concentration: 1–30 kg/m^3

The objective of the filtration step is the separation of the microbial biomass and other particulates from the antibiotic solution. The removal of the biomass is needed for the subsequent isolation steps to function properly. Furthermore, residual colloids and dissolved polymeric contaminants in the filtrates, such as proteins and starches, can also interfere with subsequent isolation and purification steps. The contaminants form emulsions and interfacial sludge that reduce the efficiency of chromatographic and extractive processes and, thus, reduce yields. Effectiveness of removing these microcontaminants by different types of filters and centrifuges drives the selection of the solids-removal device.

The most common method of removing the microbial biomass are tangential flow (cross-flow) filtration, centrifugation, or rotary-drum vacuum filtration with added diatomaceous earth body feed. For a specific process the choice of clarifying device depends on both technical and economic factors specific to the antibiotic and its manufacturing process. In broad terms, however, the choice involves weighing the major relative advantages and disadvantages of each clarifying operation, as listed in Table 3.8.

Membrane Selection

Membranes used for antibiotic broth clarification fall into two classifications:

- Ceramic (e.g., alumina, zirconia, titania, silicon carbide)
- Polymeric (e.g., polysulfone, polyolefin, polyvinylidene difluoride)

A comparison of these two classes of membranes reveals their relative advantages and disadvantages, as listed in Table 3.9. From this table ceramic membranes are seen to be superior for most antibiotic clarifications, unless ceramic membrane life is unusually short and/or membrane cost by itself is the primary selection criteria.

Table 3.8. Comparison of Antibiotic Broth Clarification Methods

Separation Method	Advantages	Disadvantages
Centrifugation	Established process with predictable performance No disposal of spent filter aid (diatomaceous earth)	Supernatant heavily contaminated with colloids and small particles
Diatomaceous earth (DE) filters	Low capital cost Generally accepted technology	Open—allows broth contamination Disposal of waste DE expensive Yield loss 10–20%
Tangential flow filtration	Filter aid not required Complete removal of suspended solids	Higher capital cost

Table 3.9. Advantages of Ceramic and Polymeric Membranes for Antibiotic Broth Clarification

Ceramic	Polymeric
Higher flux (2–5 times)	Lower membrane cost
Broader chemical compatibility	Easier replacment
Easier cleaning	More compact
Resistant to more organic solvents	
Longer life	
Lower overall life-cycle cost	
Broader temperature range	
Better abrasion/erosion resistance	

How Tangential Flow Filtration Performs the Separation

In clarification of antibiotics using ceramic membranes, the rejection of suspended matter and colloids is virtually 100 percent because all suspended particles are larger than the diameter of the membrane's pores. The rejection of antibiotics, which are meant to pass freely through the membrane, is close to 0 percent because the pore size of the membrane is much greater than the antibiotic molecule. At times the cake layer of suspended matter next to the membrane surface offers some resistance to the passage of antibiotics, and the antibiotic rejection can rise to values as high as 30 percent. This requires additional diafiltration (washing) to maximize antibiotic yield in the filtrate, which dilutes the clarified antibiotic solutions. In Table 3.10, typical antibiotic clarification applications are presented.

Operating Conditions. When clarifying antibiotics, there is no reliable theoretical method of predicting flux; flux values need to be established empirically. In the clarification of antibiotics, the following equation has been found to describe the behavior of flux as a function of velocity past the membrane surface and the instantaneous retentate solution volume, V:

$$J = K \times U^{\phi} \times \ln[V_f/(V_0 - V)]$$

where K and ϕ are empirical constants; U is the velocity of retentate past the membrane; V_f is the volume of retentate at which the flux drops to zero; and V_0 is the starting volume of the solution being clarified.

To significantly increase the recovery of antibiotic, diafiltration is needed; in TFM it is impossible to concentrate the suspended matter to a "dry filter cake" without irreversibly plugging the filter. Consequently, water is added during the clarification processes to wash, or "diafilter," the antibiotic out of the suspension. This procedure can drive the recovery levels to more than 98 percent, at the expense of some dilution of the antibiotic.

Operating Configurations. As shown in Figure 3.20, three basic operational configurations are usable when clarifying antibiotics:

1. Batch clarification with constant-volume diafiltration, in a single filtration stage
2. Continuous multistage clarification with addition of fresh diafiltrate to each stage (cross-flow diafiltration)
3. Continuous multistage clarification with diafiltrate addition to the last stage and diafiltrate flow from the last stage to previous stages (countercurrent diafiltration)

Table 3.10. Typical Antibiotic Applications

Application Name	Membrane Type	Pore Size (μm)	Feed Channel (mm)	Cross-flow Velocity (m/sec)	Typical Recovery (%)	Typical Flux (ℓ/m²/h)
Penicillin V	Ceramic	0.2	6	7	97	80
Erythromycin	Ceramic	0.2	3	5	95	60
Everinomycin	Ceramic	0.2	6	6	100	75
Oxytetracycline	Ceramic	0.2	4	5	95	40
Bacitracin	Ceramic	0.2	4	5	90	65
Semduramicin	Ceramic	0.2	6	8	100	85
Cephalosporin C	Ceramic	0.2	4	4	90	50
Terramycin	Ceramic	0.2	4	5	95	110

Figure 3.20. Operating configurations for antibiotic clarification.

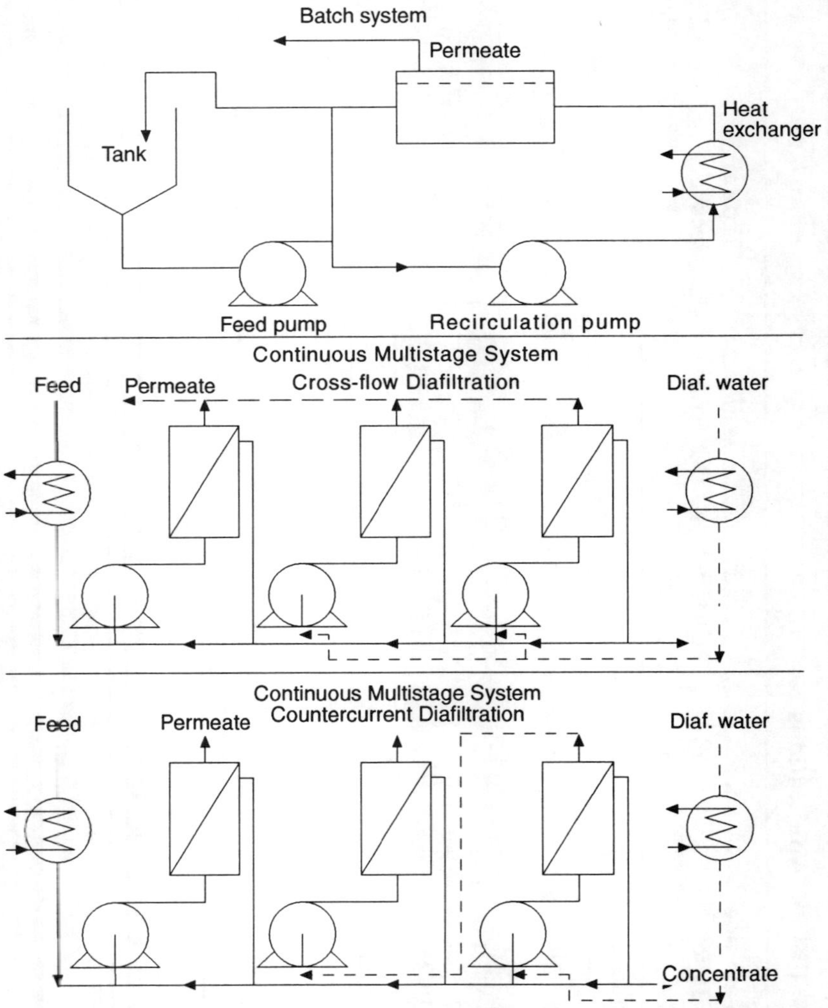

Table 3.11 outlines the advantages and disadvantages of operating configurations.

Important Control Variables. The productivity of tangential flow filters when clarifying antibiotics is not predictable on theoretical grounds and must be measured as a function of operating and design parameters. The following are the most important independent variables to evaluate.

Table 3.11. General Difference Among Process Options

Operating Mode	Dilution	Residence Time	Membrane Area	Cost
Batch	Low	Longer	Small	Low
Continuous cross-flow	High	Short	Large	Medium
Continuous countercurrent	Lowest	Short	Very large	High

In general, no significant increases in flux are obtained above six bar. Operating below three bar is to be avoided, because it results in a needless lowering of flux. Figure 3.21 presents a typical relationship between process flux and transmembrane pressure.

The circulation velocity is chosen on the basis of minimizing costs. Typical circulation velocities used in this application range from 3 m/sec–6 m/sec.

Backflow of permeate (backflush) can help to reduce fouling, thus increasing flux. Figure 3.22 presents the effect of backflushing on long-term flux decay.

The effects on filter productivity of two significant design parameters also need to be evaluated. First, pore size influences both the clarity of the permeate produced and the flux. As a general rule, increasing the pore size above 0.5 μm does not result in higher fluxes. Decreasing the pore size below 0.2 μm does not improve colloid removal. Thus, for clarification purposes only pore sizes between 0.2 μm and 0.5 μm need to be considered. To remove soluble large polymeric contaminants as well, a 0.005 μm (approximately 50–100 kDa NMWL [nominal molecular weight limits]) UF membrane can be used.

Second, feed-channel diameter has major effects on flux and circulation requirements. While no specific guidelines can be given, two factors are important in choosing the feed-channel diameter. The feed-channel diameter should be greater than 10 times the size of the largest particles to be filtered to avoid occasional plugging of the filter's flow channels. Furthermore, larger feed-channel diameters should be used with viscous liquids.

Typical tubular filters are available in the following standard diameters: 2.0 mm, 3.0 mm, 4.0 mm, and 6.0 mm; special intermediate sizes can be manufactured. Table 3.12 provides guidelines on the proper choice of tube diameter as a function of liquid viscosity.

Figure 3.21. Flux vs. transmembrane pressure.

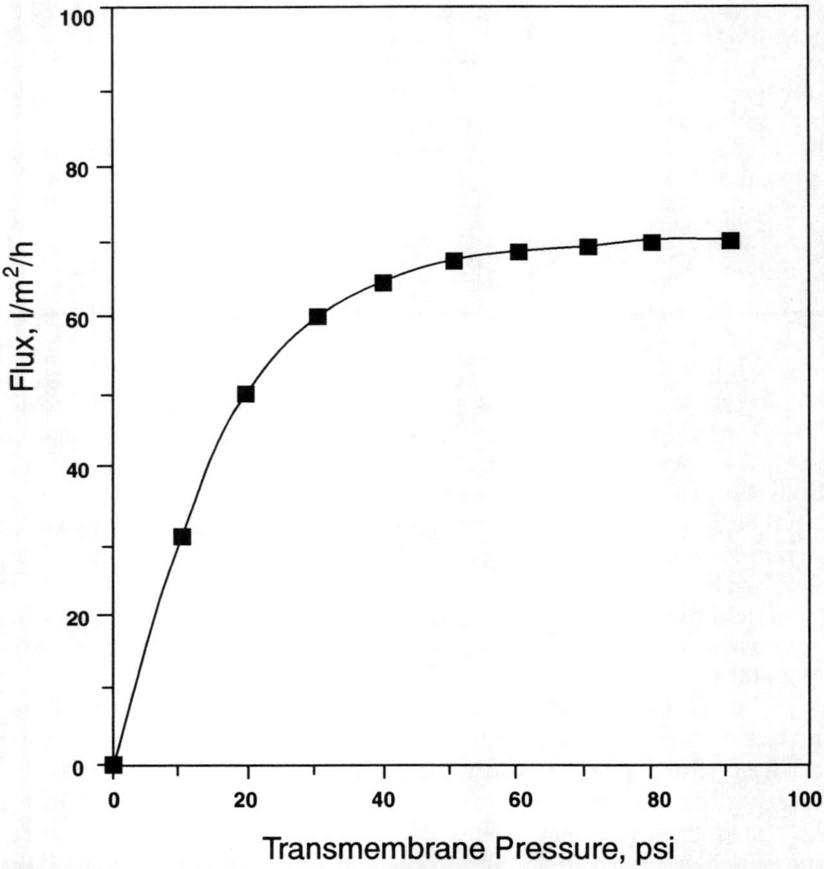

Cleaning. A typical operating cycle is 20 hours. At the end of the process cycle, the system is cleaned using the following chemicals:

- 0.25 N caustic solution and 500 mg/ℓ sodium hypochlorite at 50°C for 30 minutes
- 0.25 N nitric acid at 25°C for 20 minutes

Clarification of and Cell Removal from Biological Cultures

Separation Objective

Tangential flow filtration has been used for the recovery and separation of cells from fermentation broths (Defrise and Gekas 1988). Cells

Figure 3.22. Effect of backflush on process flux.

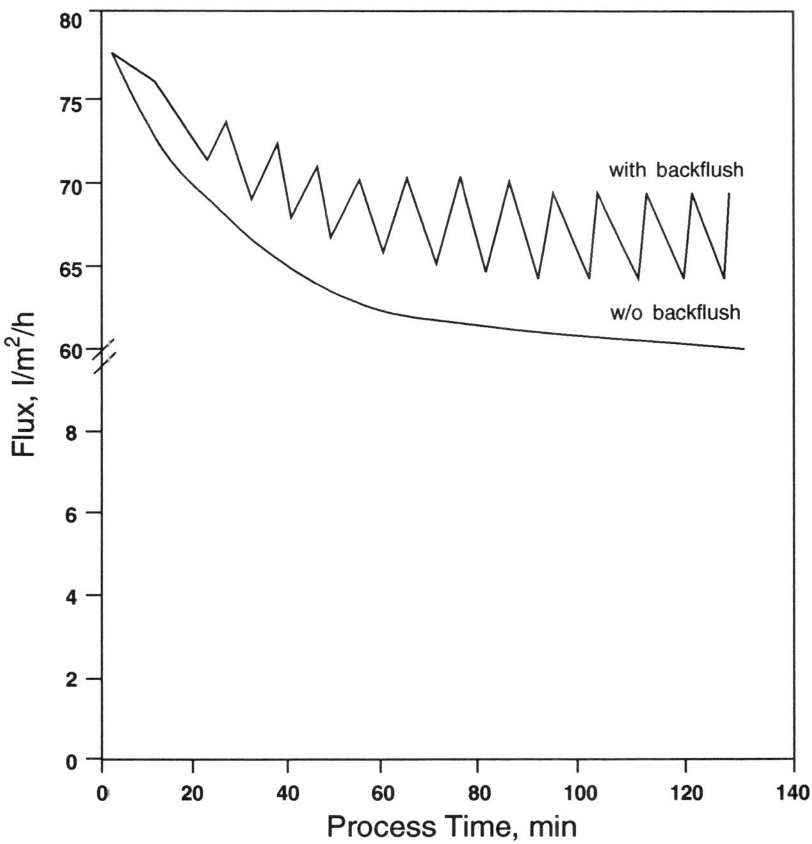

cultured in fermenters may be bacterial, fungal (including yeast), insect, or mammalian cells. The cell-separation operations performed by TFF can have different objectives.

- Cell concentration during harvesting prior to other operations to be performed on the cells, as occurs if the end product of interest is intracellular.
- Removal of whole cells to clarify the broth, as occurs if the end product of interest is excreted by the cells into the broth.
- Changing the solution conditions surrounding the cells by diafiltration (washing), as occurs in perfusion and cell recycling operations that can increase fermenter productivity.

Table 3.12. Guide to Selection of Feed-Channel Diameter

Effective Viscosity (*cP*)	Feed-Channel Diameter (mm)
<3	2
<5	3
<15	4
<20	6

Note: Effective viscosity is defined as shear stress divided by shear rate.

How Tangential Flow Filtration Performs the Separation

In these cases the TFF membrane retains whole cells because its pores are too small to permit cells to enter. Only the solvent and those solutes small enough to pass through the pores are able to penetrate and be removed in the permeate. By selecting a pore size smaller than the cells, yet as large as needed to pass the soluble components, a clean, washed, and concentrated suspension of cells can be produced as the retentate and a highly clarified solution can be produced as the permeate.

During this type of microporous TFF, a cake layer of retained cells and other particles builds up on the membrane surface. The thickness, compressibility, and particle-size distribution of this cake layer controls the passage of most soluble components and must be carefully managed by the TFF system. In addition, the inevitable presence of this cake layer ensures that TFF devices cannot separate different-sized insoluble particles from cells or one type of cell from another. This section describes how microporous TFF devices should be operated to permit high cell recoveries and/or high recovery of clarified solution from whole-cell broths.

Comparison of Tangential Flow Filtration to Traditional Methods of Cell Harvesting

Cell harvesting has been traditionally performed by centrifugation or deadend filtration (filter presses, rotary vacuum filters). In centrifugation the separation is based on density differences. The small density difference between most cells and their fermentation broths makes centrifugation difficult. Both capital expenditure and running costs are high. A portion of the cells ends up suspended in the supernatant (overflow), reducing cell recovery. Tangential flow filtration is also preferable from a safety and contamination perspective, as TFF does not generate any aerosols during cell harvesting. The aerosol containment by centrifuges becomes very important when harvesting pathogenic and recombinant organisms.

In deadend filtration devices the filtrate flow rate decreases as the cell deposits accumulate on the filter medium. In TFF operations the buildup is minimized by the sweeping action of the recirculating flow. Also, the use of filter aids in deadend filtration makes these filters less attractive, as the filter aid can contaminate the cell suspension.

Issues and Solutions

Cell Harvesting. During cell harvesting the retentate is recirculated to the feed tank. If the permeate contains the product of interest (e.g., extracellular proteins), it will be collected and subsequently concentrated by UF. If the product of interest is retained within the cells, the concentrated cell suspension can be lysed by various methods after the cell harvesting operation. Cell harvesting operations have been performed successfully both in laboratory scale and industrial scale by the use of TFF (Henry and Allred 1972; Vaheri et al. 1979; Reid and Adlam 1974).

During cell harvesting the concentration of the cell suspension increases with time since the permeate is not returned to the feed tank. The flux does not decrease linearly with the logarithm of the cell concentration, as would be expected from the gel-polarization theory (Michaels 1968). Instead, the fluxes follow a sigmoid curve (Kroner et al. 1984), decreasing sharply at first, followed by a stable region where the fluxes show a small decrease with concentration. At a certain cell concentration the fluxes fall until the permeate flow ceases.

This sigmoid behavior can be explained in terms of the formation of an initial fouling and/or cake layer. The stability of the flux afterward seems to indicate that concentration polarization does not play a major role during the TFF of cells. This agrees with the abnormally high fluxes, termed *flux paradox*, observed during the filtration of colloidal solutions (Porter 1972). Since the viscosity of cell suspensions increases drastically after a certain concentration (18–20 percent), the final drop in the flux can be produced by the high viscosities. The response of the flux to various factors is not clearly understood and some results reported in the literature conflict. The following is a description of the different factors affecting the fluxes:

- **Membrane pore-size rating:** The membrane must be selected to retain the cells in suspension. Workers in the field have experimented with different pore sizes to obtain the highest flux. Some of these results are cited below:

1. During the concentration of *Bordetella pertussis,* Vaheri et al. (1979) have tried 1000 kDa NMWL UF and 0.22 μm microporous membranes. The microporous membrane produced higher fluxes.

2. Le et al. (1984a) used 0.2 μm, 0.45 μm, and 0.6 μm microporous membranes during the harvesting of *Erwini carotovora*. The 0.45 μm membrane resulted in the highest fluxes.

3. Gatenholm et al. (1988) used various ultrafiltration (100 and 300 kDa NMWL) and MF (0.2 μm) membranes for the harvesting of *E. coli*. The MF membranes showed a considerable drop in flux in the first few minutes of operation. The flux decline in the UF membranes was slower.

4. Some think that the drastic flux drop observed with large-pore microfilters is due to the lodging of smaller cells in the pores of the filter (Patel et al. 1987). Since the pores of ultrafilters are much smaller, the cells are removed from the surface of the filter by the shear forces of the recirculating flow. The response of the flux to the membrane rating will depend on the particular cell suspension. Experiments should be performed by the TFF user to determine the optimum membrane pore rating.

- **Tangential flow rates:** Tangential flow rates increase the mass transfer of cells from the membrane surface and reduce cake buildup. Generally, the permeate fluxes increase with increasing tangential flow rates. One should avoid, however, damage caused to shear-sensitive cell lines (e.g., many animal cells) by high shear generated at high tangential flow rates. In addition, high tangential flow rates may lead to high TMPs in some regions of the TFF devices, which can counteract the flux increase, as discussed in the next paragraph.

- **Transmembrane pressure:** Initially, increasing the TMP increases the fluxes. But high TMPs accentuate fouling and accelerate the flux decline (Patel et al. 1987). The impact of TMPs on cell viability is discussed at the end of this section.

- **Fermentation broth:** The composition of the fermentation broth can affect the flux decline. A component frequently found in fermentation media that contributes to fouling is an antifoam. Most commonly used antifoams include polypropylene glycols, copolymers of polypropylene and polyethylene glycols, and silicone oil emulsions. These substances are hydrophobic and adsorb to the surface of some membranes, causing flux decline. In addition, under some conditions the antifoams may come out of solution, forming insoluble micelles that can sequester and retain normally soluble components of the broth as well as greatly increase the fouling tendency of the broth. If an antifoam is necessary, one with minimum fouling properties should be selected and its concentration should be minimized. The pH and ionic strength of the broth can also have an effect on the flux decay by altering the solubility of antifoams and other solutes in the broth and the adsorption of solutes to the membrane.

Cell Washing. As the permeate leaves the system during cell wash operations, a new solution is simultaneously added to the feed tank. This allows solution conditions such as pH or ionic strength to be changed and enables small soluble solutes to be removed from the cells. The concentration of the removed solute can be expressed as a function of the wash volume by the expression:

$$C = C_0 \times \exp(V_d/V_0)$$

where C is the final concentration of the solute, C_0 is the initial concentration, V_d is the wash volume, and V_0 is the system volume.

Usually cell harvesting and cell wash operations are combined. The cell suspension is first concentrated and then washed. The concentration should be performed to a point slightly less than that value where a rapid flux decline takes place (Tutunjian 1984).

Cell Recycling. Tangential flow filtration can improve fermenter productivity by continuously removing inhibitory metabolic products (and excreted end products of interest) from the fermentation broth. In this operation the fermentation broth is recirculated through the TFF device. If the permeate contains the product (extracellular), it is collected and sent for further downstream processing. Permeate containing solely waste products can be discarded. The concentrated cell suspension is returned to the fermenter and fresh nutrient broth is added to the fermenter to maintain a constant volume (Michaels and Matson 1985; Bibal et al. 1991; de la Broise et al. 1992).

Cell Viability. The effect of shear and TMP on cells has been investigated by various workers (Le et al. 1984b; Maiorella et al. 1991). In addition, as shown in membrane plasmapheresis (Malchesky et al. 1981), cells that are very shear sensitive (e.g., erythrocytes and many other animal cell lines) may be ruptured at TMPs as low as 7 kPa when these pressures force the cells packed onto the membrane surface to extrude into the mouths of the membrane pores. Microbial and yeast cells are quite sturdy and are not damaged by the flow rates or filtration in most TFF devices. Maiorella et al. (1991) have defined a region of operation to prevent the lysis of the more delicate mammalian cells. They recommend keeping the shear rate less than 3000 sec^{-1} and limiting the TMP. As discussed previously, for mammalian cell harvesting a mode of operation has been suggested where the permeate flow is controlled at a constant rate and the TMP is kept low (Millipore 1986). In this operation the viability of mammalian cells is maintained by the use of a second pump on the permeate side. This mode can also be used to obtain high average fluxes during cell harvesting.

Lysed Cell Separations: Clarification and Purification of Intracellular Proteins

Separation Objectives

All microbial-based recombinant proteins are processed and purified through a combination of extraction, solubilization, membrane filtration, and chromatographic separation. Several (5–15) process steps can be used to purify a recombinant protein derived from bacterial or yeast cells. Lysate clarification represents one of the early steps of purification.

In microbial hosts the recombinant protein of interest can be expressed as a soluble protein within the cell (or in the cell membrane) or as an insoluble refractile body or inclusion body within the cell. For heterologous proteins that are expressed intracellularly, the host microbes are mechanically lysed by high-pressure homogenization.

The resulting suspension is composed of insoluble and soluble components. The insoluble materials can comprise colloids, such as cell-wall polysaccharides, lipids, or aggregates formed during lysis. The soluble components can be macromolecules, such as nucleic acids, homologous proteins, and other low-molecular-weight components (salts, peptides, and sugars).

The general objective of lysate clarification is to produce a particulate- or colloid-free solution appropriate for loading onto the primary chromatography column. Since packed-bed ion-exchange chromatography (IXC) is commonly used for initial and intermediate purifications, these protein solutions must be clarified. Fouling of the columns can diminish protein yield and limit the useful life of the column.

The specific separation objective will vary, depending on whether the protein is insoluble or soluble. For recombinant proteins that are soluble, the objective is to clarify the lysate suspension to remove the particulate and colloidal material.

For inclusion-body proteins the separation objective may be twofold. The first objective is purification of the inclusion-body suspension by the elimination of soluble nonproduct proteins (and other constituents) via diafiltration. The inclusion body is retained with the suspended cell debris. Soluble nonproduct proteins are eliminated via diafiltration.

The second objective is clarification of the recombinant protein after solubilization of the inclusion body. The inclusion body is isolated from the crude lysate (typically by high-speed centrifugation) and then solubilized with urea or guanidine. The now-soluble protein is then clarified from colloids and aggregates formed during the solubilization process.

Single-stage membrane-based separations have poor resolution; they cannot separate proteins whose molecular dimensions differ by less than a factor of five without substantial yield loss. For this reason, TFF cannot compete with multistage IXC or affinity chromatographic procedures for fractionation of protein mixtures. Macromolecular separation techniques are discussed later in this chapter. The clarification of proteins from bacterial lysates with high yield is thus a persistent challenge for purification engineers and scientists.

How Tangential Flow Filtration Performs the Separation

As with any filtration process that removes suspended material, the separation is based on size differences. In principle, the clarification operation is simple; the suspended material larger than the pore size is retained and the soluble components that pass through the membrane are recovered in the clear permeate. In practice, however, the application is one of the more difficult TFF separations and can be likened to the separation of macromolecular mixtures by TFF. There are, however, demonstrated examples where TFF is a viable process for lysate clarification on a process-scale level (Ju et al. 1991; Kellaris et al. 1991; Myers et al. 1990). Typical operating conditions and membrane performance for lysate processing are given in Table 3.13.

Tangential flow filtration system performance for this application can be enhanced by the proper selection of liquid conditions, membrane type, operating pressures, cross-flow rates, and system configuration. These aspects are discussed below.

Issues Related to Recombinant Protein Clarification

During membrane filtration of lysates or any protein solution containing colloids, the protein or macromolecule of interest often will not pass freely through the membrane. For example, a 22 kDa protein in a

Table 3.13. Lysate Clarification by Tangential Flow Filtration: Typical Performance Data

Device Type	Membrane Type	Protein MW (Daltons)	Protein Retention (%)	Process Flux (ℓ/m²/h)	Transmembrane Pressure (kPa)
Plate and frame open channel	0.22 μm hydrophilic	70,000	50	15	100
Plate and frame open channel	0.22 μm hydrophilic	200,000	60	10	100
Flat sheet	300,000 UF	12,000	30	20	70
Screened channel	300,000 UF regenerated cellulose	20,000	40	25	30

lysate suspension may show 50–70 percent retention by a 0.22 μm microporous membrane. This observation is common to many TFF lysate processes. The challenge is to maximize mass flux (grams of protein/m^2/hr) through the membrane.

High retention of soluble macromolecules is the result of two basic mechanisms:

1. Increase in apparent membrane retention properties
2. Increase in the apparent protein size

As retained particulates and/or colloids accumulate at the membrane surface, a "dynamic membrane" or polarization layer is formed. This porous accumulated layer consolidates under the TMP, reducing its porosity and impairing protein passage. This phenomena is analogous to the formation of filter cake in deadend filtration. Because the particulates are relatively small, amorphous, and deformable, the polarization layer is characteristically dense and has low hydraulic permeability. This "cake layer" formation leads to low flux and high retention of macrosolutes. In fact, at high enough TMP, retention of proteins in the range of 20–200 kDa can exceed 90 percent with microporous membranes.

The strategy to enhance protein passage in this situation is to choose hydraulic operating conditions that minimize deposition and consolidation of particulates. By controlling the flux or convective transport of solids to the membrane surface, the formation of a restrictive cake layer can be minimized. Direct measurements have been taken that correlate protein retention with increasing TMP. This is thought to be related to the pressure-dependent compaction of the accumulated material at the membrane surface (Forman et al. 1990).

Protein retention is shown to increase nearly linearly with TMP; therefore, operation at low TMP is beneficial. In general, flux in the range of 5–15 ℓ/m^2/hr, or TMP in the range 10–30 kPa are operating conditions that have been shown to result in optimal mass flux (g/m^2/hr).

Increased cross-flow velocity or turbulence in the feed side channel will reduce cake-layer thickness. It would seem appropriate to operate at high cross-flow. However, excessive cross-flow is counterproductive; high inlet pressure (therefore high inlet TMP) results from high cross-flow in conventional cross-flow devices. The inlet pressure will increase exponentially (exponent >1.0) with cross-flow, while the sweeping effect on boundary-layer resistance increases with an exponent much less than 1.0. Therefore, any benefit of greater turbulence or mixing of the boundary layer is countered by the high inlet TMP.

The effect of TMP can be somewhat uncoupled from cross-flow velocity by controlling membrane flux with a secondary permeate pump.

This technique has worked well for many types of protein clarification.

Because many membrane devices are available, with a wide variety of feed-channel geometries, specific cross-flow recommendations are difficult to state. In general, cross-flow should be chosen such that the feed-side pressure drop (pressure inlet to pressure outlet) for a given device does not exceed three times the operating TMP.

Effects of Membrane Fouling

Membrane fouling by adsorption of lysate components to the membrane surface and pore walls or by penetration of colloids into the membrane matrix may also contribute to increased retention characteristics. Membranes with low, nonspecific, adsorptive binding characteristics are best suited for this application. These include loose (>300,000 kDa NMWL) regenerated cellulose UF membranes or hydrophilic microporous membranes.

There is direct evidence that the topography of the membrane surface may also influence both passage and flux for lysate clarification (Le et al. 1984a). The benefit of smooth surfaces, such as loose UF membranes or open asymmetric pores, is not clear. Passage and flux may be influenced more by the size distribution and nature of the retained particulates than by the membrane properties themselves.

The effects of fouling can be minimized by the selection of a pore that minimizes particle penetration and by the selection of low protein binding membranes that minimize adsorptive fouling.

Increase in Effective Size of Protein

The protein of interest may be associated with a larger insoluble component in the liquid and is, therefore, retained with the larger component. The insoluble component could be cell-wall fragments, nucleic acids, lipids, or other cell debris. Association of intracellular enzymes with cellular debris has been demonstrated by successive centrifugation and resuspension (Le et al. 1984a). Adsorption of protein to cellular debris, as with any other adsorptive media, will be a function of pH, ionic strength, temperature, and nature of the debris (e.g., particle size). In the evaluation of TFF for lysate clarification, these aspects should be investigated.

Many recombinant proteins can polymerize via disulfide linkages or aggregate in certain ionic environments. High levels of cross-linking or aggregation will increase the effective molecular weight of the protein. The addition of reducing agents and the elimination of oxygen during processing is useful for proteins with cysteine or methionine residues. The addition of surfactants to reduce intramolecular hydrophobic interactions may also show some benefit. Reducing the ionic strength of the solution has been shown to reduce aggregation of cellular debris to inclusion bodies (Dorin 1989).

The mechanisms described imply that the chemistry of the solution and the stability of the protein are important. The strategy to enhance protein passage, therefore, would be to reduce the potential for protein association with other lysate constituents. Efforts should be made to keep the protein in its most soluble and unassociated or unaggregated state. Because each protein/lysate system will behave differently, the aspects of solution conditions should be investigated in each case.

Yield Parameters and System Design. Since the protein of interest is partially retained, yield can be increased by diafiltration. Typically, the liquid is concentrated, then diafiltered. The yield of permeable protein across a batch lysate clarification process is calculated by the following equation:

$$Y = 1 - [X^{(R-1)} \times e^{-N(1-R)}]$$

where X = volumetric concentration factor (initial volume/concentrated volume); R = retention coefficient = (1 − [permeate concentration/feed concentration]); and N = number of wash volumes (wash volume/feed volume).

This equation describes a batch process consisting of concentration followed by constant volume diafiltration. An analysis of the yield equation indicates that even with retention coefficients of 50–60 percent, yields of 90–95 percent are efficiently reached.

As a result of diafiltration, the total permeate volume produced will be greater than the initial feed volume. However, this volume increase can easily be counteracted by including a secondary UF step to concentrate the clarified liquid. These two TFF steps can be integrated, as shown in Figure 3.23. The secondary UF system concentrates the protein prior to chromatographic column loading and also allows the protein-free permeate to be recycled as diafiltration liquid for the clarification step.

Membrane Area Requirements. Membrane area requirements will be a function of the yield objective, protein retention, concentration factor, and process flux. For a goal of 90 percent yield, membrane area can be estimated by using a specific flux of 0.2–0.4 m^2hr/liter of initial lysate. For example, to achieve a 90 percent yield, 25 liters of lysate can be processed in 2 hours with approximately 4 m^2.

Figure 3.23. Integrated clarification and ultrafiltration for lysate processing.

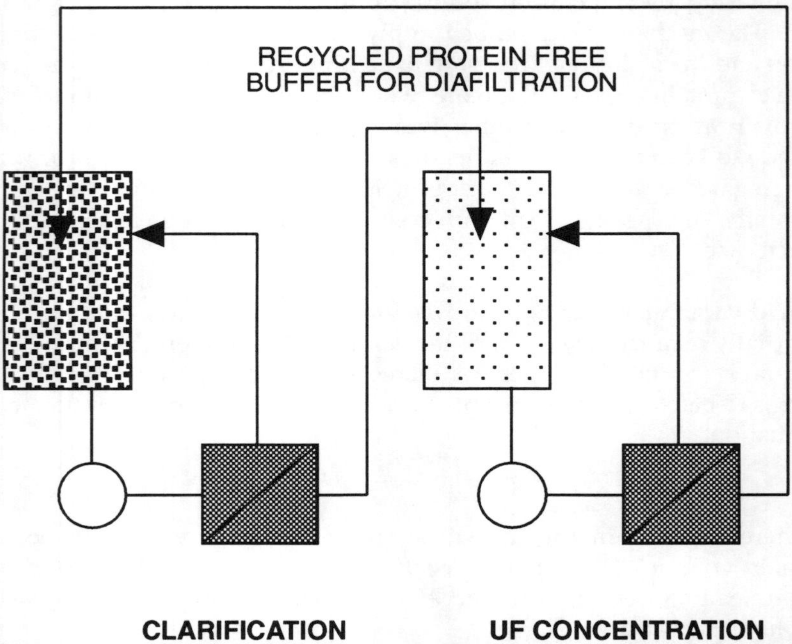

Conclusions

Tangential flow processing for clarifying soluble protein from lysate suspensions is a viable process. Various degrees of success will be obtained due to the complexity of lysate liquids. There are, however, strategies that will increase TFF performance. These strategies rely on choosing the correct membrane and device type, operating the membrane device at the appropriate TMP and cross-flow velocity, and establishing the appropriate solution conditions. These parameters can be easily evaluated on a small scale to verify the applicability of TFF for a particular lysate stream.

Precolumn Clarification

Separation Objective

During the downstream processing of biological products, the fermentation broth is processed through different unit operations, such as cell harvesting, homogenization, concentration, and chromatography. The formation of colloidal precipitates in the initial stages of the process can have adverse effects on the downstream operations of chromatography columns. Tangential flow filtration can be used to protect the columns from plugging, thus lengthening the loading cycle time and increasing the column capacity.

Issues and Solutions

The dynamics of fine clarification are similar to lysate clarification, as discussed previously. The fluxes and the passage of the product through the filter depend on the fine-particle-size distribution and the pore-size distribution of the membrane. If the pore-size distribution of the membrane is too wide, fine particles can lodge in the pores and drastically reduce both the flux and the product passage. Ultrafilters rated to pass the desired molecules or tight microfilters should be used for clarification. High cross-flow rates should be used to sweep the surface of the membrane. High TMPs should be avoided to prevent the penetration of particles into the membrane pores and to minimize consolidation of the cake into an impermeable layer.

Macromolecular Concentration and Diafiltration

Separation Objective

Ultrafiltration can be used to concentrate and diafilter biological molecules. During concentration the volume of the solution is reduced by removing solvent, thus increasing the concentration of the macromolecule. As permeate exits the TFF system during diafiltration, a new solution of different composition is added to replace it. Diafiltration permits desalting and buffer exchange of solutions containing many types of macromolecules.

How Tangential Flow Filtration Performs the Separation

Ultrafiltration membranes retain macromolecules primarily by size exclusion; macromolecules cannot enter or pass through the membrane pores and are thus kept in the retentate. Much smaller molecules (e.g., water, small organic molecules, salts) easily pass through the membrane pores and can be separated from the macromolecules.

The ability of UF membranes to discriminate among macromolecules of similar molecular dimensions is extremely poor. Thus, single-stage TFF systems cannot efficiently separate mixtures of macromolecules that differ less than fivefold in Stokes radius (effective diameter in solution). For these types of fractionations, multistage size-exclusion methods like gel-permeation chromatography are more efficient and cost-effective. Separation of macromolecular mixtures by TFF is discussed in detail later in this chapter.

Macromolecular concentration and diafiltration is perhaps the most widely used application of TFF—primarily with UF membranes. A wide variety of macromolecules is concentrated and diafiltered by UF membranes at every scale, from the laboratory to the manufacturing plant. Each macromolecular solution is processed with a unique composition and a unique separation objective, so it is not useful to provide a list of specific examples here. Instead, this section outlines key principles of concentration and diafiltration with UF membranes that are generally applicable. Papers describing examples of macromolecular concentration and diafiltration experiments are cited below as examples.

Issues and Solutions

Concentration. Concentration operations should be performed with a membrane that demonstrates maximum retention for the macromolecule of interest. During the operation a permeate volume V_p is produced. If the initial volume of the system is denoted by V_i, the final volume V_f of the system is as follows:

$$V_f = V_i - V_p$$

The ratio of the initial volume to the final volume is called the concentration factor, C_f. The final concentration C of the macromolecule can be determined by the equation

$$C = C_i \times (C_f)^R$$

where C_i is the initial concentration of the macromolecule and R is the rejection coefficient of the ultrafilter (Cheryan 1986).

The protein yield Y can be determined from the following expression (Tutunjian 1985):

$$Y = (C_f)^{(R-1)}$$

During the concentration operation the permeate flux drops as the solution becomes more concentrated. The process can be characterized by a time average flux. Most TFF systems are sized based upon this average flux.

Since its inception, TFF has been used to concentrate biologicals. Charm and Lai (1971) compared UF systems for the concentration of catalase solutions. Mitra and Fillmore (1980) concentrated prothrombin complex.

Diafiltration. Macromolecules may be diafiltered to eliminate lower-molecular-weight contaminants. Multiple ways to diafilter macromolecular solutions during TFF exist, such as batch, fed-batch, and constant volume. The most efficient method in terms of minimizing time and volume of diafiltrate solution is constant-volume diafiltration, so this method is most widely used.

In constant-volume diafiltration the system volume is kept constant by the simultaneous addition of the displacing solution to the feed tank. The final concentration C of the removed microsolute can then be expressed as a function of the diafiltrate volume by the following expression:

$$C = C_0 \times \exp(-V_d/V_0 \times [1-R])$$

where R is the rejection of the microsolute by the membrane, C is the final concentration of the microsolute, C_0 is the initial concentration, V_d is the wash volume, and V_0 is the system volume (Tutunjian 1985).

Diafiltration is widely used to purify macromolecules. Cooper and Booth (1979) diafiltered polymeric dyes successfully. Veide et al. (1989) recovered β-galactosidase from polyethylene glycol by diafiltration.

In many applications concentration and diafiltration operations are combined. A macromolecular solution is first concentrated to a certain level and subsequently diafiltered at that concentration. Ng et al. (1976) determined the macromolecule concentration that optimizes the processing time for diafiltration in a combined concentration-diafiltration operation. The optimum concentration C^* at which processing time is minimized can be expressed as

$$C^* = C_g/e$$

where e is the natural logarithm base (2.71828 . . .) and C_g is the concentration, derived from a plot of flux vs. log (concentration), where flux is extrapolated to zero. Cooper and Booth (1978) extended this relation to other applications.

A wide selection of UF membranes, filters, and systems is commercially available for the concentration and diafiltration of macromolecular solutions. Vendors of UF equipment should be consulted to ensure the best match of the membrane, filter, and system to the specific application and separation objective, especially with regard to

- Chemical compatibility and biocompatibility
- Fouling behavior of the membrane
- Cleaning, depyrogenation, and sanitization or sterilization of the filter system and filter

Virus Concentration and Diafiltration

Separation Objective

In the manufacture of viral vaccines and viral antigens (i.e., for diagnostics), TFF permits whole viruses and parts of the viral coat to be concentrated and washed (diafiltered) to remove low-molecular-weight impurities. In this TFF application, the virus or viral antigens are biopolymeric solutes. The membrane chosen should retain the whole virus or viral antigens, while allowing water, salts, and solutes much smaller than the virus or viral antigens to pass.

Because whole viruses are typically greater than 25 nm in diameter, UF membranes with NMWL greater than 100 kDa are usually employed. These UF membranes concentrate whole viruses with little or no loss (>99 percent virus retention). Ultrafiltration membranes with lower NMWL ratings may be required to concentrate viral antigens that are smaller than the whole viruses. To retain viral antigens, the membranes may sometimes have 10 kDa or 30 kDa NMWL ratings. In all cases the product (virus or antigen) is in the retentate from the TFF system, while the waste solution is in the permeate.

How Tangential Flow Filtration Performs the Separation

Ultrafiltration membranes separate solutes by size exclusion. During the UF of viruses the virus particles or viral antigens themselves are too large to enter and pass through the pores of the UF membrane and are retained in the retentate solution, where they are concentrated. Water, salts, and other small molecules (e.g., formalin or other solvents used to inactivate viruses) are removed through the membrane pores. Examples of virus concentration and related applications are shown in Tables 3.14 and 3.15.

Issues and Solutions

Virus concentration should be clearly differentiated from the application of virus removal from drugs. In virus concentration the product is the virus or viral antigen, whereas in virus removal the product is a virus-free drug. To obtain a high yield of virus or viral antigen during virus concentration, a membrane need only provide >99 percent retention of the virus or antigen. However, this level of virus retention would not assure a virus-free permeate. Virus removal is discussed later in this chapter.

Virus concentration in buffers or aqueous solutions with low protein contents is straightforward and similar to the concentration of macromolecules by UF, as discussed previously. The UF membrane should be selected, according to the manufacturer's recommendation, to provide >99 percent retention of the virus or viral antigen. The device should be selected, again according to the manufacturer's recommendation, to be able to concentrate the virus suspension or antigen solution from the starting volume to the concentrated volume in the desired time. Most viral vaccines and viral suspensions are prepared

Table 3.14. Examples of Virus Concentration by Tangential Flow Filtration

Virus Application	Device	NMWL, Daltons	Typical Conditions (cross-flow, pressure)	Typical Fluxes ($\ell/m^2/h$)	Reference
Polio	Cassette	100,000–300,000	8–12 $\ell/min/m^2$ 34–69 kPa	20–40	Millipore 1987
Influenza	Cassette	1,000,000		30–40	Vaheri et al. 1981
Epstein-Barr	Cassette	1,000,000	12 $\ell/min/m^2$ 83 kPa	22	Shibley et al. 1980
Measles	Cassette		12 $\ell/min/m^2$ 103 kPa	22	Bellini et al. 1979
Arenavirus	Cassette	1,000,000	12 $\ell/min/m^2$ 69 kPa	32	Gangemi et al. 1977

Table 3.15. Viral Vaccines Concentrated at Clinical or Production Scale by Tangential Flow Filtration

Polio	Influenza	Hepatitis A
Mumps	Feline leukemia	Rabies
Hepatitis B	Measles	Rubella
HIV	Foot and Mouth disease	

by rapidly filtering small volumes to preserve antigen activity and minimize product loss. Thus, the most frequently selected TFF devices for this application are low-working-volume, small-area devices with high filtration rates, such as narrow-channel, flat-sheet devices; small-area, narrow-channel, spiral-wound devices; and small-area, hollow fine-fiber devices.

A typical process includes concentrating the feed viral suspension or antigen solution to its desired maximum concentration (determined by solubility and stability), followed by constant-volume diafiltration (if desired) to remove salts, inactivation agents, or other low-molecular-weight impurities. A process incorporating 10× concentration followed by 4–5 volumes of diafiltration would remove 99.9 percent of salts and other small (nonproteinaceous) solutes in most cases.

Membranes and TFF devices should be compatible with all chemicals in the virus suspension or antigen solution. Most UF membranes and devices are thoroughly compatible with aqueous solutions of viruses or viral antigens, but the presence of an inactivating organic solvent may limit membrane and device choices. For example, the presence of formalin permits the use of most commercially available UF membranes and devices, but the presence of phenol or acetone might limit the choice to devices containing regenerated cellulose membranes and polypropylene support structures.

Because virus suspensions and viral antigen solutions always contain proteins and may contain other biopolymers, these streams will foul hydrophobic UF membranes and reduce the fluxes possible with these membranes. This fouling is not serious enough in most cases to prevent UF with hydrophobic membranes like polysulfone membranes. In fact, protein fouling of the membrane may increase the retention of the virus or antigen. If fouling results in a filtration rate too low to be economically attractive, the filtration rate can be increased with hydrophilic UF membranes (e.g., regenerated cellulose, cellulose acetate, and modified polysulfones) shown to be low protein binding and low fouling.

When concentrating whole viruses or viral antigens, yield losses result from (in order of importance)

- Losses due to passage of the virus or antigen through the filter
- Losses due to failure to flush the concentrated virus or antigen out of the TFF system at the end of the batch cycle
- Losses due to adsorption

Losses due to virus or antigen passage are minimized by selecting a membrane with >99.5 percent retention of the desired virus or antigen. In most cases empirical retention measurements must be obtained to confirm correct membrane selection. Some traditional asymmetric, thin-skinned UF membranes with high-molecular-weight-limit ratings tend to have a variable incidence of microdefects, leading to reduced or variable retention from filter to filter. If this problem occurs, retention can sometimes be improved without sacrificing flux by selecting high-NMWL membranes whose structures lack subsurface macrovoids ("void-free" UF membranes).

Losses due to insufficient flushing are easily minimized through a variety of techniques. For example, at the end of the concentration-diafiltration procedure, drain the system as thoroughly as possible to recover a maximum amount of retentate. Strict observance of the U.S. Department of Agriculture's 3A sanitary equipment design rules in TFF system design can ensure the maximal drainability of systems.

Another way to minimize losses is to fill the system to its minimum working volume (i.e., the minimum volume needed to operate the system with no air introduction or foam generation) with a buffer compatible with the retentate solution. Instead of fresh buffer, an equal volume of permeate recently removed from the system may be used. Circulate this volume of solution through the feed/retentate side of the TFF system for 3–5 minutes, with the permeate flow restricted, to resuspend virus or antigen deposited on the membrane surfaces. Do not shut off the permeate flow when recirculating buffer, as the resulting Starling flow within some TFF devices (permeate flow in reverse across the membrane where feed-side pressure is less than average permeate side pressure) may rupture some membranes. Stop the recirculation pump for 3–5 minutes to let the virus or antigen diffuse from the boundary layers of the TFF system into the bulk solution, then resume recirculation for 1–2 minutes.

A third way to minimize losses due to insufficient flushing is to pump out or drain the flush solution and add it to the retentate.

As for adsorptive losses of virus or antigen, if the total amount of virus or antigen in the feed is small (e.g., <0.2 g/ℓ), the losses due to adsorption to the membrane are significant only if no other proteins are present to compete with the virus or antigen for binding to the membrane. Antigen or virus losses shown to be caused by adsorption to the membrane can be minimized in most cases with hydrophilic UF membranes shown to be low protein binding.

Tangential flow filtration systems for concentrating viruses or antigens are best cleaned with 0.1 N NaOH at 45°C for 1 hour to hydrolyze proteins and other biopolymers, saponify fats, and destroy bacterial endotoxins (if any). Most UF membranes cannot be readily steam-sterilized without causing changes in their pore size distributions and retention properties, which may change the separation achieved in successive uses of the filter; such changes may complicate process validatability. Chemical sanitization of the TFF system is satisfactory for preventing contamination of subsequent batches. To sanitize the system, after flushing the NaOH from the system, it should be rinsed for 15–30 minutes with 50 ppm free chlorine (as NaOCl) adjusted to pH 7.0, which is highly virucidal and bactericidal. An alternative sanitant is 200 ppm peracetic acid, sold in commercial preparations for this purpose.

The TFF system, membrane, and filter device should be stored between uses according to the manufacturer's recommendations. Commonly used storage solutions include 2–3 percent formalin, 0.1 percent quaternary ammonium salt solutions, 0.1 N NaOH, or phosphoric acid (pH = 2), all of which prevent growth of microorganisms in the system.

Macromolecular Separations (High-Resolution Tangential Flow Filtration)

Separation Objective

Tangential flow filtration can concentrate macromolecules in aqueous solutions and remove salts and other small molecules from these solutions. However, unlike chromatographic separations, which are high-resolution multistage separations, most economical TFF processes involve only a single separating stage. For this reason conventional TFF systems cannot fractionate mixtures of macromolecules in a complex solution as completely or efficiently as chromatography, especially if the macromolecules are of similar size.

This limitation appears as the inability of UF membranes to separate macromolecular solutes differing in molecular diameter by as much as a factor of 5–10×. However, the resolution achievable with single-stage TFF separations can be maximized. These methods are referred to as high-resolution tangential flow (HRTF) filtration.

In HRTF filtration the objective is most frequently to separate, in a single-stage TFF process, two macromolecular species that differ in molecular diameter (not necessarily molecular weight) by at least a factor of 3–5×. The larger macromolecule is concentrated in the retentate, while the smaller macromolecule is removed in the permeate. Diafiltration permits maximum removal of the smaller macromolecule from the larger macromolecule, but results in dilution of the permeate solution.

An alternate but similar objective for HRTF filtration is less specific, involving the removal of a mixture of smaller macromolecules from a mixture of larger macromolecules—in other words, a crude fractionation of a polydisperse solution into two polydisperse solutions, one of larger macromolecules in the retentate and one of smaller macromolecules in the permeate. Again, due to the inherent limitations of UF membranes and devices, both solutions will contain solutes in the intermediate size range while being enhanced in either the larger or smaller molecules.

How Tangential Flow Filtration Performs the Separation

As in simpler macromolecular concentration applications, the UF membrane retains larger macromolecules by size exclusion. Molecules larger than the largest pores in the membrane are retained approximately 100 percent. Molecules smaller than the smallest pores in the membrane are not retained. Intermediate-sized molecules are partially retained, depending on the membrane's pore-size distribution and chemical interaction with the solutes (e.g., adsorption), plus the amount of concentration polarization and/or gel-layer formation occurring in the TFF device.

An example of the retention behavior of UF membranes challenged with polydisperse mixtures of macromolecules is shown in Figure 3.24, in which a selection of UF membranes is challenged under low-polarization conditions with a polydisperse mixture of dextrans. This challenge test is in fact used for quality control of some UF membranes because of its ability to discern variations in the pore-size distribution and retention performance of UF membranes. The curves in Figure 3.24 are calculated from the concentrations of many different-sized dextran molecules in the permeate and retentate from the UF membrane device. To illustrate the benefit of the HRTF filtration technique, Table 3.16 provides experimental results of a very difficult protein/protein separation—IgM from IgG and albumin in plasma—which cannot be achieved with traditional single-pump UF systems.

Macromolecular separation by HRTF filtration remains a novel application of UF membranes, one that can benefit from improvements in UF membranes, TFF devices, and operating conditions. In the example the flux that ensures a low and reproducible retention of IgG is extremely low, so this application is generally not practical to

Figure 3.24. Polydisperse dextran retention profiles of a family of UF membranes (Millipore 1991).

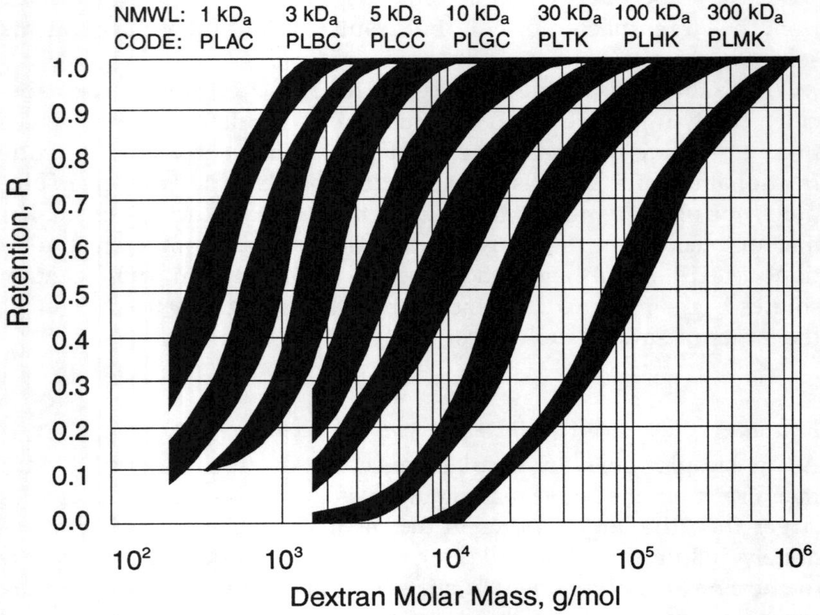

perform by UF alone. For these reasons, there are few commercial applications of HRTF techniques today. Examples of macromolecular separations that can be conducted or enhanced with HRTF techniques are listed in Table 3.17. Most HRTF applications in commercial operation or under development are confidential, so no operating data can be reported here.

Issues and Solutions

The limited resolution permitted by UF membranes in macromolecular separations arises from two interacting causes:

1. **The retention profile of UF membranes.** All UF membranes demonstrate rejection profiles having the sigmoid shapes shown in Figure 3.24. This rejection curve demonstrates a limit to how well a UF membrane can discriminate among macromolecules of similar size or molecular weight.

Table 3.16. Example—Separation of IgM from IgG and Serum Albumin in Plasma (Millipore 1992b)

Device	NMWL (Daltons)	Flux (ℓ/m²/h)	Retention of Protein		
			IgM	**IgG**	**Albumin**
Plate/frame	300,000	0.78	100%	24%	0%
Plate/frame	300,000	4.0	100%	45%	0%
Plate/frame	300,000	7.8	100%	60%	10%

Table 3.17. HRTF Application Examples

Removal of PEGs from proteins
Removal of serum or allantoic proteins from viruses
Crude fractionation of protein mixtures by removal of small proteins from much larger proteins

2. **Polarization and fouling**—the accumulation at the membrane surface during filtration of macromolecules. The accumulated layer is usually a combination of macromolecules adsorbed onto the membrane surface and within its pores (fouling) and macromolecules concentrated above the membrane surface during filtration (concentration polarization). As shown in Figure 3.25, polarization and fouling reduce the membrane's effective pore size and permeability, causing the fouled membrane to retain solutes that might pass through the unfouled, unpolarized membrane.

In the absence of any fouling or polarization, the pore-size distribution of a membrane determines its maximum resolving capability. Any fouling or polarization further diminishes this resolving capability. Thus, to obtain the best performance from tangential flow filters in clarifying macromolecular solutions or in separating large macromolecules from small macromolecules in solution, both fouling and concentration polarization must be minimized.

To minimize fouling, a membrane material that minimizes adsorption of the macromolecular solutes must be chosen. For protein

Figure 3.25. Effect of retained and adsorbed species (fouling) on the effective pore size of UF membranes (Millipore 1991).

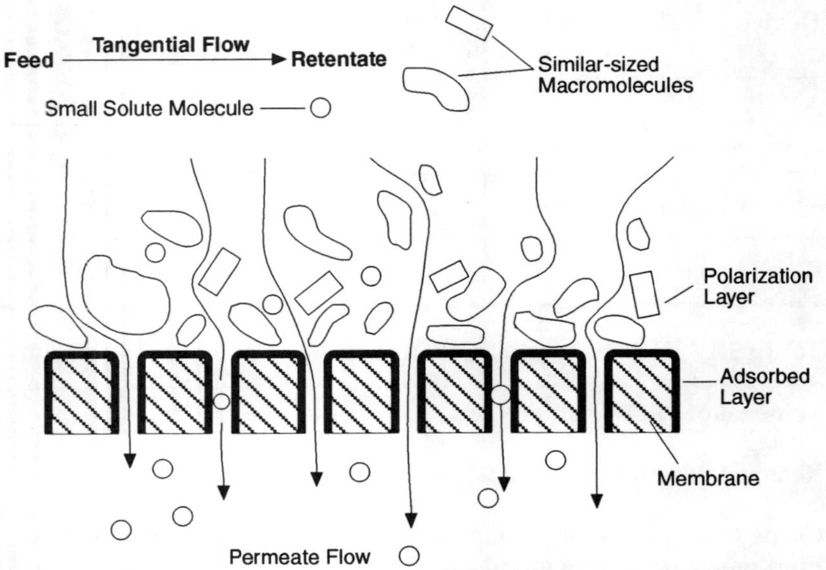

solutions this requires the use of low-protein-binding membranes. Commonly found UF membrane materials with very low protein adsorption include regenerated cellulose and cellulose acetate.

The pore-size distributions of these low-protein-binding membranes are changed very little by irreversible adsorption of proteins during filtration. Protein-binding data for these membranes and others are shown in Table 3.18.

Once fouling has been minimized by using a low-protein-binding membrane, polarization must be minimized to improve separation resolution. Conventional single-pump TFF systems are designed to increase flux instead of separation resolution. While reducing polarization, high tangential flow rates create high inlet pressure in the feed channel of the filter, as shown in Figure 3.26. High tangential flow rates control polarization at higher TMPs to yield greater filtration rates. Higher inlet TMP, however, causes polarization to be greater at the inlet to the filter than at the retentate outlet. Thus, single-pump TFF systems provide high volumetric flux but do not minimize polarization across the entire membrane surface in the TFF system.

To minimize polarization throughout the filter, control of TMP and flux must be decoupled from the tangential flow rate. Flux is most easily and reproducibly controlled by metering permeate out of the TFF

Table 3.18. Adsorption of Proteins to Various Membrane Materials (g/m²) (Millipore 1992a)

Membrane Type	Bovine Serum Albumin	IgG
Cellulose	0.34	0.32
Cellulose acetate	0.50	n/a
Polyvinylpyrrolidone-polyethersulfone blend	0.82–1.26	0.78
Polyethersulfone	0.73–1.81	1.58–1.90

Figure 3.26. Controlling cross-flow rate and transmembrane pressure in TFF devices (reproduced by permission of Millipore Corp.).

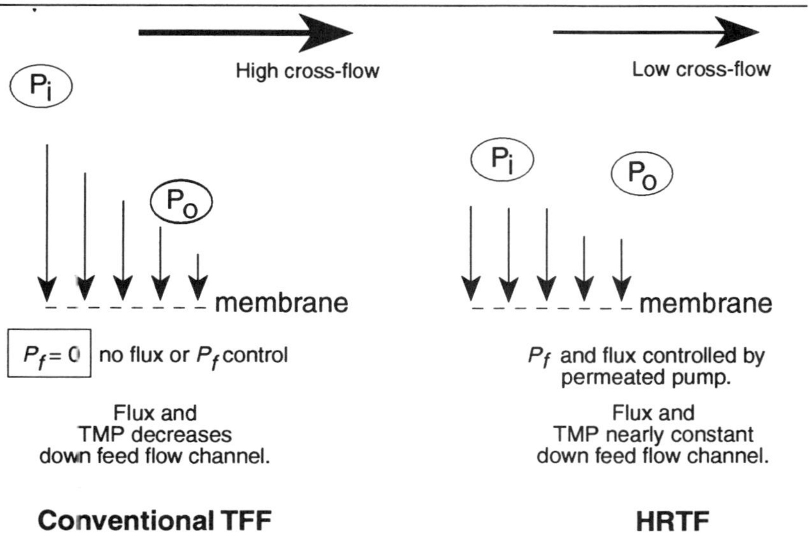

system with a second pump, as shown in Figure 3.27. With a two-pump HRTF system, tangential flow rate and feed-side pressure drop are controlled by the recirculation pump, while permeate flux and TMP are controlled separately by the permeate metering pump. Tangential flow rate and feed-side pressure drop can thus be reduced to provide more uniform TMP throughout the filter, while polarization is controlled with the permeate pump at the reduced tangential flow rate.

Figure 3.27. Two-pump TFF system.

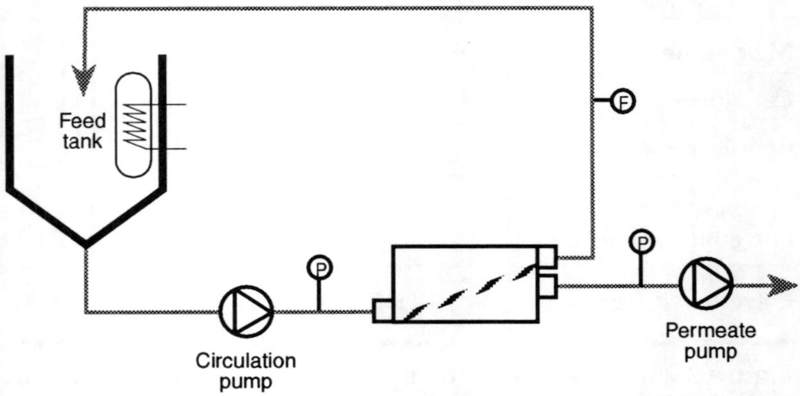

In HRTF applications there is critical flux at any specified tangential flow rate and feed concentration, above which lower-molecular-weight species are highly retained. This critical flux usually corresponds to TMPs as follows:

- ≤6.9 kPa for ≥300 kDa NMWL UF membranes
- 20.7–69 kPa for <300 kDa NMWL UF membranes

Below the critical flux, lower-molecular-weight species easily pass through the membrane. If a solute passes easily, it can be separated from the retained species quickly with limited diafiltration. However, if a solute has moderate-to-high retention, it either cannot be removed or removal requires a large volume of diafiltration buffer and a larger TFF system.

For this reason HRTF techniques are not intended to maximize volumetric flux but to maximize solute mass flux by significantly increasing passage of the solute through the membrane and to reduce total liquid volume to be filtered by reducing the diafiltration requirement.

Thus, the important parameter in HRTF applications is the critical flux value. Critical flux is easily determined by measuring retention of a solute as a function of permeate flux at fixed concentration and tangential flow rate. The typical relationship of these two variables is shown in Figure 3.28(a), which shows the critical flux value as the inflection point in this plot, where retention starts to increase sharply. Figure 3.28(b) shows the plot of mass flux (volume flux × concentration × [1 – retention]) as a function of volumetric flux for the same data.

As shown in Figure 3.28, the critical flux decreases as the concentration of retained species in the feed increases. Therefore, in measuring the critical flux for an HRTF application, retention of the passing solutes as a function of flux must be measured for several concentrations of retained solutes (or insoluble particles). The lowest critical flux will occur at the highest concentration of retained species.

Recommended Conditions for HRTF Operation of Tangential Flow Filters

For UF, the best resolution for separation of large soluble macromolecules from smaller soluble macromolecules occurs when

- The large macromolecule is at least 3× larger (in diameter) than the smaller macromolecule.
- Low-protein-binding UF membranes are in use.
- The membrane is selected to retain the large macromolecule ⩾99.5 percent to minimize losses of the retained macromolecule during diafiltration and concentration.
- The flux is controlled at or below the critical flux.

For UF separations the critical flux is usually in the range of 5–20 $\ell/m^2/h$, which corresponds to a transmembrane pressure ⩽6.9 kPa.

The tangential flow rates or feed-side pressure drop conditions recommended for HRTF operation of TFF systems should be obtained from the filter manufacturer.

As discussed in the preceding sections on macromolecular concentration and virus concentration, in HRTF applications the membranes and TFF devices should be compatible with all chemicals in the feed solution. The manufacturers' guidelines on chemical compatibility usually suffice to guide selection of suitable membranes and devices.

Reproducibility of HRTF separations is much more dependent than traditional concentration applications upon the avoidance of fouling. Fouling modifies the membrane's pore-size distribution by narrowing pores and/or occluding their entries, thus changing the passage of partially retained solutes. Thus, a membrane that fouls noticeably may induce a time-dependent and difficult-to-control variation in the final composition of the retentate and permeate streams when attempting to fractionate a complex macromolecular mixture. For this reason it is far easier to avoid fouling by use of low-fouling, highly hydrophilic membranes (e.g., regenerated cellulose) than to try to compensate for fouling effects by extremely tight control of feed composition, processing time, and filtration conditions (time, temperature, cross-flow rate, and pressures).

Figure 3.28. Critical flux demonstration graph (Millipore 1992a).

(a) Retention vs. flux.

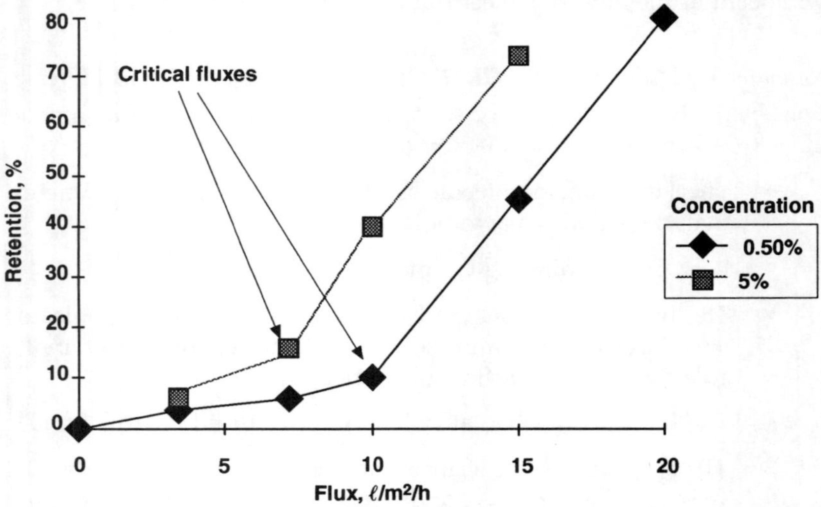

(b) Mass flux of passed component vs. volume flux.

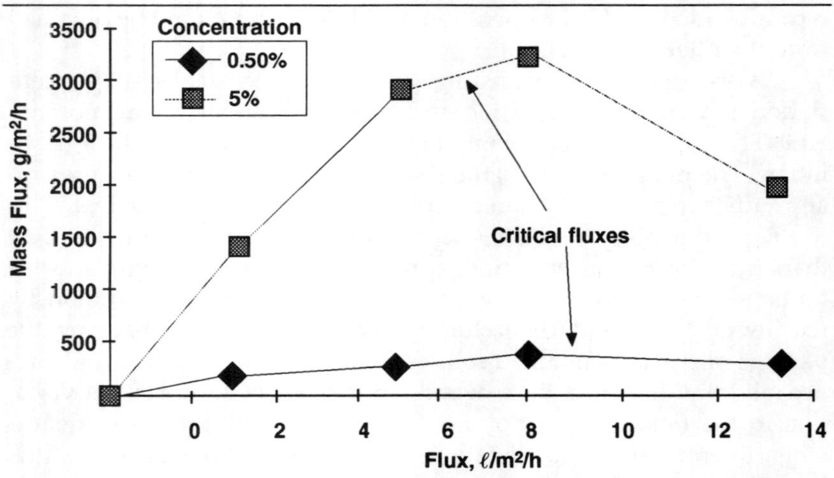

As with macromolecular concentration applications, yield losses result from the following:

- Passage of solutes desired to be retained or retention of solutes desired to pass through the membrane

- Insufficient flushing of the device to remove retentate and permeate solution
- Adsorption to the membrane

Since a low-binding membrane should be selected for all HRTF applications, adsorptive losses are usually negligible. Membranes should be selected to maximize retention of the large solutes desired to be retained at high yield. The amount of diafiltration should be adjusted to attain the desired yield of solutes through the membrane. Although increased diafiltration will dilute the smaller solutes in the permeate, these can always be reconcentrated if desired via a second UF stage containing a membrane that retains the smaller solutes. Losses due to inadequate system flushing can be avoided as described in the section on viral concentration.

HRTF filtration systems should be cleaned according to the recommendations of the vendor of the UF devices as in other UF applications.

Virus Removal from Biopharmaceuticals

Separation Objective

As the end of the century nears, blood products, immunotherapeutic drugs, infectious disease combatants, and bioactive factors are emerging as the most promising biopharmaceuticals, both financially and medically. There is, however, a long history of viral contamination of biotherapeutic products. Recently, the discovery of viruses such as hepatitis C and parvovirus B19 emphasize the vigilance required for new biotherapeutic proteins. Three segments of the industry have been significantly affected: the blood-derived products industry, particularly coagulation factors; tissue-extracted hormones; and biopharmaceuticals, notably recombinant proteins produced via mammalian cell fermentation.

The accepted practice for removing viral contaminants in these products is to include processing steps that inactivate virus; the most common practice is heat or chemical treatment. These techniques are not universally applicable; however, they require optimization for specific viruses and can only be used with heat- or chemically stabilized proteins. To augment these inactivation techniques, manufacturers rely on achieving virus removal as an indirect consequence of other downstream processing steps. These processes should be validated for their viral removal capability.

Virus-removal strategies are an outgrowth of those adopted in preparing naturally derived biologics. Complete removal of all viral contaminants, although desired, has been proven impossible with a single operation. Assurance of 12–15-log overall reduction capacity across the downstream process is generally required for parenteral protein solutions; some reports quote up to 20-log clearance. This is accomplished today using a strategy that combines physical removal with inactivation techniques; the viral clearance of individual purification operations is combined with that of at least one inactivation operation.

Other purification operations designed for protein recovery have some capability to remove virus as a byproduct of normal operation. It is accepted that the clearance achieved by complimentary operations, those based upon different physical principles, are additive and can be combined to contribute to the overall clearance.

Virus-removal technologies fall into three categories: chemical inactivation methods, physical inactivation methods, and physical removal methods. Nonspecific chemical and physical inactivation methods have the capability to remove 4 to 7 logs virus, but are limited by simultaneous protein denaturation. Concerns over protein recovery and denaturation have resulted in shorter exposure times and the use of protein stabilizers, both of which reduce the efficacy of the inactivation procedure. These concerns are paramount in the biotechnology industry, where the protein product is expensive to produce and product purity (free from potentially antigenic denatured forms and additional stabilizers) is essential. Physical removal techniques do not cause protein denaturation and do not require the addition of stabilizers.

In all applications a virus-removal system must be capable of achieving 4–6 logs virus removal, particularly of retroviruses (100 nm size) where ≥6-log removal is required. Additionally, active protein recovery must be high, greater than 90 percent and, in most cases, 95–97 percent. The performance of the virus-removal step must be validatable and scalable to production capacities. Viral clearance and protein recovery measured and validated in small-scale units must be conserved at both pilot and production scale.

Until recently, virus removal by membrane filtration has been limited by the presence of small membrane and device defects that affect integrity. As a result of these defects, membranes have been reported to reduce virus concentrations by 2–4 logs, depending on the viruses studied and the protein solutions within which they are suspended. New membrane and device technology has overcome these limitations, but a thorough examination of the performance of competitive membranes must be done. Only membrane filtration can be used in a continuous process; all others must be operated in a batch (or semicontinuous) mode.

How Tangential Flow Filtration Performs the Separation

Tangential flow filtration devices to remove virus and virus-like particles use size exclusion by UF membranes as the predominant mechanism. More recently, thin-film composite UF membranes, comprising a thin surface film on a microporous layer, have also been used for virus removal applications.

In a TFF virus-removal device separation of virus from the protein of interest is conducted by passing the feed solution over the surface of the filter membrane. The feed solution enters through one set of channels in the device; the retentate is accumulated at a different port at the opposing side and is recirculated tangentially over the membrane where virus and other larger particles are retained. The permeate stream containing the product solution with reduced viral load exits through a different port in the permeate collection section of the device.

Issues and Solutions

Material of construction of the membrane is critical to the intended application. Criteria for selecting a particular membrane type include solution pH, pore size or molecular-weight cutoff (MWCO), temperature stability of the polymer, absorptive characteristics, and the hydrophobicity of the membrane. Solution pH is critical in terms of chemical compatibility with specific polymers and protein solution stability. The temperature at which the removal system operates will also determine the membrane chosen; certain membranes perform optimally at room temperature or higher.

The hydrophobic nature of a membrane is important because the more hydrophilic the membrane, the less fouling may occur under optimized operating conditions. A low-protein-binding membrane with reproducible sieving behavior and easy cleanability is preferred for virus-removal applications. Membrane materials used in virus-removal TFF devices are cellulose acetate, polyamide, polysulfone, and polyvinylidene fluoride. Devices designed for virus removal include flat-sheet, spiral-wound, and hollow-fiber units.

Membranes currently offered in flat-plate devices include 300 kDa and 200 kDa polysulfone membranes as well as polyvinylidene difluoride (PVDF) membranes designed for virus removal from proteins smaller than 180 kDa. Characteristics of these devices include uniform liquid flow, moderate volume-to-area ratios, moderate flux performance, and moderate ease in cleaning.

Currently, several ultrafilters based on flat-sheet membranes are most often used to remove or reduce viruses in a purification train. Examples include the following (see also Tables 3.19 and 3.20):

- Omega polysulfone ultrafiltration Centrasette cassettes (200 kDa and 300 kDa) (Filtron, Northborough, MA)
- Spiral-wound ultrafilters (e.g., those provided by Amicon Division of W.R. Grace, Inc., Danvers, MA) containing membranes with 30 kDa or 100 kDa NMWL ratings
- Composite PVDF Viresolve/70 and Viresolve/180 modules (Millipore, Bedford, MA)

These devices are used in a total purification scheme that has two different objectives. The cassettes and spiral-wound ultrafilters are most often operated in a primary protein-purification step that also augments virus reduction. In contrast, the Viresolve devices are designed specifically for validatable virus removal and may be used early or late in the protein-purification train.

Also available as a specific virus-removal device is the Planova hollow-fiber device family (Asahi Chemical Industry Co. Ltd., Chiyoda-ku, Tokyo). Four types of membranes are available, with pore sizes ranging from 15–75 nm.

Virus-removal devices may be operated in either standard TFF mode (feed-recirculation pump only) or high-resolution TFF mode (feed-recirculation and permeate-metering pumps). When employing standard operation, the single pump recirculates the virus-contaminated solution over the surface of the membrane; a valve is used to control TMP. The permeate flux varies and typically decreases over the course of processing.

During standard operation there is less control over the polarization layer. As discussed earlier, operations requiring the separation of macromolecules that differ in size by less than a factor of three should employ a two-pump HRTF filtration system. This method controls the flux throughout the course of processing and improves control over the protein polarization layer for a more reproducible separation.

Various modules may be chemically sanitized or steam-sterilized, depending on the membrane type and other materials of construction. Typical chemical sanitization methods include flushing with dilute solutions of sodium hypochlorite (300 ppm), sodium hydroxide (0.1 N), alcohol (70 percent volume in water), or peracetic acid-hydrogen peroxide solutions. Manufacturers recommend sanitizing agents based on membrane compatibility as well as expected bioburden. If a cleaning step is strictly designed for postfiltration use and specifically employed for its virucidal/sporicidal strength, it need not be compatible

Table 3.19. TFF Devices Used for Virus Removal

Device Type	Manufacturer	Effective Surface Area	Material of Construction	Mode of Operation	Sanitization/ Sterilization	Application
Viresolve 70, 180 (Flat Plate Modules)	Millipore (Bedford, MA)	10 cm^2 to 10 ft^2	PVDF	HRTF filtration	Sodium hypochlorite/ autoclave	Biopharmaceuticals and plasma derivatives less than 180 kDa
YM-100 (Spiral-Wound)	Amicon (Danvers, MA)	1.0 ft^2 to 40 ft^2	cellulosic, polysulfone, polypropylene, epoxy-filled polyester	Standard TFF	None specified	Plasma derivatives <100 kDa
Planova 15, 35, 40, 75 (Hollow-Fiber)	Asahi (Tokyo, Japan)	0.001 m^2 to 1.0 m^2	cuprammonium-regenerated cellulose, polycarbonate/ polyurethane	Standard TFF	NS/autoclave	Plasma products and biopharmaceuticals
Omega 200 K, 300 K	Filtron (North-borough, MA)	up to 27.5 ft^2	polyethersulfone polyurethane	Standard TFF	None specified	Biopharmaceutical products <300 kDa

Table 3.20. Performance of Virus Removal Devices (log reduction values)

Device	Virus								
	Polio	**HCV**	**JEV**	**Sindbis**	**HBV**	**SV-40**	**Reo-3**	**HIV-1**	**MuLV**
Omega 200K	Not tested	Not tested	Not tested	Not tested	Not tested	5.5	6.0	Not tested	>4.6
Omega 300 K	3.0	Not tested	Not tested	Not tested	Not tested	<1.0	2.6	Not tested	4.0
Planova 15	5	>4	>7	>8	Not tested	>8	Not tested	Not tested	Not tested
Planova 35	0.2	4	5.5	>8	6	6.5	Not tested	>5	Not tested
Planova 40	Not tested	3.5	4.5	>8	5	Not tested	Not tested	>5	Not tested
Planova 75	Not tested	1.5	1.8	Not tested	1.9	Not tested	Not tested	>5	Not tested
Viresolve 70*	4.0	Not tested	Not tested	7.0	Not tested	6.44	6.6	8.5	>6.3
Viresolve 180*	3.1	Not tested	Not tested	6.9	Not tested	Not tested	7.6	Not tested	>6.6

*Viresolve results are representative of worst case process conditions.

with the membrane. Those designated for prefiltration use must fit both criteria.

Depending on the membrane type, sterilization by autoclaving is an option. Some polysulfone devices (such as the Omega-series devices) are not autoclavable and require chemical sanitization. Other devices, such as the Viresolve and Planova, are designed to withstand a single autoclave sterilization cycle.

Small Molecule Concentration and Desalting by Reverse Osmosis and Nanofiltration

Separation Objective

Tangential flow filtration using membranes with NMWL in the range of 100–3000 is often referred to as reverse osmosis or nanofiltration. Reverse osmosis membranes are capable of retaining low-molecular-weight pharmaceutical molecules and salts. In contrast, NF membranes retain low-molecular-weight pharmaceutical molecules and small macromolecules while passing water and some salts. The objective of RO and NF, therefore, is to concentrate (and desalt, in the case of NF) aqueous solutions of pharmaceutical molecules, such as antibiotics and small polypeptides. Removal of organic solvents from these types of solutions is discussed later in this chapter.

How Tangential Flow Filtration Performs the Separation

Reverse-osmosis and NF membranes are cast from polymers such as the following:

- Polysulfone or polyethersulfone (sometimes sulfonated)
- Regenerated cellulose (NF only)
- Cellulose acetate
- Polyamides (often as thin-film composites on polysulfone UF supports)

These membranes are designed with average pore sizes and pore-size distributions low enough to retain pharmaceutical molecules in the range of 100–3000 Daltons. Membranes with NMWL ratings above 1000 Daltons ("tight" ultrafilters) typically pass most or all dissolved inorganic salts at high levels, while membranes with 300–1000 Daltons NMWL ratings (nanofilters) pass monovalent salts in the range of 20–60 percent but retain multivalent salts at levels of 80–100 percent. Membranes with NMWL ratings from 100–300 Daltons (RO membranes) retain all salts and pass only water.

These membranes retain uncharged and/or larger-molecular-weight pharmaceutical molecules primarily by a size-exclusion mechanism. Salt retention, where it occurs, may be due to a combination of Donnan (charge-repulsion) exclusion, size exclusion, and low solubility in the membrane matrix. Examples of concentration and desalting of small pharmaceutical molecules are provided in Table 3.21.

Issues and Solutions

All these membranes effectively concentrate retained solutes, as none retain water. However, membranes with lower NMWL ratings are less permeable and thus require operation at higher TMPs to produce acceptable filtration rates. In addition, for small solutes the TMP will have to exceed the osmotic pressure difference between the feed and filtrate to induce flow through the membrane. If salts are retained along with pharmaceutical molecules, the combined osmotic pressure of all retained solutes will increase as the solution is concentrated, eventually equaling the TMP, thereby stopping filtration. Therefore, there is a practical limit to the maximum concentration achievable by this method, set by the osmotic pressure difference between the retentate and filtrate. For this reason RO and NF are normally conducted at higher pressures than UF—most often 1.4–4.2 MPa (200–600 lb/in.^{2}).

The most cost-effective and widely sold tangential flow filter geometry designed for operation at these pressures is the spiral-wound

Table 3.21. Small Pharmaceutical Molecules Concentrated and Desalted by Reverse Osmosis and Nanofiltration

Retained Solute	Retention	Passed Solutes	Retention	Pressure, MPa (lb/in.^{2})	Average Flux (ℓ/m^2/h)
Penicillin G, K salt	>99%	NaCl	<50%	2.76 (400)	90
Aspartame	>95%	NaCl	<50%	2.76 (400)	65
Penicillin G, K salt	>99%	Methanol	<30%	2.76 (400)	30
6-amino-penicillanic acid	>99%	Water	0%	2.76 (400)	60

cartridge, although a limited number of flat-sheet and hollow-fiber RO/NF devices have appeared on the market over the years. Some typical operating conditions for RO and NF devices are listed in Table 3.22.

The salt retention behavior of a chosen NF membrane determines the ability of that membrane to desalt a retained solute. Membranes with higher salt retention will require much larger volumes of water to diafilter salt away from the retained solute, thereby requiring larger membrane areas and generating larger waste-filtrate volumes. The pharmaceutical product must be retained at typically >98 percent to ensure high recovery. Membranes with NMWL ratings of 1000 Daltons or higher can desalt retained products quite effectively because they do not retain salts. However, if membranes with NMWL ratings below 1000 Daltons are required to retain the pharmaceutical product, these cannot remove multivalent salts (e.g., Ca, Mg, SO_4, PO_4) from these products without excessive area and diafiltration volume requirements. Therefore, desalting of molecules below 1200 Daltons usually requires an ionic separation method such as ion exchange, continuous electrodeionization, or electrodialysis.

Maximum yield of the retained pharmaceutical product may be ensured by the following techniques:

- Selecting a membrane with maximum retention of the product (>99 percent if possible)
- Avoiding scaling during operation
- Designing the RO/NF system for maximum drainability
- Flushing the system carefully with the minimum required volume of water after draining and adding this flush to the retentate

The chemical compatibility of spiral-wound RO/NF devices is limited by all wetted materials of construction in the cartridge—usually including membrane, substrate, adhesive, feed and permeate carrier screens, center tube, housing, antitelescoping disk, gaskets, and outerwrap (if any). Most of these devices are designed to filter aqueous solutions only and are susceptible to damage (swelling, extraction, dissolution, or softening) if exposed to organic solvents. Low concentrations (<50 percent) of ethanol and methanol have been successfully diafiltered from aqueous mixtures for short periods of time without damage to polyamide thin-film composite membranes. Filtration of other organic solvents is discussed in the next section of this chapter.

Reverse osmosis and NF membranes are also susceptible to hydrolytic attack at extreme pH; they lose retention if exposed to acid or alkali. For example, the continuous operating pH ranges of RO/NF membrane materials are typically the following:

Table 3.22. Typical Operating Conditions for RO/NF Devices

NaCl Retention, %	Diameter, cm (in.)	Cross-flow Rate, ℓ/min	Pressure, MPa (lb/in.^{2})
RO Membranes Polyamide thin-film composites			
>99	10 (4)	40	3.5 (500)
>99	20 (8)	150	3.5 (500)
98.9	10 (4)	40	5.5 (800)
98	20 (8)	150	2.9 (420)
96	10 (4)	40	1.6 (225)
96	20 (8)	150	1.6 (225)
Cellulose acetate			
>99	10 (4)	40	5.5 (800)
>99	15 (6)	90	5.5 (800)
>95	10 (4)	40	3.5 (500)
>95	20 (8)	150	3.5 (500)
>94	10 (4)	40	2.8 (400)
>94	15 (6)	90	2.8 (400)
Sulfonated polysulfone			
95	10 (4)	40	3.5 (500)
95	20 (8)	150	3.5 (500)
Aromatic polyamide			
>98.5	10 (4)	26	5.5 (800)
>98.5	20 (8)	80	5.5 (800)

Continued on next page

Continued from previous page

NaCl Retention, %	Diameter, cm (in.)	Cross-flow Rate, ℓ/min	Pressure, MPa (lb/in.^{2})
NF Membranes			
Polyamide thin-film composite			
95	10 (4)	40	1.0 (150)
95	20 (8)	150	1.0 (150)
90	10 (4)	40	1.0 (150)
90	20 (8)	150	1.0 (150)
Cellulose acetate			
>90	10 (4)	40	2.1 (300)
>90	15 (6)	90	2.1 (300)
0	10 (4)	40	0.7 (100)
0	15 (6)	90	0.7 (100)
Regenerated cellulose			
0	10 (4)	60	0.7 (100)
0	10 (4)	60	0.4 (60)
Polysulfone			
0	10 (4)	60	0.7 (100)
0	15 (6)	90	0.7 (100)

- Polysulfones: pH 1–14
- Regenerated cellulose: pH 2–12
- Polyamide thin-film composites: pH 3–10
- Cellulose acetate: pH 4–7

No steamable RO/NF devices are offered at this date; they must be chemically sanitized. In addition, none of the RO/NF membrane materials except plain polysulfones are resistant to oxidants such as peroxide or hypochlorite. Sulfonated polysulfones exhibit limited hypochlorite resistance at low free-chlorine concentrations.

Fouling in RO or NF is most often due to exceeding the solubility limit of a retained species that precipitates in the high-concentration polarization (boundary) layer adjacent to the membrane. When the precipitated species is a salt, this fouling is referred to as "scaling." When scaling occurs, the flux decreases rapidly and may reach zero if the precipitated layer is impermeable. The cartridge feed channel may also become plugged if the precipitation is more widespread throughout the bulk retentate solution. Thus, to avoid scaling, the solubility limit of the least-soluble retained component should not be exceeded. In addition, prefiltration of feeds with a nominal 1–10 μm depth filter is recommended to avoid plugging with insolubles present in the feed.

Cleaning an RO/NF cartridge fouled by scaling requires a cleaning solution that readily dissolves the precipitated material. For example, iron scales may be cleaned with citric acid; calcium scales may be cleaned with EDTA solutions or dilute acids.

The choice of cleaning and sanitizing agents for RO/NF devices is dictated by the chemical compatibility of the membrane. Consult the RO/NF manufacturer for specific cleaning and sanitization procedures.

Organic Solvent Removal and Exchange by Ultrafiltration or Reverse Osmosis

Separation Objective

Some pharmaceutical compounds are chemically modified or purified in nonaqueous solvents. Traditionally, the removal of an organic solvent from a pharmaceutical compound has been conducted by crystallization and/or drying, which generally involves evaporation of the volatile solvent to remove it from the nonvolatile drug compound. Unfortunately, for many organic macromolecular drugs, the energy input required for evaporation often destabilizes the drug and decreases its yield or inactivates it. This evaporation problem, combined with high energy and time requirements, sometimes prevents the use of organic solvents in the preparation of some drugs, especially certain peptides, antibiotics, proteins, and polysaccharides.

Fortunately, with the recent introduction of UF and RO membrane and device materials with broader solvent compatibility, membrane separations are slowly gaining acceptance as an alternative to evaporation for removal and/or exchange of organic solvents in drug solutions. In these applications the membrane is selected to retain the drug while the solvent—and sometimes other small solutes like salts—pass through the membrane. The drug may thus be concentrated by removing the solvent. Alternatively, the solvent carrying the drug may be exchanged by diafiltering with the new solvent, which often may be an aqueous buffer if the solvent is to be completely removed from the drug solution. In all cases the drug is in the retentate while the solvent is removed in the permeate.

How Tangential Flow Filtration Performs the Separation

Both UF and RO membranes operate primarily by size exclusion; the membrane's pores must be small enough to retain the drug. The much smaller solvent molecules pass through the pores unhindered. Reverse-osmosis membranes also work in part through Donnan rejection of charged species, permitting them to retain ionic species by size exclusion and by charge repulsion at the membrane surface; thus RO membranes will retain small charged species, especially salts, at much higher levels than small uncharged species, such as many solvents.

Examples of solvent-removal applications are listed in Table 3.23. Due to the developmental nature of these applications, performance data are not available.

Issues and Solutions

The key issue affecting solvent-removal applications is the lifetime of the membrane device when in the solvent mixture. Most UF and RO membrane materials are incompatible with many or any organic solvents. Symptoms of chemical incompatibility include the following:

- Rapid, irreversible loss of membrane retention and integrity
- Rapid, irreversible loss of or increase in process and water flux
- Softening and mechanical failure of the device
- Extraction of organic compounds and polymeric byproducts from the device into the solvent

It is thus mandatory to verify the compatibility of all materials in the membrane device—membrane, support structures, screens, elastomers, and adhesives—with the solvent mixtures to be filtered.

Generally, regenerated cellulose membranes (*not* cellulose esters) are compatible with organic solvents such as alcohols, nitriles, esters, ketones, aldehydes, and hydrocarbons. However, if these membranes are supported by polypropylene structures (as most commercial cellulose membranes are), the polypropylene eliminates compatibility with

Table 3.23. Examples of Solvent Removal by Ultrafiltration and Reverse Osmosis

Application	Types of Membranes Used
Removing HPLC solvents (e.g., acetonitrile, alcohols) from polypeptides after chromatographic purification	Regenerated cellulose UF 1–5 kDa NMWL
Depyrogenating solvents used in purification or synthesis	Regenerated cellulose UF 10 kDa NMWL
Removing lower alcohols from antibiotics	Polyamide thin-film composite RO/NF 100–400 kDa NMWL

chlorinated hydrocarbons and aliphatic and aromatic hydrocarbons. Presence of polyurethane and/or epoxy adhesives or polysulfone membranes or support parts eliminates compatibility with most polar solvents except lower alcohols. The best UF devices available for solvent removal from macromolecular solutes, therefore, are plate-and-frame devices containing regenerated cellulose membranes and only polypropylene support parts, usually with ethylene-propylene dimer (EPDM) or similar elastomeric gaskets. An example of this device is Millipore Corporation's (Bedford, MA) Prostak-UF module.

Most RO membranes are either polyamide thin-film composites supported by polysulfone parts, cellulose ester membranes, or polysulfone membranes. Of these membranes the cellulose esters have the least solvent compatibility. The polysulfone RO membranes are compatible only with dilute solutions of lower alcohols. The thin-film composites, having polysulfone supports, are similarly susceptible to damage by most solvents besides lower alcohols, but they have shown the capability to survive extended exposures to methanol and ethanol at concentrations below 20–30 percent. These severe restrictions on solvent compatibility of commercially available RO membranes have so far prevented the commercialization of solvent removal RO processes. However, new RO membrane materials under development in Europe and Israel show promise for compatibility with broader families of solvents if methods of fabricating devices without using solvent-sensitive materials can be devised.

The first solvent-removal applications under development are UF applications using regenerated cellulose membranes. These membranes demonstrate reversible swelling behavior when transferred from aqueous to nonaqueous solvent environments. This reversible behavior affects process flux and partial retention properties of these membranes in predictable ways. For example, when a regenerated cellulose membrane is wetted with an organic solvent, its permeability may be only 20–40 percent of its permeability when wetted with water. The retention of a solute only partially retained when in water may be increased when in solvent for the same reason. This reversible change in flux should be measured for the actual solution being filtered and should be factored into system sizing and process times.

Because most solutes from which solvents are being removed or exchanged are at relatively high purity, partial retention of impurities is rarely a consideration. The membrane should be selected to ensure high retention of the drug both in solvent and in water or aqueous buffer. Solvent, salt, and water retention by these UF membranes is very low in aqueous or nonaqueous environments. If partial retention of an impurity becomes an issue during solvent removal via UF, diafiltration using an aqueous buffer rather than with solvent will often enhance removal of the impurity.

Because most solvent-removal applications involve drugs with molecular weights below 10 kDa, "tight" UF membranes are most often chosen. These membranes have low permeabilities associated with their high retention of small solutes and, therefore, must be typically operated at 310–620 kPa TMP to obtain economically attractive filtration rates. In addition, due to the small size of these solutes, they exert a small, but significant, osmotic pressure that must be overcome during concentration. These pressures are slightly higher than those at which most traditional macromolecular concentration applications are conducted.

When RO membranes can be used for solvent removal or exchange, the smaller size of the solutes significantly increases their osmotic pressure; the increased retention performance requires a lower permeability. As a result, RO devices are usually operated at 1.34–4.14 MPa TMPs to obtain economical filtration rates.

As the drug is always in the retentate from these devices, yield losses result from

- Losses due to passage of the drug through the membrane
- Losses due to not flushing the retentate solution out of the filter at the end of the batch cycle

Adsorptive losses in these applications are normally insignificant due to the hydrophilic nature of the membranes and the tendency of the solution conditions to interfere with adsorption. To maximize yields, therefore, the membrane must show high retention of the drug (usually >99.5 percent) and stability in the solvents being filtered for an acceptable lifetime. Flushing the TFF device with buffer, solvent, or retained permeate at the end of the batch cycle and recombining the retentate flush with the retentate, helps to maximize recovery of the drug from the device. The flushing method described in the virus concentration section is also suitable for this application.

Cleaning solutions should be compatible with the device materials of construction and be able to dissolve, decompose, and remove solutes retained in the device. Cellulose membranes may be repeatedly cleaned with 0.1 N NaOH or 0.1 N H_3PO_4 solutions, but they are rapidly damaged by cleaning with strong oxidants like chlorine (as NaOCl). Polyamide and cellulose ester RO membranes have more restricted pH ranges and may be damaged by acid and base cleaners; therefore, vendors should be contacted for the recommended cleaning agents for these membranes. If the device is compatible with the solvent in which the drug is first introduced into the TFF system, that solvent may be used to clean residuals out of the device after filtration is complete.

Cellulose UF membranes can be sanitized with 0.1 N NaOH or 0.1 N H_3PO_4 without damage, but chlorine sanitization, even at low concentrations (30–50 ppm), may shorten membrane lifetime with repeated use. Peracetic acid (at 200 ppm) is safer to use for cellulose membranes. The solvent being filtered by the UF device may also be used as a sanitant if it is bactericidal at the concentration at which it is filtered.

Sanitization of RO membranes is more difficult because they are not acid- or alkali-compatible and are susceptible to damage by oxidants. Possible sanitants for RO membranes include 20 percent ethanol, 0.1 percent quaternary ammonium salt solutions, and 1 percent sodium metabisulfite, although the choice of sanitant should be confirmed with the vendor of the membrane and device.

The best storage agents for cellulose UF membranes are solutions without extreme pH, such as 2–3 percent formalin, 20–40 percent ethanol or isopropanol, or 0.1 percent quaternary ammonium salt solutions. Most cellulose membranes can be stored safely in 0.1 N NaOH or 0.1 N H_3PO_4 with some reduction in effective lifetime. Reverse-osmosis membranes are much less solvent, acid, or alkali resistant, however, and should be stored in solutions recommended by the vendors. Examples of RO storage agents include 1 percent sodium metabisulfite and 0.1 percent quaternary ammonium salt solutions.

Endotoxin Removal by Ultrafiltration

Endotoxins are agents that cause fever (pyrogens) when administered intravenously to man or certain experimental animals.

Lipopolysaccharides (LPSs) from the outer wall of gram-negative bacteria are the principal group of pyrogens. These endotoxins are released when the bacterial cell is destroyed. The size of the bacterial endotoxin is a function of its state of aggregation in relation to the solution chemistry, as presented in Table 3.24.

The amount of endotoxin required to generate a pyrogenic reaction is on the order of 0.1 ng per kg of body weight, equivalent to approximately 1 EU (endotoxin unit, a unit in which endotoxin data are commonly reported).

Separation Objective

Mechanisms to remove endotoxins include both adsorption and size exclusion, depending on the properties of the solution. The objective of endotoxin removal by UF is to remove endotoxin molecules from the process stream via size exclusion (i.e., retaining the endotoxin molecules while passing the solutes and solvent of interest). More important, considering the minute amount of endotoxin present in most solutions, UF will successfully remove the endotoxin if

- The ultrafilter is compatible with the process solution to avoid the introduction of extractables that may cause other adverse reactions.

Table 3.24. Aggregation State of Bacterial Pyrogens as a Function of Solution Chemistry

Solution	Endotoxin Aggregation Form	Size
Water	Vesicle	Submicron
Saline	Vesicle	Submicron
$MgCl_2$	Vesicle	Submicron
EDTA*	Micelle	300–1000 kDa
Na Cholate	Micelle	100–700 kDa
Cholate and EDTA	Subunit	10–20 kDa

*EDTA = Ethylenediaminetetraacetic acid

- The operating procedures ensure that the system can be regenerated to a endotoxin-free state and maintained in that state between batches.

Some applications of UF systems in endotoxin removal include the following:

- Water for injection (WFI), 20 percent glucose, 30 percent dechloramphenicol succinate, and 4 percent amino acid solutions (Zimmerman et al. 1976)
- Carbencillin and sodium cephalothin solutions (Nelsen 1978)
- 10 percent amino acid, 30 percent penicillin derivative (Karbachsch et al. 1984)

Filtration performance and endotoxin reduction for some of these applications, as well as some unpublished data, are provided in Table 3.25.

Issues and Solutions

Choosing the Proper Membrane. Endotoxin removal is often of similar importance to sterile filtration in the production of parenteral drugs. However, there are no prescribed standards for endotoxin removal via UF. The USP XXI (1985) chapter 1211 requires demonstration of a three-log reduction for endotoxin removal, in reference to the final rinse water from containers. Another reference for product endotoxin levels would be that consistent for USP WFI, which is to be less than 0.25 EU/ml.

Given that the endotoxin molecules and active subunits can vary in size from submicron particles to approximately 50 kDa soluble polymers, nearly all UF membranes will remove pyrogens to varying degrees. The selection of the most appropriate membrane will depend on the solution to be processed and the performance of the membrane. Table 3.26 provides relative retention characteristics of a variety of membranes. The solutions used in Table 3.26 cover the varying aggregation states possible for LPSs.

The 10 kDa NMWL ultrafilter satisfies the required log reduction of endotoxin in all cases. However, when the solution being treated is altered due to retention of other desired components, UF membranes with higher NMWLs must be used to ensure adequate product passage. These less-retentive UF membranes also yield lower log reductions of endotoxin in single-stage operation. Thus, to obtain the aforementioned three-log reduction and/or 0.25 EU/ml level, more open UF membranes may have to be operated as several stages in series to provide the desired endotoxin removal and product yield.

Table 3.25. Endotoxin Removal Application Data

Solution	Membrane MWCO	Solution Permeability ($\ell/m^2/h/bar$)	Endotoxin Log Reduction	Source
WFI	10,000	27	>3	Zimmerman et al. 1976
4% amino acid	10,000	17	>3	Zimmerman et al. 1976
10% amino acid	10,000	25	>>1	Karbachsch et al. 1984
10% amino acid	20,000	113	>>1	Karbachsch et al. 1984
Penicillin derivative, 30%	20,000	45	>3	Karbachsch et al. 1984
Contrast agent (35%)	10,000	3.4	>3	Millipore 1989a
Contrast agent (35%)	30,000	8.7	>2	Millipore 1989a
Contrast agent (35%)	100,000	13.0	>2	Millipore 1989a

Table 3.26. Log Reduction of Pyrogens after 1,000 ng/ml Challenge

Solution	0.2 µm	0.025 µm	1,000,000 NMWL	1000,000 NMWL	10,000 NMWL
Water	0	>4	>4	>4	>4
0.9% NaCl	0	>4	>4	>4	>4
$MgCl_2$	0	>4	>4	>4	
EDTA		0	>4	>4	
Na Cholate			0	2	>4
Na Cholate and EDTA			0	0	>4

Actual retention of endotoxin by a UF device can be tested at reduced scale using the actual product solution to determine log reduction values and to validate large-scale endotoxin-removal processes using ultrafilters.

System Design. Given that the nature of the endotoxin present is retained by UF, successful application of UF to endotoxin removal is dependent on design and operation of the system.

System design is generally application dependent. The primary criteria combine maximum retention of endotoxin with required product yield. The retention of endotoxin provided in Table 3.26 was greater than 99.99 percent; this retention of endotoxin is an instantaneous value; however, it is not the overall retention. Table 3.27 shows the overall retention for various TFF system designs, based upon 99.99 percent instantaneous retention of endotoxin. With the constraint of no dilution, the resultant yield and actual endotoxin reduction are calculated. In the scenarios represented in Table 3.27, a process volume, V, is reduced to the system minimum volume (V_0) which is then continuously washed with a volume (V_w) of solvent (water, buffer) equal to that held in the system.

Contrast agents are generally nonionic compounds that range in size from 300–1500 Daltons. These products are generally handled in high concentration and are generally highly viscous. Given the molecular size, the high concentration of product, and the high viscosity of contrast agents, operation at high TMP can lead to elevated product

Table 3.27. Effect of System Design on Apparent Endotoxin Retention

System Design (Endotoxin Retention = 99.99%)	System Minimum Working Volume*	Yield	Apparent Reduction**
Batch	0.05 V	98.16	99.96% (3.8 LRV)
Feed and bleed	0.01 V	99.64	99.94% (3.4 LRV)
Continuous (single stage)	0.01 V	99.64	99.8% (2.8 LRV)

*V = initial feed volume
**LRV = log reduction value

retention. Product retention appears as deteriorating process flux and altered product formulation as the volume is reduced and retained component concentrations increase. Contrast agent UF is characterized by the relatively high cost of contrast agents, the high yields required, and restrictions on allowable dilution. A batch concentration of the feed to minimum volume followed by minimal diafiltration provides the highest overall log reduction of endotoxin with high product yield.

Antibiotics and other therapeutics are sometimes formulated for parenteral use, in which case they are often ultrafiltered in the course of bulk manufacture. Generally, the objective in these applications is not to provide validatable endotoxin removal, but to control endotoxin burden. Because this process step may be far removed from the formulation area and because ease of operation is more important than maximizing endotoxin removal, single-stage continuous operation incorporating diafiltration for yield improvement is preferred.

Chemical Compatibility. Some of these products may be dissolved in organic solvents or in aqueous solutions containing residual organic solvents. In either case the solvent presents a chemical compatibility issue for all wetted materials in the TFF device and system. While in many cases the compatibility issues can be eliminated with careful membrane and device selection, deviations from the UF device's specifications for water permeability and integrity need to be understood, noted, and documented for later validation and/or quality control exercises.

Cleaning. To define a suitable cleaning procedure, the potential foulants in the TFF system must be identified. In the aforementioned applications the foulants would be the retained endotoxin (LPS), residual product, and inorganic impurities found in the process water. The next section of this chapter provides strategies for cleaning and sanitization. Endotoxin reduction itself can be accomplished through the use of acid or base.

Storage. The storage solutions for an ultrafilter system vary, depending on the device material of construction and suitability of the chemical to the specific operation. In general, 0.1 N NaOH is preferred because it degrades all endotoxins and is easily removed by rinsing with WFI to leave a pyrogen-free system. The quality of rinse water used is critical. If the storage solution is validated as effective in keeping a system endotoxin-free, rinsing with WFI prior to processing can provide a suitable endotoxin-free TFF system.

PRINCIPLES OF TANGENTIAL FLOW FILTRATION MAINTENANCE

Maintenance of TFF membranes, devices, and systems should be conducted on a regular, planned schedule to ensure the following:

- Each filtration run provides the same separation within an acceptable time period.
- Prior runs do not leave solutes in the system to contaminate subsequent runs.
- Wetted materials of construction are adequately preserved and protected from microbial or chemical attack and are replaced when no longer functional.

System maintenance for TFF should comprise the following activities:

- Flushing
- Cleaning
- Sanitization or sterilization
- Depyrogenation (for systems filtering injectable drugs)

- Storage (between runs)
- Planned hardware maintenance

Cleaning should be conducted immediately after flushing the process retentate out of the TFF system, before sanitization or any separate depyrogenation steps. Depyrogenation is the implementation of a cleaning step with an agent selected for its specific ability to degrade and remove endogenous and exogenous pyrogens (i.e., bacterial endotoxins possibly introduced into the TFF system in the process feed or by accidental bacterial contamination of the system). Some frequently used cleaning agents (e.g., acids and alkalis) are also effective depyrogenation agents, permitting cleaning and depyrogenation as a single step.

If a separate depyrogenation step needs to be conducted, the order of maintenance steps should be as follows:

1. Flush out process retentate.
2. Confirm membrane integrity (optional).
3. Clean.
4. Flush out cleaning agent.
5. Confirm membrane cleaning efficacy (usually by measuring clean water permeability of the cleaned TFF device and comparing against the permeability of the same device when new).
6. Sanitize.
7. Flush out sanitizing agent.
8. Depyrogenate.
9. Flush out depyrogenation agent.
10. Confirm membrane integrity (recommended).
11. Instill storage agent into system or start next filtration run.

If the cleaning agent is an effective depyrogenation agent, steps 8 and 9 can be eliminated.

This section describes these steps and their objectives in general. Specific procedures for each activity should be obtained from the vendor of the TFF system.

Flushing

A TFF system is flushed to remove the bulk of the process liquid or residual solution prior to a maintenance step. Thus, flushing a system at the end of a TFF operation permits improved recovery of the pharmaceutical solution from the system and reduced cleaning solution volume requirements and cleaning time by minimizing the amount of residual solutes in the system prior to cleaning.

After any maintenance step, effective flushing is required to remove the maintenance agent (i.e., cleaning or sanitizing agent) from the system and to prevent contamination of the process feed by any residual cleaning, sanitizing, or storage agent.

Effective flushing requires pumping the flush solution or solvent once through the system with flow removed from both the retentate and permeate ports. When flushing out solutes retained by the membrane, the following protocol works best.

1. At the end of the concentration/diafiltration procedure, drain the system as thoroughly as possible to recover a maximum amount of retentate. Strict observance of the U.S. Department of Agriculture's 3A sanitary equipment design rules in TFF system design can help to ensure the maximal drainability of systems.

2. Fill the system to its minimum working volume (i.e., the minimum volume needed to operate the system with no air introduction or foam generation) with a buffer or solvent compatible with the retentate solution. Instead of fresh buffer, an equal volume of permeate recently removed from the system may be used. Circulate this volume of solution through the feed/retentate side of the TFF system for 3–5 minutes at minimum TMP, to resuspend virus or antigen deposited on the membrane surfaces. Do not shut off the permeate flow when recirculating buffer, as the resulting Starling flow (permeate flow in reverse across the membrane where feed-side pressure is less than average permeate-side pressure) within some TFF devices may rupture some membranes. Stop the recirculation pump for 3–5 minutes to let virus or antigen diffuse from the boundary layers of the TFF system into the bulk solution, then resume recirculation for 1–2 minutes.

3. Pump out or drain this flush solution and add it to the retentate (to recover the solutes) or discard.

After performing the procedure for recovering retained solutes, and when flushing solutions not retained by the membrane (e.g., cleaning and storage solutions), the following protocol is suggested.

1. Drain the system as thoroughly as possible to remove the maximum amount of solution.
2. Fill the system with a buffer or solvent compatible with the retentate solution. Pump this solution once through the feed/retentate side of the TFF system for 3–5 minutes with the permeate flow completely open and the retentate flow restricted to generate a TMP in the range approximating the operating pressure for the filtration step. Do not recycle the flush solution in this step. This action forces flush solution through the membrane to completely flush both retentate and permeate channels and piping. This flush solution should be directed to the drain.

The actual volumes of flush solutions required in both of the above procedures are dependent upon the membrane area installed and the working volume of the specific TFF system. System vendors can supply guidelines.

Cleaning and Depyrogenation

System and membrane cleaning is an essential step to regenerate the wetted surfaces of the TFF system.

- Efficient productivity is maintained by counteracting the effects of membrane fouling.
- A clean and sanitary system is maintained by removing microorganisms and their metabolites.

Since TFF devices are designed to operate over many cycles, for lifetimes of a year or longer, cleaning methods should be consistent, time-efficient procedures that will not limit the useful life of the membrane or device.

Cleaning agent selection depends on

- The type of foulant to be removed from the system
- The chemical compatibility of the wetted materials of construction of the membrane, device, and system
- The difficulty in confirming removal of the cleaning agent from the system and the dangers of contaminating the drug product with residual cleaning chemicals

Cleaning agents that double as depyrogenation agents are alkaline solutions at $pH > 12$ and acidic solutions at $pH < 2$.

Many process liquids are complex mixtures of organic and inorganic compounds that may require more than one type of cleaning agent and more than one cleaning step. To select an effective cleaning procedure, the process liquid being filtered must be well characterized chemically. Based on the foulants to be removed from the system and the vendor's recommendation of cleaning agents compatible with the system's and device's wetted materials, an effective cleaning procedure can then be devised.

Compatibility of the cleaning agent with the process application itself is also a very important consideration when processing drugs. For example, an enzyme-based cleaner may be very effective at removing protein residuals from a system, but if traces of the enzyme are left behind after cleaning, they may be hard to measure and yet may risk contaminating the drug with a reactive or sensitizing protein. Most cleaning agents selected for drug filtration systems have easy and reliable sensitivity measurement and verification of removal. Most frequently used cleaning agents include the following:

- NaOH solutions for removal of proteinaceous and lipid foulants (confirm removal by pH measurement)
- H_3PO_4 solutions for removal of inorganic foulants (confirm removal by pH measurement)
- NaOCl solutions for removal of bacterial and polysaccharide foulants (confirm removal by colorimetric or conductivity measurement)

Commercial membrane-cleaning preparations are also available from such companies as Henkel (Plymouth Meeting, PA), Ecolab (St. Paul, MN) and Calgon (St. Louis, MO). Some of these are suitable for cleaning drug-filtration systems, while others are intended for cleaning food- and beverage-filtration systems and may not be as easy to analyze. Examples of general classes of cleaning agents and their ranges of efficacy are listed in Table 3.28.

The effectiveness of cleaning agents is enhanced by cleaning at elevated temperatures, for longer periods of time, and at higher cleaning-solution concentration. The most essential factors for effective cleaning are maintaining a recommended cleaning solution concentration, which is often restricted by the chemical compatibility of the wetted materials of construction in the TFF system, and maintaining an elevated temperature, typically in the range of 40–50°C. Cleaning TFF systems at room temperature or lower is frequently ineffective and requires greatly extended cleaning times.

Mixing the cleaning solution within the system and wetting all surfaces is also essential for effective and complete cleaning and depyrogenation. For this reason the vendor's recommended cross-flow

Table 3.28. General Classification of Cleaning Agents (Millipore 1989)

Cleaning Agent	Examples	Mode of Action	Foulant Removed
Alkalies	NaOH	High pH hydrolysis	Biopolymers
		Solubilization	Fats
		Saponification	Proteins Starches Endotoxins
Acids	H_3PO_4	Solubilization	Inorganic salts
	HNO_3	Low pH hydrolysis	Nucleic acids Endotoxins
Surfactants	Sodium dodecyl sulfate	Wetting	Biopolymers
	Triton X-100®	Emulsifying	Fats
		Suspending	Oils
		Dispersing	Proteins Insoluble particulates
Oxidizers	NaOCl	Oxidative degradation	Biopolymers Proteins Polysaccharides
Enzymes	Proteases Amylases Lipases	Enzymatic digestion	Biopolymers Proteins Polysaccharides
Solvents	Alcohols	Solubilization	Oils Fats Lipids Grease

rates and operating pressures should be maintained throughout the cleaning procedure, with the solution circulating throughout the system and wetting all systems normally in contact with process solutions.

Finally, the quality of the water in the cleaning solution can affect cleaning efficacy. Poor-quality water can introduce additional and more difficult-to-remove foulants, such as silica, humic acids, and iron oxides. Some water-borne contaminants may also react with cleaning agents that can deposit onto membrane surfaces. The quality of water used for maintaining TFF systems should therefore satisfy the guidelines in Table 3.29.

Sanitization

Most TFF membranes and devices are incompatible with steam sterilization or autoclaving. The elevated temperatures required for heat

Table 3.29. Recommended Water Quality Guidelines for TFF Maintenance (Millipore 1989)

Fe	<0.05 ppm
Mn	<0.05 ppm
Al	<0.5 ppm
Reactive silica	<2 ppm
Colloidal silica	nil
Ca, Mg	<25 ppm
Turbidity	<1.0 JTU
SDI_{15} (fouling index)	preferably <3.0
Particulate matter	None
Oil, grease, etc.	None

Reverse-osmosis permeate or WFI is recommended wherever possible, although distilled and filtered deionized water may be used if the former are not available.

sterilization either damage or modify the retention properties of the membranes and may soften or destroy other materials of construction in the filter devices. Therefore, many TFF devices must be chemically sanitized instead. Sanitization reduces bioburden inadvertently introduced into the system to a level low enough for product contamination or degradation to be averted during normal processing cycles. Unlike sterilization, sanitization procedures do not eliminate bioburden completely. They reduce bioburden by lower log factors than sterilization methods, which typically are designed for a greater than seven-log reduction.

Sanitization can be achieved by a variety of chemical or thermal mechanisms, depending on compatibility with the membrane, device, and system design, so the vendor should be consulted. Examples of sanitizing agents and conditions are shown in Table 3.30.

As with cleaning and depyrogenation, sanitizing agents should be circulated through the TFF system at the vendor's recommended cross-flow rates and pressure conditions to ensure complete wetting of all system surfaces.

Sterilization

Sterilization of TFF systems is most often accomplished by autoclaving or steaming the systems in place. By raising all system surfaces to at least 121°C for 15–60 minutes, greater than seven-log reduction of bioburden (including spores) is ensured (Block 1991), meeting the definition of sterilization. However, most commercially available UF and

Table 3.30. Typical Sanitizing Agents Effective Against Bacteria, Fungi, and Many Viruses (Block 1991)

Agent	Conditions
Chlorine (as NaOCl)	pH 7.0, 30–50 ppm, 20–50°C, 10–20 min
Peracetic acid	100–300 ppm, 20°C, 10–20 min
Formaldehyde	1–3%, 20°C, 5–10 min
NaOH	pH 12–13, 40–50°C, 10–60 min
Ethanol or isopropanol	40–70%, 20°C, 10–30 min
Hot water	80°C, 1 hr (consult vendor for temperature limits of device)

RO membranes cannot withstand the sterilization temperatures and must be chemically sanitized as noted earlier. Nonetheless, a number of commercial TFF devices and systems containing microporous membranes can withstand repeated steam-in-place (SIP) or autoclaving cycles without membrane or device damage, permitting these systems to be sterilized instead of chemically sanitized.

Systems intended for SIP must contain high-point vents (to remove all air) and low-point drains with steam traps (to remove all condensate during steaming). All materials of construction and all instruments must be able to withstand repeated long-term exposures to 120–130°C saturated steam and temperature cycling from this temperature to cold-room temperatures without damage or loss of calibration. Because all wetted surfaces of a TFF system must reach 121°C for at least 15 minutes to ensure sterilization, typical SIP cycles may require at least one hour of steam flow into the system to ensure enough time at sterilization temperature.

One other complication associated with SIP involves the dewetting of hydrophobic microporous membranes (e.g., polypropylene, hydrophobic polyvinylidene difluoride (PVDF), and polysulfone membranes) during the steaming and cooling procedure. If the membrane is dried out by the steaming, no flow through the membrane will occur until it is rewetted with water. Dewetting is observed as a drop in water flux of the membrane devices after steaming. Since dewetting often occurs with hydrophobic membranes, a rewetting procedure must be available to correct it. Hydrophilic membranes do not require rewetting.

Because microporous membranes require high water-intrusion pressures to rewet them with pure water—typically from 310 kPa to 1.34 MPa—rewetting usually requires that a solution of water plus a surfactant or wetting enhancer be circulated through the membrane. Typical rewetting agents vary from buffer solutions to 70 percent ethanol or isopropanol. The rewetting solution is circulated through the system at TMPs of 310–520 kPa until a strong, stable permeate flow is established. The TMP is then reduced and the rewetting solution is recirculated through the system for 10 minutes to ensure complete wetting of both the retentate and permeate sides of the membrane. Finally, the rewetting solution is drained from the system and flushed out with buffer or water to complete the procedure.

Integrity Testing

A TFF system is effective at conducting a separation only if the membrane and device are installed into the system in such a manner as to prevent feed flow from bypassing the membrane into the permeate.

For this reason the integrity of the membrane and TFF module should be verified regularly. Integrity can be confirmed directly by analyzing the feed and permeate, calculating the retention of specific solutes expected to be retained, then comparing the actual retention against specified retention for the process. However, this direct method requires placing process solution into the system, so if integrity is compromised, product may be lost into the permeate before the problem is identified.

Some vendors have developed nonintrusive, nondestructive integrity test methods that can be implemented before process feeds are introduced into the TFF system. The most frequently recommended method for microporous TFF and UF modules measures the airflow through the fully wetted membranes in the installed TFF devices. Because air is only transported through the wetted membranes by diffusion through the water in the pores, airflow through integral devices is always very slow. When a membrane, device, or seal is damaged, however, the damaged area does not contain water under the test pressure and permits convective airflow at a much higher rate (typically 2–5× the diffusive rate). This diffusional airflow test is sufficiently sensitive to identify single submicron defects in small areas of the membrane and multiple defects in large-area systems that can lead to measurable solute passage through an otherwise fully retentive membrane.

Diffusional airflow rates vary with the pressure, temperature, and composition of the solution in the membrane pores. Most integrity tests for UF and microporous membrane devices are designed for operation at 34–138 kPa, 20–25°C and with pure water wetting the membrane.

While the integrity and pore size of many microporous membranes can also be confirmed by a bubble-point test, the design of most microporous TFF devices does not permit a bubble-point test to be conducted. Many devices physically obscure the permeate side of all membrane sheets so that the first bubble cannot be observed. In addition, the water bubble points of most microporous membranes occur at pressures in excess of the design maximum pressures of many microporous TFF devices. Use of a pure-solvent bubble point may lower the test pressure, but may not be chemically compatible with the TFF device's materials of construction. Therefore, bubble-point tests are rarely appropriate for confirming the integrity of TFF devices.

For RO devices and membranes, a direct-retention challenge with an appropriate salt solution—most often NaCl or $MgSO_4$—confirms both the integrity and the agreement with specified retention of the device. Conductivity measurements of the retentate and permeate provide a means of measuring salt retention.

Integrity test methods and specifications should be obtained from the TFF device's vendor.

Storage

When a TFF system will be shut down for prolonged periods, the devices should be stored properly to maintain membrane properties and to limit or prevent bacteriological growth. Most TFF membranes can be easily stored within their respective systems for periods up to 15 days, but for longer periods the filters should be removed from the system and stored separately in a container of a suitable storage solution. Tight-sealing polypropylene containers are suitable for storing filters outside of systems.

Some storage agents are bacteriostatic and suitable for storage periods of up to one week, while others are bactericidal and permit storage periods of up to one year. Some storage agents may cause corrosion of system hardware during long storage periods, forcing the filters to be stored outside the system, while others are relatively innocuous to both hardware and filter materials.

To prepare a TFF system for storage, the system should be thoroughly flushed with water after cleaning, depyrogenation, and sanitization. Then the storage solution should be introduced and circulated to fill the system completely. All air should be vented from the system so that all internal surfaces are wetted with the storage agent.

If the filters will be stored outside the system, the storage solution should first be circulated through the filters in the system, then drained by gravity from the system and filters. The filters are then removed from the system, immersed in storage solution in suitable containers, sealed, and placed in the storage location. Storage at 4–10°C can further enhance the bacteriostatic effects of the storage agent.

Table 3.31 lists some typical storage agents for TFF devices. Consult your TFF vendor for specific recommendations for a particular device and membrane.

PRINCIPLES OF TANGENTIAL FLOW FILTRATION VALIDATION AND QUALIFICATION

Process validation is a requirement of the U.S. Food and Drug Administration's (FDA) *Current Good Manufacturing Practices Regulations for Finished Pharmaceuticals,* CFR Title 21 parts 210 and 211 (cGMP). These regulations and the FDA's *General Principles of Process*

Table 3.31. Typical TFF Storage Agents

0.05–0.5 N NaOH (pH 12–14)
0.05–0.1 N H_3PO_4 (pH 1–2)
1–3% formalin
0.1% quaternary ammonium salt solutions
1–2% sodium metabisulfite
20–70% ethanol or isopropanol
5% NaCl (for short storage periods)
0.05% sodium azide

Validation (U.S. FDA 1987) identify key information and experimental evidence as necessary to demonstrate control and reproducibility of pharmaceutical manufacturing and purification processes.

Because each TFF system in a pharmaceutical process is committed to a process objective unique to that process and to the product being manufactured, no generic qualification and validation protocols are suitable for TFF steps. Instead, the elements of validation for each TFF system and process step must be developed by the user with or without assistance from the system vendor. This section outlines the types of information that may be obtained to support TFF validation, but it cannot specify a method of validation.

Subprocess Definition

The subprocess definition is a description of the function of the TFF system and process step within the overall pharmaceutical manufacturing process. This item should become part of the overall process description, which helps to support the process specifications, flow schematics, and mass and energy balances. The TFF subprocess definition may include the following (this is not a comprehensive list):

- Objective of the TFF step
- Where the feed comes from
- What vendor, membrane, filter, and system is used

- Maintenance procedure overview (e.g., cleaning regimen, etc.)
- What is done with the retentate and with the permeate

System Specification

The TFF system specification is a detailed description of the TFF system (hardware and filters), which is sufficient to define the installed equipment. A system specification should cover both which components are in the system and how the system should perform. A system specification may include the following (again, not a comprehensive list):

- Feed, retentate, and permeate composition specifications and allowable variations
- Step process flow schematic
- Step mass and energy balances
- Cycle-time specifications
- Operating conditions and allowable variations
- List of key components (instruments, pumps, controls, filters; sometimes with vendors and part numbers)

Often, the draft system specification is a key part of the request for quotation submitted by a pharmaceutical company to a TFF system vendor prior to purchasing a system. The vendor's final quotation will then contain this specification plus any agreed-upon amendments. When the vendor supplies the system, the specifications are usually finalized to agree with the actual scope of the system supplied and are supplemented by component vendor information describing components in detail. This combined system-design package (drawings plus literature plus specifications) becomes the nucleus of the installed system's specification.

Vendor Qualification

Pharmaceutical companies are expected to provide information demonstrating that their vendors are qualified to manufacture and supply equipment suitable for processing pharmaceuticals. This information may consist of the following:

- Description of the vendor's product testing, which ensures that the products are suitable for contact with pharmaceuticals (e.g., toxicity testing of materials)

- Description of the vendor's proof of performance claims for the product (e.g., retention testing of product)
- Description of the vendor's quality control (QC) and quality assurance (QA) system (to assure that products will consistently satisfy performance claims)
- Description of the vendor's financial stability

Furthermore, the cGMPs require that pharmaceutical companies audit the qualifications of their vendors to supply products for specified uses. Audits reinforce and clarify vendor qualification as well as identify manufacturing and QC issues requiring correction.

The information required for vendor qualification may often be obtained by contacting the vendor's QA Department. Some vendors also assemble product-specific validation data-packages containing results of product testing that was conducted to demonstrate suitability for processing pharmaceuticals and to establish and support product performance claims.

Installation Qualification and Operational Qualification

Installation qualification (IQ) and its subset, operational qualification (OQ), are experimental protocols implemented to ensure that a TFF system is correctly installed and operates mechanically as specified in the system specification. Installation qualification reviews and confirms all aspects of how the system is installed and connected into the utilities and process lines in the process plant.

Operational qualification operates the TFF system in the absence of process feed—most often with water in the system—to demonstrate that all mechanical components, controls, alarms, and instruments perform as specified. It also includes the following verifications:

- Instrument-calibration procedures (measurement qualification)
- Maintenance procedures
- Personnel training
- Written and approved standard operating procedures
- Written and approved manufacturing work orders

Most components of IQ and OQ protocols must be specified and executed by the pharmaceutical manufacturer. Selected items in the IQ and OQ protocols, however, can be jointly executed by the pharmaceutical manufacturer and the TFF system vendor during the vendor's system QC testing prior to shipping the system for final installation. In this case the pharmaceutical manufacturer's representative attends the vendor's QC tests at the vendor's site and verifies completion of those OQ and IQ items that are also part of the TFF system's QC tests. Examples of items common to both vendor tests and OQ/IQ protocols are listed in Table 3.32.

In addition to these items, the pharmaceutical manufacturer may adapt the vendor's general filter and system operating maintenance recommendations into the Standard Operating and Maintenance Procedures for the TFF system.

Table 3.32. Items Common to Vendor TFF System Quality Control Tests and Operational/Installation Qualification Protocols

Installation Qualification
Equipment description
System specifications
List of materials in product contact
List of materials not in product contact
Description of utility requirements for TFF system
TFF system operating manuals
Critical instrumentation calibration procedures used (if any)
Critical instrumentation calibration certification data (if any requested prior to shipment)
Recommended spare parts list
Recommended lubricants list
Engineering drawings and installation drawings (if any)
Operational Qualification
Verification of specified operation of pumps, valves, instruments
Verification of leak-proof operation of system
Verification of specified operation of control software and hardware and alarms (if any)

Performance Qualification

Performance qualification (PQ) requires experimental protocols designed to demonstrate that the actual TFF process, operated on actual process solutions, satisfies the performance specification for the process step. These protocols are unique to each process and product and involve actual operation of the entire process on the installed plant equipment. Multiple experiments must usually be conducted to verify that the process, when operated within the specified operational conditions, produces a drug satisfying the product specification. Experiments may also be conducted under conditions exceeding the acceptable control bands to verify that the process, when properly operated, is robust and that the specified control conditions in fact ensure that the product will satisfy its specification. The definition of which experiments must be included in PQ protocols, and how many times each experiment must be repeated for statistical reliability, is made by the pharmaceutical manufacturer.

System and filter vendors for TFF cannot normally assist PQ in a significant manner, as PQ tests are process specific and often confidential. In most cases the TFF vendor can, at best, offer advice on certain types of PQ tests to conduct for a specific objective. For example, when PQ requires (as it often does) proof that a cleaning regimen is effective at removing foulants from the membrane, the TFF vendor can sometimes help the pharmaceutical manufacturer design experimental protocols that satisfy this objective.

Key PQ objectives of TFF systems often include the items shown in Table 3.33. This table also indicates which of these items a TFF vendor may assist the pharmaceutical manufacturer to assemble.

Table 3.33. Key Objectives of Validation of Tangential Flow Filtration Systems (Millipore 1990; Michaels 1991)

1. Chemical Compatibility	
General:	Demonstrate that all materials of wetted surfaces should not corrode, extract, swell, weaken, or react with feeds, flush, cleaning, sanitizing, depyrogenating and storage solutions. If steam sterilization is intended, these materials should not be affected by steam.
Membrane:	Same as general[a]
Device:	Same as general[a]
System:	Same as general

Continued on next page

Continued from previous page

2. Integrity	
General:	Demonstrate that the seals within the system and between the system and the environment remain as barriers to the solutions in the system.
Membrane:	Membrane must remain attached and intact.
Device:	Module must be installed properly and be strong enough to withstand the operating conditions. Modules should have some easily conducted, nondestructive integrity test.[a]
System:	The system should have pressure and temperature capability to prevent leaks. It should be hydrotested.[a]
3. Retention	
General:	Demonstrate that the system has constant and defined retention and passage of macromolecules.
Membrane:	The membrane should maintain retention performance. The pharmaceutical manufacturer should supply a definition of performance at recommended operating conditions (tangential flow rate, pressure, and temperature).
Device:	The module design should minimize the effect of polarization on retention when operated as specified by the module manufacturer.[a]
System:	The system may employ techniques, such as permeate flow control, to minimize polarization and control retention.
4. Product Recovery	
General:	Demonstrate that the product is removed from the system at the conclusion of the process run.
Membrane:	Evaluate binding by the membrane.[b]
Device:	Evaluate the ability to flush and drain the modules.[b] Unrecoverable volume should be reproducible and controllable.

Continued on next page

Continued from previous page

System:	Unrecoverable volume should be reproducible and controllable. Evaluate flush out.
5. Cleanability	
General:	Demonstrate that contaminants are removed to permit reuse. Restore original and/or reproducible performance.
Membrane:	Demonstrate ability to remove foulants.[b]
Device:	Demonstrate ability to remove entrained contaminants.[b]
System:	Demonstrate properties consistent with sanitary design.
6. Depyrogenation	
General:	Demonstrate ability to remove and destroy endotoxins.
Membrane:	Desorb and destroy endotoxins.[b]
Device:	Demonstrate ability to flush out or destroy.[b]
System:	Demonstrate ability to flush out or destroy.
7A. Sanitizability	
General:	Demonstrate ability to lower microbiological population to reduce contamination during a process run.
Membrane:	Demonstrate compatibility with effective sanitants.[a]
Device:	Demonstrate ability to get sanitant on both sides of the membrane and throughout the module.[b]
System:	Demonstrate sanitary design with no section isolated from flow.
7B. Sterilizability (not all membranes can be sterilized; many can only be sanitized–consult with the manufacturer)	
General:	Demonstrate ability to sterilize the TFF system.

Continued on next page

Continued from previous page

Membrane:	Demonstrate that the membrane can withstand repeated sterilization temperatures, radiation, or chemicals.[a]
Device:	Demonstrate penetration of sterilizing agent throughout the module.[b]
System:	Demonstrate penetration of sterilizing agent throughout the system.
8. Flushability	
General:	Demonstrate the ability to remove cleaning, storage, sanitizing, and depyrogenating chemicals from the system.
Membrane:	Demonstrate that there is no irreversible adsorption of the chemicals to the membrane surface.[b]
Device:	Demonstrate that the areas upstream, downstream, and in the seal areas are flushed.[b]
System:	Demonstrate sanitary design with no areas isolated from flow. Assure that flush rate is high enough to avoid channeling.
9. Maintenance of Activity	
General:	Demonstrate that the product maintains activity after processing in the system.
Membrane:	Demonstrate that there is predictable and controllable loss due to irreversible adsorption to the membrane.
Device:	Demonstrate that the materials in the module are compatible with the products.[b]
System:	Demonstrate that the pumps, tanks, and valves do not affect activity of the product.
10. Biocompatibility	
General:	Conduct USP Class VI testing of wetted polymeric materials or use accepted metals and nonplastic materials (e.g. 316SS)

Continued on next page

Continued from previous page

Membrane:	Same as general.[a]
Device:	Same as general.[a]
System:	Same as general.

[a]Many TFF vendors can supply data to satisfy these objectives.

[b]Many TFF vendors can supply experimental protocols that can be modified and adapted into PQ protocols to satisfy these objectives. In many cases the protocols are accompanied by data proving that the objective is attainable for a generic application.

REFERENCES

Bellini, W. J., A. Trudgett, and D. E. McFarlin. 1979. Purification of measles virus with preservation of infectivity and antigenicity. *J. Gen. Virol.* 43:633–639.

Bibal, B., Y. Vayssier, G. Goma, and A. Pareilleux. 1991. High-concentration cultivation of *Lactococcus cremoris* in a cell-recycle reactor. *Biotechnol. Bioeng.* 37 (8):746–754.

Blatt, W. F., A. Dravid, A. S. Michaels, and L. N. Nelsen. 1970. Solute polarization and cake formation in membrane ultrafiltration: Causes, consequences, and control techniques. In *Membrane Science and Technology*, edited by J. E. Flinn. New York: Plenum Press.

Block, S. S., ed. 1991. *Disinfection, sterilization and preservation*, 4th ed. Philadelphia: Passim, Lea and Febiger.

de la Broise, D., M. Noiseux, B. Massie, and R. Lemieux. 1992. Hybridoma perfusion systems: A comparison study. *Biotechnol. Bioeng.* 40 (1):24–32.

Charm, S. E., and C. J. Lai. 1971. Comparison of ultrafiltration systems for concentration of biologicals. *Biotechnol. Bioeng.* 13 (20):185–202.

Cheryan, M. 1986. *Ultrafiltration handbook.* Lancaster, PA: Technomic Press.

Cooper, A. R., and R. G. Booth. 1979. Ultrafiltration of synthetic polymers. 2. Purification of polymeric dyes by ultrafiltration. *J. Appl. Poly. Sci.* 23 (5):1373–1384.

Cooper, A. R., and R. G. Booth. 1978. Ultrafiltration of synthetic polymers: 1. Optimzation of solvent flux during diafiltration. *Sep. Sci. and Technol.* 13 (8):735–744.

Cowieson, D. R., and P. M. Gallagher. 1991. Recent developments in formation and applications of metal mesh/ceramic composite membranes. *Key Eng. Mate.* 61-62:23–32.

Defrise, D., and V. Gekas. 1988. Microfiltration membranes and the problem of microbial adhesion. *Process Biochem.* 23:105–116.

Dorin, G. J. 1989. Patent. *Process for Recovering Refractile Bodies Containing Heterologous Proteins from Microbial Hosts: Biochemical Separation or Recovery.* U.S. Patent no. 5,248,769. International Patent no. WO 8912092.

Forman, S. M., E. R. DeBarnardez, R. S. Feldberg, and R. W. Swartz. 1990. Cross flow filtration for the separation of inclusion bodies from soluble proteins in recombinant *E. coli* cell lysate. *J. Membr. Sci.* 48 (2-3):263–279.

Gangemi, J. D., E. V. Connell, B. G. Mahlandt, and G. A. Eddy. 1977. Arenavirus concentration by molecular filtration. *Appl. Environ. Microb.* 34 (3):330–332.

Gatenholm, P., S. Paterson, A. G. Fane, and J. D. Fell. 1988. Performance of synthetic membranes during cell harvesting of *Escherichia Coli*. *Process Biochem.* 23 (3):79–81.

Henry, J. D., and R. C. Allred. 1972. *Dev. Indust. Microb.* 12:177.

Ho, W. S. Winston, and K. K. Sirkar, ed. 1992. *Membrane handbook.* New York: Van Nostrand Reinhold.

Ju, S. C., T. C. Wu, and S. H. Pai. 1991. *Abstracts of Papers of the American Chemical Society*, 202nd meeting, Part 1, BIOT session 48.

Karbachsch, M., G. Steinhoff, and H. Monzavi. 1984. Depyrogenation of solutions of low molecular weight substances by ultrafiltration. *Drugs Made in Germany* 27 (72):74–76.

Kellaris, P. A., C. A. Beard, P. D. Wyman, G. S. Gray, T. E. Maione, and J. A. Myers. 1991. Large-scale purification of a recombinant platelet factor-IV from a cleavable fusion protein. *Abstracts of Papers of the American Chemical Society*, 202nd Meeting, Part 1, BIOT session 50.

Kroner, K. H., H. Schutte, H. Hustedt, and M.-R. Kula. 1984. Cross-flow filtration in the downstream processing of enzymes. *Process Biochem.* 19 (2):67–74.

Le, M. S., L. B. Spark, and P. S. Ward. 1984(a). The separation of aryl acylamidase by cross flow microfiltration and the significance of enzyme/cell debris interaction. *J. Membr. Sci.* 21:219–232.

Le, M. S., L. B. Spark., P. S. Ward, and N. Ladwa. 1984(b). Microbial asparaginase recovery by membrane processes. *J. Membr. Sci.* 21:307–319.

Maiorella, B., G. Dorin, A. Carion, and D. Harano. 1991. Cross-flow microfiltration of animal cells. *Biotechnol. Bioeng.* 37 (2):121–126.

Malchesky, P. S., A. Werynski, Y. Asanuma, I. Zawicki, J. W. Smith, K. Kayashima, H. Guland, and T. Nosém. 1981. Clinical operation of Asahï plasma separators. *Artif. Organs* 5 (suppl.):113–116.

Michaels, A. S. 1968. Separation technique for the chemical process industry. *Chem. Eng. Prog.* 64 (12):31–43.

Michaels, A. S., and S. L. Matson. 1985. Membranes in biotechnology: State of the art. *Desalination* 53:231–258.

Michaels, S. L. 1991. Validation of tangential flow filtration systems. *J. Paren. Sci. and Technol.* 45 (5):218–223.

Millipore Corporation. 1992(a). Using Millipore high-resolution tangential flow systems to improve the resolution of TFF separations. *Tangents,* No. 9, Part 1, lit. no. SD207. Bedford, MA: Millipore Corporation.

Millipore Corporation. 1992(b). Using Millipore high-resolution tangential flow systems to improve the resolution of TFF separations. *Tangents,* No. 9, Part 2, lit. no. TB007. Bedford, MA: Millipore Corporation.

Millipore Corporation. 1990. *Validation of tangential flow filtration (TFF) systems,* lit. no. TB052. Bedford, MA: Millipore Corporation.

Millipore Corporation. 1989(a). Unpublished experimental results.

Millipore Corporation. 1989(b). *Maintenance guide for tangential flow filters and systems,* lit. no. P17511. Bedford, MA: Millipore Corporation.

Millipore Corporation. 1987. Unpublished data.

Millipore Corporation. 1986. *Downstream process update,* No. 2. Bedford, MA: Millipore Corporation.

Mitra, G., and K. J. Fillmore. 1980. Ultrafiltration of prothrombin complex. In *Ultrafiltration membranes and applications,* edited by A. R. Cooper. New York: Plenum Press.

Müller, J., and K. H. Kroner. 1992. Rotational shear filtration of biosuspensions—Influence of Taylor vortices on filtration performance. In *Proceedings of the Conference on Engineering of Membrane Processes,* 13–15 May, Garmisch-Partenkirchen, Germany.

Myers, J. A., C. A. Beard, S. Almeda, D. J. Peters, and O. J. Konarkowski. 1990. Large scale purification of HIV-1 envelope proteins GP160 and GP120 expressed in baculovirus. *J. Cell. Biochem.* supplement 14D:38.

Nelsen, L. N. 1978. Removal of pyrogens from parenteral solutions by ultrafiltration. *Pharma. Technol.* 2 (5):46–49,80.

Ng, P. K., J. L. Lundblad, and G. Mitra. 1976. *Sep. Sci. and Technol.* 11:499–502.

Patel, P. N., M. A. Mehaia, and M. Cheryan. 1987. Cross-flow membrane filtration of yeast suspensions. *J. Biotechnol.* 5:1–16.

Porter, M. C. 1972. Concentration polarization with membrane ultrafiltration. *Indust. Eng. Chem.* 11 (3):234–248.

Reid, D. L., and C. Adlam. 1974. Large-scale harvesting and concentration of bacteria by tangential flow filtration. *J. Appl. Bacter.* 41 (2):321ff.

Shibley, G. P., M. Manousos, K. Munch, I. Zelljadt, L. Fisher, S. Mayyasi, K. Harewood, R. Stevens, and K. E. Jensen. 1980. New method for large scale growth and concentration of the Epstein-Barr viruses. *Appl. Environ. Microb.* 40:1044–1048.

Siegel, M. H., A. Condemine, C. W. Robinson, C. Fee, T. Z. Fahidy, and B. R. Glick. 1992. Electrically enhanced protein fractionation and separation using constant and pulsed-field cross-flow ultrafiltration. *Proceedings of the Conference on Engineering of Membrane Processes*. 13–15 May, Garmisch-Partenkirchen, Germany.

Tutunjian, R. S. 1985. Ultrafiltration processes in biotechnology. In *Comprehensive biotechnology*, edited by M. Moo-Young. Oxford, U.K.: Pergamon Press.

Tutunjian, R. S. 1984. Cell separations with hollow fiber membranes. *Dev. Indust. Microb.* 25:415–435.

U.S. Food and Drug Administration. 1987. *Guideline on general principles of process validation.* Rockville, MD: Center for Biologics Evaluation and Research (Center for Drugs and Biologics).

United States Pharmacopeial Convention, 1985. *United States Pharmacopeia, XXI.* Rockville, MD: U.S. Pharmacopeial Convention. Chapter 1211.

Vaheri, A., G. Gazzei, and G. Genna. 1979. Tangential flow filtration of *Bordetella pertussis* submersed cultures. *Experientia* 35:1535–1536.

Vaheri, A., G. Gazzei, R. Botti, V. Pellegrini, A. Corradeschi, and D Soldateschi. 1981. One-day purification of influenza A and B vaccines using molecular filtration and other physical methods. *Microbiologica*, 4:403–412.

Vandamme, E. J. 1984. *Biotechnology of industrial antibiotics*. New York: Marcel Dekker.

van Reis, R., E. M. Clabb, J. W. Kaspari, and S. E. Builder. 1992. High-performance tangential flow filtration. *Engineering Foundation Conference on Recovery of Biological Products*, VI, Abstract 9.1, 20–25 Sept., Interlaken, Switzerland.

Veide, A., T. Lindback, and S. Enfors. 1989. Recovery of beta-galactosidase from a poly(ethylene glycol) solution by diafiltration. *Enzyme Microb. Technol.* 11:744–751.

Wijmans, J. G., S. Nakao, J. W. A. van den Berg, F. R. Troelstra, and C. A. Smolders. 1985. Hydrodynamic resistance of concentration polarization boundary layers in ultrafiltration. *J. Membr. Sci.* 22:117.

Zimmerman, G., D. Kruger, and H. Woog. 1976. Pyrogen elimination from parenteral medicines by means of ultrafiltration. *Drugs Made in Germany* 19:123–128.

Zydney, A. L. 1992. Protein fouling during membrane microfiltration. *Engineering Foundation Conference on Recovery of Biological Products IV*, Abstract 9.3, 20–25 Sept., Interlaken, Switzerland.

RECOMMENDED READING

Abramson, D., D. Butler, and S. Chrai. 1981. Depyrogenation of a parenteral solution by ultrafiltration. *J. Paren. Sci. Technol.* 35 (1):3–7.

Devereux, N., and M. Hoare. 1986. Membrane separation of protein precipitants. *Biotechnol. Bioeng.* 28:422–431.

Fane, A. G., and J. M. Radovich. 1990. Membrane systems. In *Separation processes in biotechnology*, edited by J. A. Asenjo. New York: Marcel Dekker, Inc.

Gabler, F. R. 1987. Pyrogens and the depyrogenation of solutions with ultrafiltration membranes. In *Filtration in the pharmaceutical industry*, edited by T. Meltzer. New York: Marcel Dekker.

Ingham, K. C., T. F. Busby, Y. Shahlestrom, and F. Castino. 1980. Separation of macromolecules by ultrafiltration: Influence of protein adsorption, protein-protein interactions, and concentration polarization. In *Ultrafiltration membranes and applications*, edited by A. R. Cooper. New York: Plenum Press.

Lange, E. 1983. *Molecular and macromolecular sieving by asymmetric ultrafiltration membranes*. Ph.D. thesis, Stanford University.

Millipore Corporation. 1991. *Selecting a membrane for concentration, desalting and buffer exchange of macromolecules (Ultrafiltration)*, lit. no. AB002. Bedford, MA: Millipore Corporation.

Nilsson, J. L. 1990. Protein fouling of UF membranes: Causes and consequences. *J. Membr. Sci.* 52:121.

Opong, W. S., and A. L. Zydney. 1991. Hydraulic permeability of protein layers deposited during ultrafiltration. *J. Coll. and Int. Sci.* 142:41.

Reihanian, H., C. R. Robertson, and A. S. Michaels. 1983. Mechanisms of polarization and fouling of ultrafiltration membranes by proteins. *J. Membr. Sci.* 16:237.

Ruth, B. F. 1946. Correlating filtration theory with industrial practice. *Indust. Eng. Chem.* 38:564.

Shettigar, U. R., H. Mann, and A. Gressner. 1982. Continuous enrichment of albumin in relation to globulins in plasma. *Artif. Organs* 6:163.

Sieberth, H. G., W. M. Glockner, and H. Kierdorf. 1983. Cascade filtration. In *Plasmapheresis,* edited by Y. Nosé, H. E. Cambric, and S. Matsubara. New York: Raven Press.

Trettin, D. R., and M. R. Doshi. 1980. Ultrafiltration in an unstirred batch cell. *Indust. Eng. Chem.* 19:189–194.

Vilker, V. L, C. K. Colton, and K. A. Smith. 1981. Concentration polarization in protein ultrafiltration, II: theoretical and experimental study of albumin ultrafiltered in an unstirred cell. *A.I.Ch.E. Journal* 27:637–645.

Wolber, P., and M. Dosmar. 1987. Depyrogenation of pharmaceutical solutions by ultrafiltration: Aspects of validation. *Pharma. Technol.,* 11 (9):38–46.

Zahka, J., and T. J. Leahy. 1985. Practical aspects of tangential flow filtration in cell separations. *ACS Symposium Series*, pp. 51–69.

4

Supercritical Fluid Extraction/ Chromatography

Kent M. Payne, Ph.D.
Jerry W. King, Ph.D.

The use of supercritcal fluids to effect separations has received increased attention in the last 15 years. This is due to the attendant ability of a supercritical fluid to offer selective solubilities by adjusting the pressure and temperature as well as chemical composition of the fluid. Preparative and process-scale supercritical separation techniques offer many advantages over similar liquid technologies (Table 4.1). Processing plants which use supercritical carbon dioxide (CO_2) as a solvent have the advantage of leaving no detectable chemical residues in the end products. Commercial grade CO_2 is also one of the cheapest solvents available. As such, supercritical processes developed for the pharmaceutical industry using CO_2 could have similar success, to those in the food industry, in obtaining FDA approval.

The use of supercritical CO_2 has been demonstrated on analytical, preparative, and process scales for the separation, via extraction or chromatography, of high value products. Due to large capital investment and complex technology, supercritical fluid separation processes are usually most cost-effective when the isolated material is of a high value. Pharmaceuticals that fall into the above category are excellent candidates for supercritical fluid extraction (SFE) and supercritical fluid chromatography (SFC).

The use of supercritical fluids for process-scale separation in pharmaceutical applications has yet to be realized in commercial

Table 4.1. Supercritical Fluids

Advantages	Disadvantages
Applicable to low temperature extraction of high boiling or temperature sensitive compounds.	High pressure process, management and operators may be used to processes operated at near ambient conditions.
Extract is readily removed from the solvent.	Plant costs are comparatively high.
Solvent is easily recovered for for reuse.	Supercritical fluid plants require high pressure vessels and pipework.
Reduced loss of heat-sensitive components.	Sophisticated process control systems are adviseable.
Solvent is approved as food ingredient.	Maintenance costs could be relatively high as a consequence of sophistication.
No toxic solvent residues are contributed from the solvent.	Full chemical engineering and design data are not yet mature.
High purity of commercial carbon dioxide available.	Many variants are already patented.
Readily available and comparatively inexpensive solvent.	
Some extractions could not be made by other techniques.	
Very pure products in relatively few steps.	

applications. It is hoped that through examination of some of the theoretical aspects of supercritical fluids, along with the current commercially viable applications, that some possibilities can be suggested for the pharmaceutical industry. In addition, several studies have been conducted successfully on pharmaceutically related compounds involving the modeling and economics of process scale systems. Review of these applications should provide a guideline as to the utility of supercritical fluid techniques, for potential exploitation by the pharmaceutical industry.

THEORY

Definition of a Supercritical Fluid

A supercritical fluid is obtained when a substance is brought to temperature and pressure above or near its critical point. Supercritical fluids are neither gases nor liquids, but possess properties of both states of matter. For example, CO_2 exists as a solid (dry ice) at -85°C and sublimes directly to the gas phase at ambient conditions. Hence, liquid CO_2 exists only under pressure and between -56°C and 31°C. However, at even higher pressures (≥72.9 atm) and above the critical temperature (31.3°C), CO_2 exists in its supercritical state.

In the supercritical state properties such as density, viscosity, and vapor-liquid equilibrium ratios become strongly temperature dependent at a given pressure. Subtle variations in temperature and pressure can produce wide variations of density, particularly in the region of the critical point. This affords the ability to control solubility of solutes in the supercritical fluid. The control of solute solubility in a supercritical fluid, by manipulating pressure and temperature, is the basis of both process and analytical SFE and SFC. This is discussed in the next section.

Solubility

Pressure, Temperature, and Density

The solvating power of supercritical fluids is highly dependent on the density of the fluid. High fluid density in the supercritical fluid is achieved as a result of the high pressures created by the system rather than by intermolecular attraction as with liquids. Supercritical fluids, in general, exhibit lower viscosities and permit greater solute diffusivities, than those associated with liquids. In addition, no surface tension is associated with supercritical fluids, permitting greater wettability of sample matrices. By varying the temperature and pressure on the fluid, the density of the supercritical fluid can be controlled. This allows selective manipulation of the solvating power of the fluid. In general, higher densities lead to increased solubility.

Temperature plays a dual role in supercritical fluid separation technology, since it also influences the vapor pressure of solutes. Temperature can also play an important role in the desorption of analytes from a matrix or sorbent. Higher temperatures also decrease density and, therefore, increase mass transfer coefficients of the solutes in the fluid. In SFE enhanced mass transfer permits greater permeation of the sample matrix by the supercritical fluid, leading to more efficient extractions. For SFC the result is more frequent and rapid equilibrium of solute partitioning into and out of the stationary phase, thereby achieving higher chromatographic resolution than obtainable with liquids.

By analyzing the solubility trends for the respective components in a mixture, a set of conditions can be ascertained that can provide selective extraction or separation of the individual solutes. This often occurs in what is referred to as the crossover region. The crossover point for a single analyte occurs at the pressure where there is a change in the temperature dependence of the solubility. Below this pressure, solubility is dependent on the density of the supercritical fluid. Under these conditions a decrease in temperature leads to an increase in solubility. Above this pressure solubility is also dependent on solute vapor pressure. At this pressure, raising the temperature results in an increase in the vapor pressure and, hence, an increase in solubility. Therefore, for two solutes (being chromatographed or extracted) whose crossover regions occur at difference pressures, it becomes possible to separate the two solutes by varying the temperature or pressure under which the separation is conducted.

Cosolvent Effects

Some solutes may be quite soluble in a supercritical fluid at a given density, but when the same conditions are used in SFE or SFC, their elution may be substantially impaired due to solute-matrix interactions. This is particularly true for polar solutes which may adsorb to active surface sites in the matrix or chromatographic packing in a column. To overcome this problem, organic cosolvents (also called entrainers or modifiers) are frequently added to the supercritical fluid to enhance the solvating power over that achieved with the neat fluid. Also, those cosolvents compete with active sites in the sample matrix or chromatographic packings, thereby enhancing the extraction or chromatography of the desired components. Many low molecular weight polar organic solvents, such as ethanol, are miscible in CO_2 up to a specific mole fraction at a given pressure and temperature. For example, Brunner and Peter (1982) demonstrated that the addition of 10 percent ethanol to CO_2 increased the solubility of palm oils under supercritical conditions by a factor of 20 times as compared to neat CO_2. It should be noted that the amount of cosolvent used in conjunction with the supercritical fluid is much less than is required in conventional liquid processes.

Additional information on the physicochemical properties and behavior of supercritical fluids can be found by consulting several well known texts (Penninger 1985; McHugh 1986; Johnston 1989; Lee and Markides 1990; Bruno 1991).

APPLICATIONS OF SUPERCRITICAL FLUIDS

General Industrial Applications

Specific applications mentioned in this chapter should not be considered as an exhaustive listing either of current or potential uses of supercritical fluids, but simply as examples from which analogies can be drawn in formulating decisions about the use of supercritical fluids in pharmaceutical process separations. It is apparent that more industrial applications will be forthcoming as more physicochemical data is available for predicting solubilities of various classes of pharmaceutical compounds and as models for scaling separations to the process scale continue to improved.

Decaffeination

Probably the best-known application of supercritical fluids is for the decaffeination of coffee. The CO_2 decaffeination process was discovered and developed by Kurt Zosel and the Max Planck Institute in Germany. The process was patented in the 1970's and licensed to Kaffee HAG and General Foods (Zosel 1978). In the separation of caffeine from coffee beans, Zosel recommended extraction of the hydrated beans with CO_2 (Zosel 1981). The first commercial-scale plant to be operated in Germany (designed for the decaffeination of coffee) became operational in the late 1970s with a capacity of 30,000 metric tons/year.

Nestle, a major food company also holds patents for the extraction of the coffee aroma, using CO_2. The soluble coffee is then rearomatized just prior to packing by back addition of the volatile fraction. Far less flavor is lost, than in the conventional processing technique (Pictet et al. 1968).

The large scale decaffeination of coffee demonstrates that supercritical fluids can be used to provide selective isolation of alkaloids from a natural product. It is apparent that this would not only be applicable to the isolation of an alkaloid from coffee, but that the potential exists for applying the technology to a wide variety of other separation needs involving alkaloid, and other extraction and isolation processes within the pharmaceutical industry.

Hops

The first major plants to use CO_2 for the removal of flavor components from hops, were built in Australia and England (Clarke and Mailer 1981) to provide a product for the beer brewing industry. Initially, liquid CO_2 was used as the solvent (Wheldon 1981). A process utilizing supercritical fluid CO_2 was eventually developed in West Germany at, surprisingly, the Coal Research Institute and HAG AG (Vitzthum 1971). In 1981 a plant was commissioned for the extraction of the aroma from hops in Germany, which became operational in 1982. Plants in England and Germany are capable of processing one fifth of their respective nation's annual hop crop, whereas the Australian plant can handle over half of that country's annual production. This capability is important since the hops themselves are difficult to store in their harvested state, due to their tendency to deteriorate. Carbon dioxide derived extracts store better and are superior as flavoring agents. Supercritical fluid based hop plants are utilized for the extraction of other products, at the conclusion of the hops harvesting season. Today, there are several supercritical CO_2 based hops plants, processing products in the Yakima Valley region of the State of Washington.

A related application of SFE is the isolation of oils from brewer's grains that can be utilized as antifoaming agents in brewing processes. These oils are normally extracted and lost during the brewing process. Refortification with the supercritical fluid extract is but one example of adding a "natural" CO_2 extract back into a production process producing a foodstuff.

Other Applications

Some other references to additional applications of supercritical fluids are included in Table 4.2. These examples illustrate the wide potential that exists for supercritical fluid separations. Although many have not yet reached commercial application, they deserve serious consideration from the pharmaceutical industry as viable alternatives in ecologically compatible manufacturing processes.

Selected Pharmaceutical Applications

Supercritical Fluid Extraction

Extraction of Natural Products. Just as liquid and supercritical fluid CO_2 may be used in the extraction of foods, they may also be used to extract drugs and related precursors. Other pharmaceutically related applications are as a reaction media for chemical and enzyme catalyzed synthesis and for the purification of synthetic organic

Table 4.2. Application References

Supercritical fluids as media for enzymatic reactions	Randolph et al. (1985); Taniguchi et al. (1987a); Kamihira et al. (1987c); Russel and Beckman (1991a, 1991b)
Sterilization	Kamihira et al. (1987b); Taniguchi et al. (1987b)
Enzymatic analysis for pesticide residues	King and France (1991); Nam and King (1994)
Enzyme synthesis of aspartame precursors	Kamihira et al. (1987c)
Decaffeination of coffee	Vizthum and Hubert (1975); Asche (1980)
Production of low alcohol wines	Berger et al. (1981)
Fractionation of vegetable and cottonseed oils	Coenen and Kriegel (1984); List et al. (1984)
Extraction of spices	Avagimov et al. (1980); Brannolte (1982)
Essential oils flavors and aromas from plants	Stahl et al. (1982); Coenen and Kriegel (1984); Krukonis (1984); Caragay (1981); Moyler (1985); Williams (1981); Temelli et al. (1988)
Extraction of food colors	Degnan et al. (1991)
Regenerating lubricating oils	Coenen et al. (1980, 1982)
Enzyme catalysis to manufacture high value oils from lower value oils	Nakamura (1987)

Continued on next page

Continued from previous page

Pyrethrin insecticides from pyrethrum	Stahl and Schultz (1980); Sims (1982)
Alcohol from fermentation broths	De Filippi (1982)
Reduction of fat in potato chips	Hannigan (1981)
Regeneration of spent bleaching clays	King et al. (1992)
β-Carotene from liquid solvents	Chang and Randolph (1991)
Removal of fat and cholesterol from milk	Process for removing . . . (*The Cheese Reporter* 1989); Bhaskar et al. (1993)
Fat from meats	King et al. (1989)
Removal of nicotine and tars from tobacco	Roselius et al. (1970)
Extraction of aroma from tobacco dust	Luganskaya et al. (1967)
Dry cleaning of garments	Maffei (1971)
Extraction of paraffins and extraction of liquid fuels from coal	Williams (1981); Demitrelis et al. (1984)

chemicals. The isolation of drugs from fermentation broths may be possible in certain instances using selected supercritical fluids.

Supercritical fluid techniques are being examined in the biochemical and pharmaceutical industries as a result of the needs for separating temperature sensitive compounds using physiologically inert solvents.

The number of applications of supercritical fluid extractions has been limited, in part due to the fact that the preferred solvent,

CO_2, does not always provide the required selectivity. Biomolecules generally have low solubilities in supercritical fluids and often occur in supercritical fluids, as trace components in the presence of high concentrations of lipid coextractants. Some additional selectivity can be obtained, by using cosolvents. Further purification steps may still be required, including the removal of the cosolvent from the extract.

The following are but a few examples of the application of supercritical fluid extraction to compounds of pharmaceutical interest. Some of the reported studies are concerned with solubility studies in supercritical fluids, or in analytical SFE and SFC studies on compounds of pharmaceutical interest. These studies, although not strictly process in scale or scope, suggest potential applications in the pharmaceutical industry and invite further research investigations.

Tocopherol concentrates are utilized in various applications as antioxidants and vitamin supplements. The recovery and purification of tocopherols from natural substances, such as vegetable oils, has been accomplished using SFE, coupled with preparative scale SFC (King 1993a). In a two-step process the tocopherols were isolated and transferred from the SFE stage, and deposited at the head of a preparative column, containing silica sorbent. The tocopherols were then eluted from the preparative column via SFC, using supercritical CO_2 in order to isolate the enriched alpha, beta, gamma, and delta fractions. Recoveries of approximately 75 percent were reported, based on the original concentration in the starting oil. A schematic depiction of the device used in these experiments is given in Figure 4.1. Enrichment factors are shown in Table 4.3. Results demonstrated that the additional enrichment afforded by SFC is significant, and achieves a result, not possible by SFE alone.

Taxol (MW = 853.9) is a natural product, exhibiting anti-cancer properties, which can been isolated from *Taxus brevifolia* (Northwestern pacific yew tree) as well as other botanicals of the *Taxus* genus. It has been used in the treatment of breast, ovarian, and other forms of cancer. Harvesting taxol from *Taxus brevifolia* is somewhat problematic (Borman 1991). Since taxol occurs in small amounts in the bark of *Taxus brevifolia* [50 to 100 mg/kg bark (Zuer 1988)], it is estimated that the bark of two 60-year-old trees would be required for the treatment of a single patient (Caruana 1991). Obviously, extraction of taxol from such a natural source is not a long term solution. Synthesis of taxol has been demonstrated using a compound derived from extracts of the leaves of *Taxus brevifolia* as a precursor (Denis et al. 1988). More recently, Chun et al. (1994) described the extraction of taxol, baccatin III and 10-deacetyl baccatin III used as precursors of an efficient semi-synthetic method for producing taxol. The leaves of the tree (or

Figure 4.1. Schematic of SFE/SFC used for the production of tocopherol concentrates. (Source: King et al. (1993a) with permission)

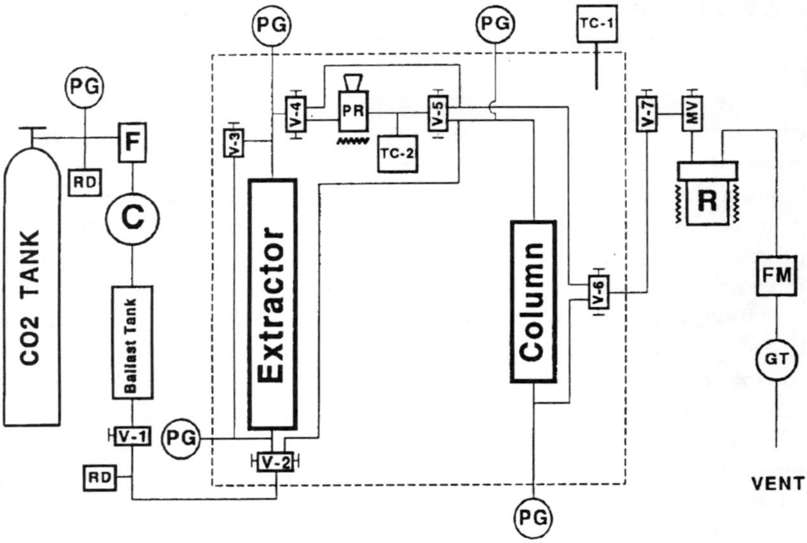

Table 4.3. Enrichment of Tocopherols from Soybean Flakes

	SFE	SFE/SFC
alpha	4.33	12.1
beta	1.83	2.4
gamma	3.94	15.0
delta	3.75	30.8

Source: King et al. 1993a with permission.

possibly plant tissue culture) being a renewable resource may prove to be an alternative solution (Witherup et al. 1990; Caruana 1991).

The application of SFE for the extraction of taxol from the *Taxus brevifolia* bark has been investigated (Jennings et al. 1992). Ground bark was alternately extracted with neat supercritical CO_2 and CO_2 with ethanol as a cosolvent. The experiments were conducted isothermally at 318 K and the amount of taxol extracted from the bark was shown to increase with increasing pressure for both neat supercritical CO_2 and supercritical CO_2/ethanol. In both cases, taxol was more selectively recovered from the bark using supercritical extraction compared to the liquid extraction method. The CO_2/ethanol mixtures were the most selective of all, providing the highest percentage of taxol per weight of total extractants. Although the supercritical fluid extraction conditions were not optimized in this study, the ability of supercritical fluid extraction to provide a purer extract than that obtained with liquid extraction was demonstrated.

In addition, coextracted materials that were present in the supercritical fluid extract were more amenable to separation by liquid chromatography than the coextractants recovered via liquid extraction. This is of significance when one considers the economics for the total isolation of natural products. Recently, (King 1993b) has demonstrated that 85 percent recovery of taxol is possible from yew heartwood via SFE.

Monocrotaline is valuable as a precursor in the production of the semisynthetic pyrrolizidine alkaloids for use as anticancer agents (Gelbaumm et al. 1982). It can be obtained from the seeds of *Crotalaria spectabilis.* The conventional method for isolating monocrotaline from *Crotalaria spectabilis* is quite difficult and expensive. Schaeffer et al. (1989a, 1989b) have demonstrated the extraction of monocrotaline from *Crotalaria spectabilis* using supercritical CO_2 with ethanol as a cosolvent.

By conducting experiments in the temperature-solubility crossover region, an attempt was made to find a set of conditions that would allow the selective separation of coextracted lipids from the monocrotaline (Schaeffer et al. 1988). A 22 percent increase in monocrotaline purity was obtained, however, the monocrotaline content of the total extracted mass was still only 49 percent.

Additional experiments were conducted (Shaeffer 1989b) to evaluate the feasibility of conducting SFE in conjunction with a bed of cation-exchange resin used for selectively trapping the alkaloid moieties in the presence of supercritical CO_2. By incorporating the ion-exchange column, the extracted lipid components can be removed by washing, leaving the monocrotaline behind for selective elution from the column (Gelbaumm 1982).

To verify this, supercritical CO_2 modified with ethanol, and with water, was used in separate experiments, to extract *Crotalaria spectabilis* seeds. The experiments were conducted as follows; the seeds were extracted with the supercritical fluid, the solute-laden supercritical effluent was passed (still at system temperature and pressure) through the cation-exchange column (in the H^+ form). The column effluent was then depressurized into a collection vessel, leaving the monocrotaline adsorbed onto the cation-exchange resin. Some lipid material was also trapped on the resin, however, the cation-exchange column was then washed with ethanol:water 95:5 (v:v), which removed the adsorbed lipid materials. No monocrotaline was detected in the ethanol wash. The cation-exchange column was then washed with 1 N NH_4OH, thereby releasing the adsorbed monocrotaline.

Using both ethanol and the water modified supercritical CO_2, monocrotaline purities of >95 percent were obtained. No lipid material was detected in the adsorbed fraction wash.

The significance of the above process lies in the applicability not only for the specific isolation of monocrotaline but to a wide variety of compounds from plants and other matrices that could be treated with SFE. It follows that this process could also be extended to the use of an anion-exchange resin for the removal of various acidic drugs from supercritical fluids.

The use of SFE in combination with ion-exchange columns for trapping extracted analytes was used to design a prospective process for the production of monocrotaline on the kilogram scale (Dyken et al. 1990). Computer modeling was performed to study the production process, utilizing data from Schaeffer's previous work (Schaeffer 1989b). Information was obtained related to predicted production levels, and an economic analysis was performed. It was concluded that SFE in conjunction with ion-exchange, affords a low-energy, moderate temperature recovery process, that provides an end product of high purity. The resulting information suggested that the final extracted product must have a high value in order for supercritical processing to be cost-effective.

Vindoline and catharanthine are used as precursors in the biosynthetic pathways of high value dimeric indole alkaloids, such as vinblastine and vincristine, that are used in chemotherapeutic treatment of acute leukemia and Hodgkin's disease (Lee et al. 1992). However, Lee and coworkers, using only CO_2, were able to demonstrate the successful extraction of both alkaloids. It seems clear that further studies including the utilization of modifiers would certainly result in similar results to other studies where alkaloid substances were successfully isolated. In fact, other alkaloids that have been isolated by SFE, include morphine and quinine (Ndiomu and Simpson 1988).

Antibiotics. *Penicillin.* Antibiotics present a unique challenge for SFE due to their structural attributes (i.e., high molecular weight, high polarity, low volatility, and thermal lability). The β-lactam antibiotics represent the largest group of commercially available antimicrobial agents, of which penicillin's are one chemical type. Penicillin V is the most widely used of the penicillin derivatives. The classical separation and purification process for antibiotics includes a series of solvent extractions and precipitations culminating with crystallization of the product (Florey et al. 1949). Additional separation techniques such as ion-exchange, metal complexation, and chromatographic processes may be necessary for compounds which are difficult to crystallize. These purification processes can take up to 60 processing steps and account for as much as 80 percent of the expense of an antibiotic production operation (Ko et al. 1991). Supercritical fluid extraction could reduce the cost of the separation and purification process, by possibly extracting the antibiotic directly from the fermentation broth or by purifying the solid precipitate.

Polycyclic Ethers. The solubility of the sodium salts of the polycyclic ether class of veterinary antibiotics (Figure 4.2) have been examined (Maxwell et al. 1992). The purpose of the study was to see if SFE could be used in the analysis of incurred antibiotic residues in edible tissues. The solubility data obtained however is of value for separation and purification purposes in pharmaceutical production. The solubility of four polyether antibiotics (lasalocid, monensin, narasin, and salinomycin) were measured in neat CO_2 and cosolvent modified CO_2

Figure 4.2. The structure of the four polycyclic ether antibiotics and their equilibrium solubilities at 80°C in pure and modified CO_2. (Source: Maxwell et al. (1992) with permission)

	Molecular Weight Sodium Salt	Equilibrium Solubility (mol/ℓ) at 80°C, 400 bar
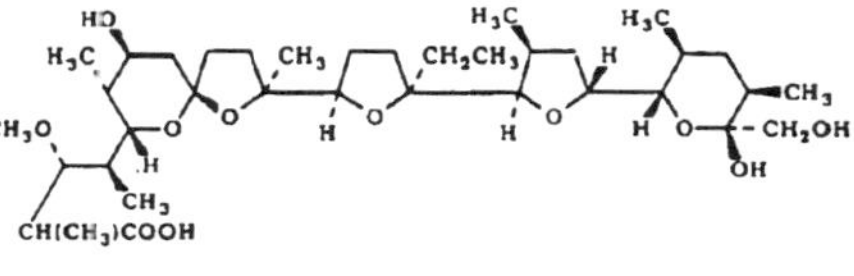 Lasalocid	$C_{34}H_{53}O_5Na$—612	2.3×10^{-4} (CO_2/1% MeOH)
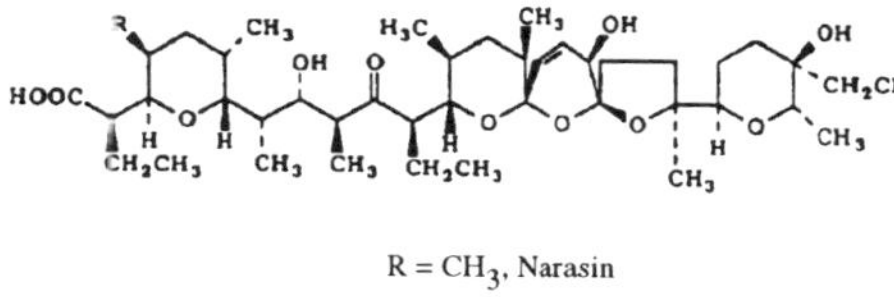 Monensin	$C_{36}H_{61}O_{11}Na$—692	5.7×10^{-5}
R = CH_3, Narasin	$C_{43}H_{71}O_{11}Na$—786	1.4×10^{-3}
R = H, Salinomycin	$C_{42}H_{69}O_{11}Na$—772	7.4×10^{-3}

(methanol and water) (Table 4.4). Attempts to correlate the molecular structure with solubility trends were frustrating due to lack of physicochemical data on the solutes.

Table 4.4. Solubility Parameters for the Polycyclic Ether Antibiotics in Modified and Pure Carbon Dioxide

		Solubility	
Pressure (bar)	CO_2 Density (mol/ℓ)	Mol/ℓ	Mole Fraction
	Lasalocid		
CO_2 + 1% MeOH/80°C			
152	10.07	7.00×10^{-5}	6.95×10^{-6}
207	13.60	5.00×10^{-5}	3.68×10^{-6}
279	16.44	6.55×10^{-5}	3.98×10^{-6}
344	18.17	1.17×10^{-4}	6.44×10^{-6}
382	18.93	1.37×10^{-4}	7.22×10^{-6}
390	19.08	2.34×10^{-4}	1.23×10^{-5}
	Monensin		
CO_2/80°C			
201	13.60	1.23×10^{-5}	9.04×10^{-7}
277	16.40	1.57×10^{-5}	9.57×10^{-7}
309	17.15	3.39×10^{-5}	19.77×10^{-7}
408	18.85	5.71×10^{-5}	30.29×10^{-7}
CO_2 + 1% MeOH/80°C			
185	12.36	7.59×10^{-6}	6.14×10^{-7}
243	15.22	4.35×10^{-5}	2.86×10^{-6}
316	17.48	1.20×10^{-4}	6.87×10^{-6}
402	19.30	2.08×10^{-4}	1.08×10^{-5}
CO_2 + H_2O[1]/80°C			
182	12.01	7.68×10^{-6}	6.35×10^{-7}
242	14.49	8.02×10^{-6}	5.55×10^{-7}
316	16.30	1.43×10^{-5}	8.77×10^{-7}
401	17.71	2.54×10^{-5}	1.43×10^{-6}

Continued on next page

Continued from previous page

		Solubility	
Pressure (bar)	CO_2 Density (mol/ℓ)	Mol/ℓ	Mole Fraction
Narasin			
CO_2/60°C			
195	16.30	2.6×10^{-4}	1.59×10^{-5}
206	16.65	3.3×10^{-4}	1.98×10^{-5}
264	18.20	4.3×10^{-4}	2.36×10^{-5}
274	18.40	4.4×10^{-4}	2.39×10^{-5}
332	19.35	7.0×10^{-4}	3.62×10^{-5}
401	20.25	1.12×10^{-4}	5.53×10^{-5}
CO_2/70°C			
175	13.65	3.10×10^{-4}	2.27×10^{-5}
241	16.48	4.80×10^{-4}	2.91×10^{-5}
314	18.18	8.40×10^{-4}	4.62×10^{-5}
405	19.55	1.26×10^{-3}	6.44×10^{-5}
CO_2/80°C			
141	8.80	1.61×10^{-4}	1.82×10^{-5}
217	14.35	2.69×10^{-4}	1.87×10^{-5}
276	16.35	5.31×10^{-4}	3.24×10^{-5}
364	18.15	1.35×10^{-3}	7.43×10^{-5}
CO_2 + H_2O[1]/60°C			
181	14.81	3.9×10^{-4}	2.63×10^{-5}
184	14.94	5.7×10^{-4}	3.82×10^{-5}
243	16.88	2.23×10^{-3}	1.32×10^{-4}
311	18.34	4.48×10^{-3}	2.44×10^{-4}
401	19.72	6.29×10^{-3}	3.19×10^{-4}
Salinomycin			
CO_2/70°C			
185	14.23	2.42×10^{-4}	1.70×10^{-5}
249	16.73	9.86×10^{-4}	5.39×10^{-5}
318	18.25	2.37×10^{-3}	1.30×10^{-4}
391	19.38	2.92×10^{-3}	1.51×10^{-4}

Continued on next page

Continued from previous page

Pressure (bar)	CO_2 Density (mol/ℓ)	Solubility Mol/ℓ	Solubility Mole Fraction
CO_2/80°C			
142	8.90	2.32×10^{-4}	2.61×10^{-5}
202	13.60	6.35×10^{-4}	4.67×10^{-5}
256	15.80	1.77×10^{-3}	1.12×10^{-4}
273	16.30	2.05×10^{-3}	1.26×10^{-4}
318	17.35	4.71×10^{-3}	2.71×10^{-4}
395	18.65	7.36×10^{-3}	3.95×10^{-4}
$CO_2 + H_2O$[1]/60°C			
175	14.53	2.57×10^{-4}	1.77×10^{-5}
184	14.94	5.7×10^{-4}	3.82×10^{-5}
243	16.88	2.23×10^{-3}	1.32×10^{-4}
313	18.38	3.24×10^{-3}	1.76×10^{-4}
326	18.60	5.74×10^{-3}	3.09×10^{-4}
404	19.75	5.82×10^{-3}	2.95×10^{-4}

[1]Water-saturated CO_2 (experimental)
Source: Maxwell et al. (1992) with permission.

Solvent Removal from Antibiotics. The production of antibiotics as well as in the synthesis of organic drug substances often require solvents to purify the final product. Purification of the final product also requires the removal of organic solvents used in their production. Traditionally this is accomplished by vacuum drying the final product; however, traces of the solvent can be left behind in the antibiotic. Experiments were conducted by Kamihira et al. (1987a) to investigate the potential for solvent removal from antibiotics by SFE using neat CO_2. The residual amounts of organic solvents left in penicillin G and streptomycin are shown in Table 4.5.

Differences in the final residual amount of solvent remaining in the antibiotics were observed to be dependent upon the specific compound. It appeared that the results obtained were due to differences in the affinity of each antibiotic for a specific solvent. Dramatic results were shown, even when a twofold weight ratio of solvent to antibiotic was used (sufficient to immerse the antibiotic). However, supercritical CO_2 was still effective in removing the solvent.

Water was also used as a cosolvent to increase the rate and extent of solvent removal. The extraction efficiency for the solvent approximately

Table 4.5. Relative Remaining Activity and Residual Amount of Solvents after Extraction with Supercritical CO_2*

	Penicillin G, Potassium		Streptomycin Sulfate	
Solvent	**Relative Remaining Activity (%)**	**Residual Amount of solvent (g/g-dry matter)**	**Relative Remaining Activity (%)**	**Residual Amount of Solvent (g/g-dry matter)**
Methanol	94	$<1 \times 10^{-5}$	101	9×10^{-5}
Ethanol	100	2×10^{-5}	99	4×10^{-5}
Acetone	100	5×10^{-5}	98	8×10^{-5}
Isopropyl alcohol	100	6×10^{-5}	97	7×10^{-5}
Ethyl acetate	103	$<1 \times 10^{-5}$	106	$<1 \times 10^{-5}$
n-Butyl acetate	ND	0.032	ND***	ND
n-Butyl alcohol	ND	0.222	ND	ND

*Extraction conditions: 35°C, 200 atm, 2 hr
**Initial amount of solvent: 2.0–2.2 g/g-dry matter
***ND: Not Determined
Source: Kamihira et al. (1987a) with permission

doubled, compared to the use of neat CO_2. Note, that the use of a supercritical fluid resulted not only in rapid removal of the organic solvent, but produced in most cases no loss in enzymatic activity.

Supercritical Fluid Chromatography

Potential Uses. As noted previously, the greater diffusivity of solutes in a supercritical fluid allows larger chromatographic resolution to be obtained, relative to that obtained using liquid chromatographic methods. Some process scale chromatography is accomplished using the normal phase mode, hence SFC using CO_2 could be a suitable replacement for the nonpolar solvents used in this mode of chromatography. In addition, the SFC literature indicates that many compounds of pharmaceutical interest can be chromatographed (Xie et al. 1992). Considering these factors, it is worthwhile to consider several select possibilities of applying SFC to compounds of pharmaceutical interest.

Chiral Separations. Chromatographic separation of enantiomers on the preparative and process scales to isolate optically pure compounds is becoming an increasingly important (Francotte and Junker-Buchheit 1992; Blum and Kumar 1994).

The cost and time required for developing a chiral synthetic method may be prohibitive, especially early in development when the final success of the project has yet to be determined. The investment of resources to develop and implement a chromatographic separation in order to produce pure chiral compound may provide savings in time and capital. SFC, using CO_2, has the potential to be a viable and lower cost alternative to LC in preparative and process scale chiral separations.

The unique selectivity and superior chiral resolution obtained on the analytical scale by SFC compared to LC is well documented for analytical separations (Lee et al. 1990; Brugger et al. 1991; Marti et al. 1991; Xie et al. 1992). Preparative separation of several enantiomers of optically active drug substances have been recently presented (Blum and Kumar 1994). The unique selectivity offered by SFC, documented for analytical-scale separations, was also observed in preparative-scale separations as well. A schematic of typical preparative SFC equipment that was utilized is shown in Figure 4.3. Preparative-scale separations were achieved for warfarin, propranolol, and another Merck-produced drug (Figures 4.4–4.6).

Additional Pharmaceutical Applications of Supercritical Fluids

Supercritical fluids are useful in other areas of pharmaceutical manufacturing besides separation and extraction alone. Two such

Figure 4.3. Schematic of preparative scale SFC equipment used for chiral separations. (Source: Blum and Kumar (1994) with permission)

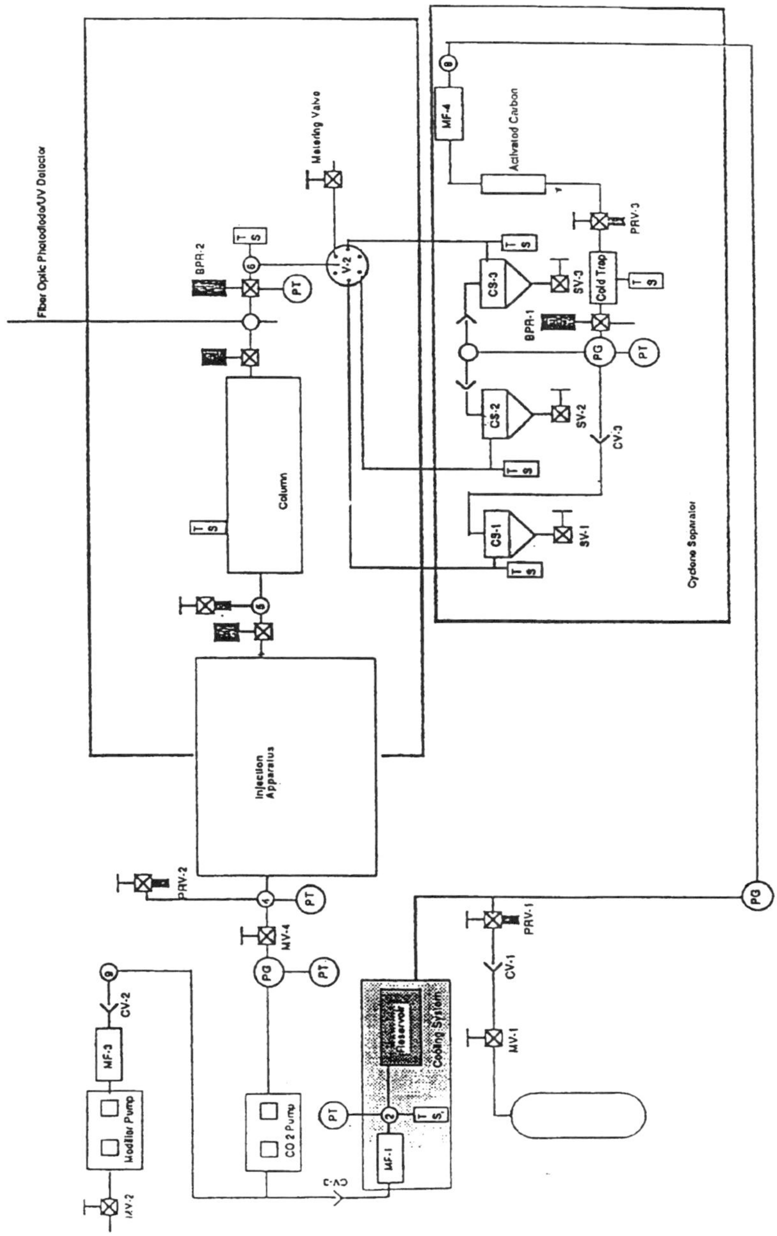

Figure 4.4. Preparative SFC of Warfarin. (Source: Blum and Kumar (1994) with permission)

Column:	Whelk-O1 - 25 × 2.54 cm
Column Temp.:	25°C
Solvent:	0.5% Acetic Acid, 25% IPA, in Carbon Dioxide
Flow:	100 ml/min
Pressure:	250 atm
Detection:	260 nm
Injection:	0.2 g

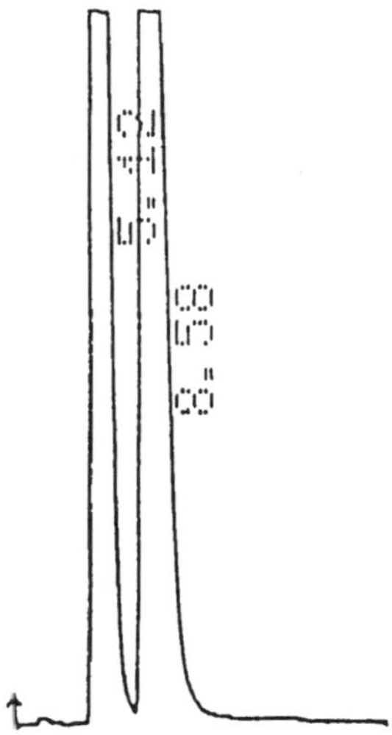

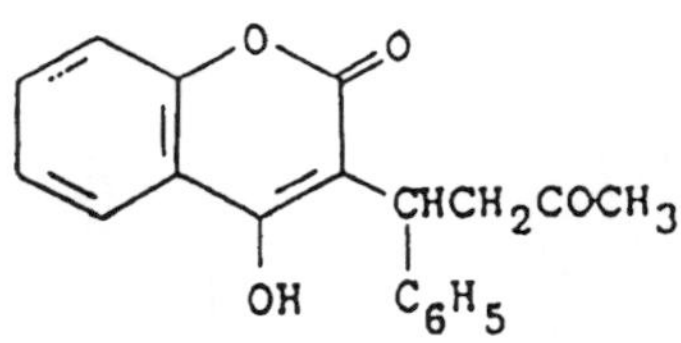

operations are described below to illustrate the diversity of supercritical fluid technology.

Micronization of Drug Delivery Particles. In order to obtain drug-loaded polymeric particles for the delivery of drug substances, processes must be developed involving their precipitation from organic solvents. In these methods, surfactants and bead suspending agents are utilized in organic solvents, that must then be removed from the microspheres, in order to allow their use. Methods currently used producing these microspheres, are spray drying and melt pressing, followed by micronization. However, both of these methods involve the application of heat which may affect the stability of drug substances.

A process known as rapid expansion of supercritical solutions (RESS) (Debenedetti et al. 1993; Srinivasan and Elliot 1992) can provide an alternative method for coprecipitating biodegradable polymers with drug substances. As noted previously, supercritical fluids exhibit no surface tension, hence they are able to "wet" particles more completely without the need for surfactants. This allows the supercritical

Figure 4.5. Preparative SFC of Propranolol. (Source: Blum and Kumar (1994) with permission)

Column: Whelk-O1 - 25 × 2.54 cm
Column Temp.: 25 °C
Solvent: 10% IPA, in Carbon Dioxide
Flow: 100 ml/min
Pressure: 250 atm
Detection: 220 nm
Injection: 0.05 g

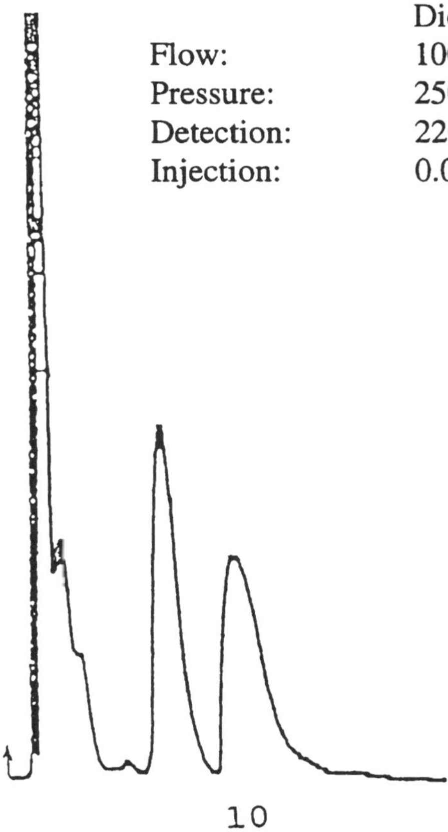

fluid to effectively infuse a dissolved drug substance into the microspheres. The drug is then precipitated in situ by reducing pressure. By selecting supercritical fluids with sufficiently low critical temperatures, this process has the capability of producing high quality, solvent-free, drug-laden microspheres for biomedical use.

Sterilization. Supercritical fluids have also been applied in the sterilization of enzymes (Kamihira et al. 1987b) and blood plasma powder (Taniguchi and Suzuki et al. 1987b), as an alternative to the use of heat, ethylene oxide, radiation, microwaves, and other methods. These

Figure 4.6. Preparative SFC of DMPC. (Source: Blum and Kumar (1994) with permission)

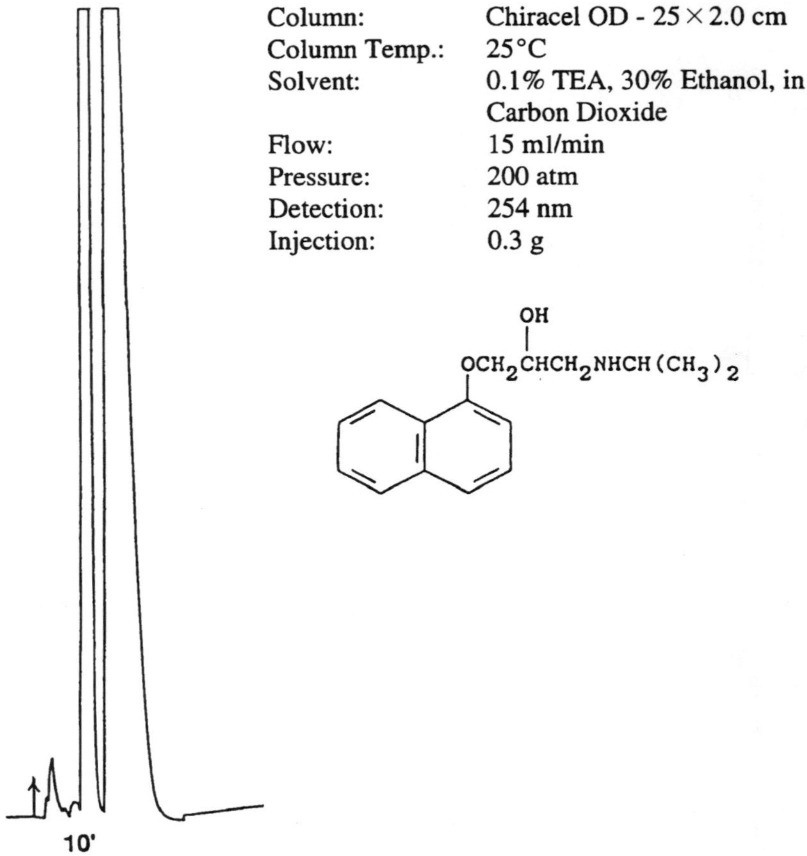

methods compromise the integrity of biologically active products, and can cause degradation of the active components. Supercritical fluid treatment is capable of accomplishing sterilization, and conserving bioactivity while not degrading thermally labile compounds.

The above study examined the influence of supercritical CO_2 under both wet and dry conditions and with/without organic cosolvents. Moist SFE conditions were shown to be more effective than dry conditions, for the sterilization of the target microorganisms. This is illustrated by the data provided in Table 4.6. The addition of organic modifier caused a dramatic improvement in the results obtained under dry conditions, while under moist conditions, a further small

Table 4.6. Sterilizing Effect of Supercritical CO_2 at 200 atm and 35°C

Microorganism	Ratio of Living Cells	
	Wet Cells*	Dry Cells*
Baker's yeast	5.4×10^{-7}	0.50
E. coli	7.2×10^{-6}	0.047
S. aureus	1.5×10^{-5}	0.037
A. niger (conidia)	1.2×10^{-5}	0.88
B. subtilis (endospore)	0.47	0.99
B. stearothermophilus (endospore)	1.07	0.80

*Water content: wet cells, 70–90%; dry cells, 2–10%
Source: Kamihira et al. (1987b) with permission

improvement was observed when a cosolvent was used in conjunction with CO_2 vs. sterilization with neat CO_2 (Table 4.7). Table 4.8 shows the effect of sterilization on the activity of enzymatic preparations. Not only was sterilization shown to be effective, but activity was essentially preserved in each case.

Morphological examination of the microorganisms after sterilization indicated that some cells had burst, but the death of the cells did not seem to be due to cell disruption. Rather it was concluded that the death of the microbial cells was due to inactivation of some enzyme via a pH decrease and/or extraction of intracellular material, such as phospholipids (provided cosolvents were used, since phospholipids are negligibly soluble in CO_2). It was concluded that sterilization by supercritical CO_2 is applicable to biologically active and heat-sensitive products and holds promise as an alternative to current methods.

SUPERCRITICAL PROCESS EQUIPMENT

Processing in high pressure extraction, can only be optimized from the thermodynamic point of view, if the relevant thermophysical properties are known (Hederer 1985). Unfortunately, there has been little impetus to experiment with SFC on this scale, due to the lack of theory or

Table 4.7. Effect of an Entrainer on Supercritical CO_2 Treatment[a]

Microorganism	Ratio of Living Cells					
	Wet cells			Dry cells		
	CO_2	CO_2 + EtOH[b]	CO_2 + AcOH[b]	CO_2	CO_2 + EtOH[b]	CO_2 + AcOH[b]
A. niger (conidia)	1.2×10^{-5}	$<2.3 \times 10^{-6}$	$<7.9 \times 10^{-6}$	0.88	1.2×10^{-6}	$<1.6 \times 10^{-5}$
B. subtilis (endospore)	1.07	0.62	0.43	0.80	0.49	0.43

[a]Treatment conditions: 200 atm, 35°C, 2 hr
[b]Ethanol (EtOH) or acetic acid (AcOH) was added to CO_2 at a weight ratio of 2% or 0.5%, respectively
Source: Kamihira et al. (1987b) with permission

Table 4.8. Sterilization of an Enzyme Preparation with Supercritical CO_2

Enzyme	Microorganism	Enzyme Activity (%)	Ratio of Living Cells
α-Amylase	*E. coli*	121	5.2×10^{-5}
	Baker's yeast	135	3.6×10^{-3}
Lipase	*E. coli*	88	8.9×10^{-5}
	Baker's yeast	78	4.7×10^{-3}

Treatment conditions: 200 atm, 35°C, 2 hr
Source: Kamihira et al. (1987b) with permission

models, whereas little can be done to model these processes, until more experimental information is available. For SFE the successful design and implementation of plants for decaffeination and hop extraction have provided the initiative for further investigation. This section will primarily concern itself with existing models for SFE processes in the literature.

Construction and Modeling

A supercritical fluid separation process according to Körner (1985) is comprised of a separation stage (with or without recycling of the supercritical fluid), a stage for the removal of the required extract, and a system for modifying pressure and/or temperature. The following questions need to be answered:

- Can thermodynamic data adequately predict solubility?
- Are laboratory tests required?
- Is a cosolvent needed?

A reputable and experienced supercritical fluid plant contractor should be engaged to be responsible for basic engineering, including design, in order to avoid accidents. It is recommended that prior to initiating construction, a contract manufacturer's pilot plant facilities be used to verify process parameters, confirm assumptions, and detect weak points. At that time the cost versus expected profit can be estimated.

Because of the diversity of products that can be extracted by SFE, production plants may vary in size (i.e., for highly valued pharmaceutical products extractors with volumes as small as 50 ℓ could be considered commercial scale). However (for example), in the removal of caffeine from coffee, the extractors would have to be several cubic meters in dimension. A point in reference is the General Foods plant in Houston, TX, which can process approximately 100 cubic meters of material (McHugh 1994).

Since most of the products to be extracted are often solids, continuous charging and discharging of the plant must be viable. Since there are no lock-hopper systems available for such high pressures, the product to be extracted must be admitted and discharged from the extractors at atmospheric pressure after extraction. Currently, most supercritical fluid processes are batch rather than continuous (Paulaitis et al. 1983). The challenge is to develop a semicontinuous mode of operation. Supercritical fluid extraction plants can operate with only one extractor, but most are equipped with several extractors to permit the batch process to be semicontinuous. Since most pharmaceutical production processes are themselves batch processes; this is not as critical as it might be in the chemical or food industries.

Some modeling as well as empirical information is available to evaluate process-scale supercritical fluid separations. For example, Eggers and Tschiersch (1980) have described the essential criteria underlying the design of a plant for the recovery of an extract from a supercritical CO_2 extraction of a natural product. Details of equipment and components as well as thermodynamic analyses and methods of optimization can be found in this reference.

Ramchandran et al. (1992) has examined the use of SFE process control, based on deviation from linear behavior for supercritical systems. Nonlinear process control (based on approximate models) uses adjustable parameters to mimic process behavior. This allows the engineer to look at the entire process in real time, using multivariable decoupling and process nonlinearities to select the appropriate control action. Computer programs have also been developed which use thermodynamic data in conjunction with mathematical models to simulate supercritical extraction, in lieu of pilot plant studies (Cesari et al. 1989).

A procedure has been developed by Colussi et al. (1992) that can be used with several equations of state for understanding the SFE process. In this procedure, the enthalpy balance along the extraction column is computed since small temperature changes in the supercritical fluid can significantly effect the resultant separation. Cubic equations of state, used for demonstrating this program, were constrained to pass through the critical point of the fluid. The resulting distortion of the isotherms near the critical point, using this program, suggest that a much more complicated thermodynamic model is required. The program was demonstrated to be reliable; since simulation by this or any other model is dependent upon the thermodynamic model chosen to describe the chemical system.

Recently, Mukhopadhyay and Raghuram Rao (1993) have used a modified covolume-dependent (CVD) mixing rule to predict the solubility of mixed solids in pure and mixed supercritical fluids. The CVD mixing rule assumes that in a dilute supercritical mixture, the probability of a molecule interacting with another depends on the fraction of the surface it can "see" of the other molecule, rather than its mole fraction. Using the Peng-Robinson equation of state (EOS), solubility predictions could be performed from pure component solubility data. Because of the complex, highly compressible, and asymmetric nature of supercritical systems, a cubic EOS is required. Peng-Robinson EOS has been shown to predict as well as more complicated perturbed hard-sphere EOS and hence was selected as the model for the above study. The required pure component properties needed are van der Waals molecular volume, dipole moment, molecular weight, and molar volume. This predictive model was shown to be effective for the preliminary screening of process design parameters. Solubilities in modified CO_2, as well as crossover separation parameters were correctly predicted using this model.

In the industrial production of a chemical commodity, separation process can account for 40–70 percent of the production cost. Supercritical fluid extraction facilities usually involve large investments. For example, a new extraction plant for decaffeination of 32 m^3/day of whole, green coffee beans can represent an investment in the range of 15 million European currency units (ECU). Nevertheless, operation costs of about 0.35 ECU/kg are acceptable (Leyers et al. 1991). There are two factors which significantly influence the investment cost for a supercritical plant.

1. The licensing royalties
2. The high pressures (1,000 to 6,000 psig) associated with supercritical fluid systems increase the capital investment significantly and only products selling for more than $1/lb are likely to justify the manufacturing cost (Humphrey 1987)

Velo et al. (1994) showed that in some cases, standard equipment can be used, thereby making supercritical fluid a more reasonable alternative to conventional processes, provided plant design is optimized.

Additional energy is required for plants that require depressurization of the extractors in order for them to be charged and discharged.

Körner (1985) described a cascading system which minimized pressure loss, thereby reducing costs due to repeated repressurizations. Similarly, more energy is required when separation of the extract from the supercritical fluid is accomplished by reducing the pressure, rather than by means of modifying temperature, using water washing systems, or adsorption traps. Since most equipment necessary is well designed and has been tested under similar conditions, no additional technical risks are involved in designing and implementing an SFE plant. Because of compression costs, it is desirable to operate at as low a pressure as possible. However, it is important to remember that generally higher densities equate with higher solute solubilities outside the crossover region, resulting in less extraction time and lower cost. Also, operation in this pressure range, can lower production costs since higher solute concentrations will result in a reduced recycle rate of supercritical fluid.

Separation of the extract from the fluid phase is desirable by using as small a decrease in pressure as possible, to minimize recompression costs. It is preferred to keep pressure constant and alter temperature, but sometimes a pressure change is necessary. By determining the region where solute solubility is most effected by pressure, recompression costs can be minimized.

One advantage of SFE over conventional extraction with liquid solvents is the ease by which the extract can be separated from the solvent. The ratio of the saturated vapor pressure of the solvent to the solute is approximately 10^3 larger for the supercritical CO_2/solute system over their liquid/solute counterparts. Therefore, it is possible to obtain a higher grade product, solvent free, in virtually a single extraction operation by SFE. Recycling the supercritical fluid can be used to improve separations. As components become concentrated in the fluid phase, they are transferred to a separator where a change in pressure and/or temperature causes a loss in solvating power and the extract is deposited. The discharge gas is recompressed and then recycled to the extraction vessel for reuse.

The National Institute of Standards and Technology (NIST) has reported an applied study which focusses on the economics of an SFE process. Called SEPSOL (Bruno 1992), it is applicable for the extraction of natural products from aqueous or wet matrices. In specific, β-carotene a high cost, and low volume chemical was justified economically. One limitation to this design, as with many other SFE designs is the batch mode of operation, resulting in additional costs compared to a continuous process. Naturally, the tendency would be to minimize the number of times the vessel is opened and maximize the length of the run and increase productivity. In the case of β-carotene, longer residence times would translate into degradation of and make for a lower quality product. Alternatively, a lock-hoper flash tank may be considered as a viable option. Collection of the carotenoid product occurs in a stabilizing oil matrix directly, and hence is suitable for commercial use. In situ process analyzers were suggested to monitor solute concentration and quality and to provide automated process control.

The above examples on the utilization of SFE in commercial processes not only demonstrate its applicability, but show that the construction and implementation of supercritical fluid plants can be done in a manner that does not attenuate the cost of the resultant process or products. Hence, the value of supercritical extracts are sufficient to accommodate the high capital outlays, while still keeping the cost per unit of product, at least low enough to maintain a competitive edge in the market. However, it is not anticipated that supercritical fluid technology will always be able to provide a sufficient economic advantage in all the process cases, to justify implementation. Although much work has been accomplished in the area of supercritical separation processes, commercial success in the future will depend upon the reliable design, simulation, and scale-up operation of actual processing plants.

Safety

The safety of supercritical fluids for process-scale separations can be rationalized on the basis that they reduce the amount of highly flammable organic solvents which are used in conventional liquid extraction processes. Even with the use of small quantities of organic solvents as cosolvents, the pressure of CO_2 reduces the possibility of a flammable explosion. There is limited, if any, toxicity associated with the use of supercritical CO_2, if adequate ventilation is provided to prevent the possibility of asphyxiation, since the safe exposure limits for CO_2 are relatively high.

Due to the concomitant high pressures associated with the utilization of supercritical fluids in relation to pilot and commercial applications, it is prudent to consider the possible dangers and appropriate precautions which should be addressed to minimize loss potential. As in any other process, hazard analysis should be performed to detect all relevant failure paths for a given system. Since many supercritical processes are new, and limited data is available from which to create reliable models, non-traditional hazard analyses has been deemed appropriate (Randhava and Calderone 1985). Some fluids such as nitrous oxide, deserve careful examination, since its use in the presence of organic solutes represents a potentially explosive hazard (Raynie 1993).

CONCLUSION

Many current applications of SFE and SFC have been discussed in this chapter and potential applications have been identified. The successful implementation of this technology in the food processing industry suggests that the pharmaceutical industry should add SFE and SFC to their arsenal. The increases in the number of patents indicates a growing interest in these versatile solvents. More physicochemical data and improved modeling schemes (Rizvi 1994) are needed to assist the engineer/scientist in implementing this technology in industry. In summary, supercritical fluids as a process tool hold great promise for future applications in the pharmaceutical industry.

REFERENCES

Avagimov, V. B., A. V. Pekhov, G. I. Kras'yanov, and B. Nad'dezhe. 1980. Carbon dioxide extracts of spices useful in sausages and canned meats. *Izvestia Vysshikh Uchebnykh Zavedenii, Pishchevaya Tekhnologiya* 4:34–36.

Asche, W. 1980. Überkritisches kohlendioxid löst trennprobleme. *Umsch. Wiss. Tech.* 80:568–569.

Berger, F., B. Cerles, and F. Sagi. 1981. Two stage extraction of wine with supercritical carbon dioxide. European patent application 77,745.

Bhaskar, A. R., S. S. H. Rizvi, and P. Harriott. 1993. Performance of a Packed column for continuous supercritical carbon dioxide processing of anhydrous milk fat. *Biotechnol. Prog.* 9:70–74.

Blum, A. M., and M. L. Kumar. 1994. The role of preparative SFC in chiral separations. Paper presented at the Fifth International Symposium on SFC and SFE, 11–14 January, in Baltimore, MD.

Borman, S. A. 1991. Scientists mobilize to increase supply of anticancer drug taxol. *Chem. Eng. News.* 69 (35):11–18.

Brannolte, H-D. 1982. Hochdruckextraktion mit CO_2—eine neue strofftrennmethode. *Seifen, Oele, Fette, Wasche.* 108:241–242.

Brugger, R., P. Krahenbuhl, A. Marti, R. Straub, and H. Arm. 1991. Chiral π-donor stationary phases with (R)-N-pivaloylnaphthyletheylamide groups for direct enantiomer separation by gas, liquid and supercritical fluid chromatography. *J. Chromatogr.* 557:163–171.

Brunner, G., and S. Peter. 1982. On the solubility of glycerides and fatty acids in compressed gases in the presence of an entrainer. *Sep. Sci. Technol.* 17:199–214.

Bruno, T. J., and J. F. Ely. 1991. *Supercritical fluid technology*. Boca Raton, Florida: CRC Press.

Bruno, T. J. 1992. Application of process analysis to a separator-solubilizer for supercritical fluid extraction. *Process Contol and Quality*. 2:59–68.

Caragay, A. B. 1981. Supercritical fluids for extraction of flavors and fragrances from natural products. *Perfumer and Flavorist*. 6 (4):43–55.

Caruana, C. M. 1991. Bioengineering looms as route for scarce drug. *Chem. Eng. Prog.* 87 (March):18–20.

Cesari, G., M. Fermeglia, I. Kikic, and M. Policastro. 1989. A computer program for the dynamic simulation of a semi-batch supercritical fluid extraction process. *Computers Chem. Engng.* 13 (10):1175–1181.

Chang, C. J., and A. D. Randolph. 1991. Separation of beta carotene mixtures precipitated from liquid solvents with high-pressure CO_2. *Biotechnol. Prog.* 7:275–278.

Chun, M-K., H-W. Shin, H. Lee, and J-R. Liu. 1994. Supercritical fluid extraction of taxol and baccatin III from needles of *taxus cuspidata*. *Biotechnol. Tech.* 8:574–550.

Clarke, B. J., and A. Mailer. 1981. How Carlton and United made hop extract history. *Brewer's Guardian*. 110 (10):13–17.

Coenen, H., and E. Kriegel. 1980. A method of regenerating lubricating oils by supercritical extraction with carbon dioxide or hydrocarbons containing 2 or 3 carbon atoms. German patent 30,38,728.

Coenen, H., R. Eggers, and E. Kriegel. 1982. The use of supercritical fluids to recover used lubricating oils. *Tech. Mitt. Krupp, Forschungsber.* 40 (1):1–11.

Colussi, I. E., M. Fermeglia, V. Gallo, and I. Kikic. 1992. Supercritical multistaged multicomponent separation: Process simulation. *Computers Chem. Engng.* 16 (1):211–224.

Debenedetti, P. G., J. W. Tom, S-D. Yeo, and G-B. Lim. 1993. Application of supercritical fluids for the production of sustained delivery devices. *J. Controlled Release*. 24:27–44.

De Filippi, R. P. 1982. CO_2 as a solvent: application to fats, oils and other materials. *Chemistry and Industry* 12:390–394.

Degnan, A. J., J. H. Von Elbe, and R. W. Hartel. 1991. Extraction of annatto seed pigment by supercritical carbon dioxide. *J. Food Sci.* 56 (6):1655–1659.

Demitrelis, D., R. Wennersten, and J. M. Prausnitz. 1984. Supercritical fluid extraction of heavy hydrocarbons using propane as a solvent. Presented at the AIChE Annual Meeting, 25–30 November, in San Fransico, CA.

Denis, J. N., A. E. Greene, D. Guenard, F. Gueritte-Voegelein, and L. Mangatal. 1988. Highly efficient, practical approach to natural taxol. *J. Am. Chem. Soc.* 110:5917–5919.

Dyken, J. J., B. L. Knutson, J. F. Morris, and J. T. Sommerfeld. 1990. Prospective design of a supercritical extraction process for the production of monocrotaline. *Proc. Biochem.* 4:47–54.

Eggers, R., and R. Tschiersch. 1980. Development and design of plants for high-pressure extraction of natural products. In *Extraction with supercritical gases*, edited by G. M. Schneider. Weinheim, Germany: Verlag Chemie.

Florey, H. W., E. Chain, N. G. Heatley, M. A. Jennings, A. G. Sanders, E. P. Abraham, F. E. Florey. 1949. *Antibiotics.* London, England: Oxford University Press.

Francotte, E., A. Junker-Buchheit. 1992. Preparative chromatographic separation of enantiomers. *J. Chromatogr.* 576:1–45.

Gelbaumm, L. T., M. M. Gordon, M. Miles, and L. H. Zalkow. 1982. Semisynthetic pyrrolizidine alkaloid antitumor agents. *J. Org. Chem.* 47 (13):2501–2504.

Hannigan, K. J. 1981. Extraction process creates low-fat potatoe crisps. *Food Eng.* 53 (7):77.

Hederer, H., and H. Heidemeyer. 1985. Process design for high-pressure extraction with supercritical gases. *Ger. Chem. Eng.* 8:112–118.

Humphrey, J. L. 1987. A guide to separation processes. *CPI Equip. Report.* 9 (September):39–40.

Jennings, D. W., H. M. Deutsch, L. H. Zalkow, and A. S. Teja. 1992. Supercritical extraction of taxol from the bark of *Taxuz brevifolia. J. Supercritical Fluids.* 5:1–6.

Johnston, K. P. 1989. New directions in supercritical fluid science and technology. In *Supercritical fluid science and technology,* edited by K. P. Johnston and J. M. L. Penninger. ACS Symposium Series 406.

Kamihira, M., M. Taniguchi, and T. Kobayashi. 1987a. Removal of organic solvent form antibiotics with supercritical carbon dioxide. *J. Ferment. Technol.* 65 (1):71–75.

Kamihira, M., M. Taniguchi, and T. Kobayashi. 1987b. Sterilization of microorganisms with supercritical carbon dioxide. *Agric. Biol. Chem.* 51 (2):407–412.

Kamihira, M., M. Taniguchi, and T. Kobayashi. 1987c. Synthesis of aspartame precursors by enzymatic reaction in supercritical carbon dioxide. *Agric. Biol. Chem.* 51 (12):3427–3428.

King, J. W., J. H. Johnson, and J. P. Friedrich. 1989. Extraction of fat tissue from meat products with supercritical carbon dioxide. *J. Agr. & Food Chem.* 37 (4):951–954.

King, J. W., and J. E. France. 1991. Supercritical fluid extraction/enzyme assay: A novel technique to screen for pesticide residues in meat. *J. Assoc. Off. Anal. Chem.* 74:1013–1016.

King, J. W., G. R. List, and J. H. Johnson. 1992. Supercritical carbon dioxide extraction of spent bleaching clays. *J. Supercrit. Fluids.* 5:38–41.

King, J. W., F. Favati, and S. Taylor. 1993a. Production of tocopherol concentrates by supercritical fluid extraction/chromatography. Presented at the American Institute of Chemical Engineers Annual Meeting, 7–12 November, in St. Luis, MO.

King, J. W. 1993b. Private communication.

Ko, M., V. Shah, P. R. Bienkowski, and H. D. Cochran. 1991. Solubility of the antibiotic penicillin V in supercritical CO_2. *J. Supercrit. Fluids*. 4:32–39.

Körner, J.-P. 1985. Design and construction of full-scale supercritical gas extraction plants. *Chem. Eng. Prog.* 4:63–66.

Krukonis, V. J. 1984. Supercritical fluid extraction-overview and specific examples in flavors applications. Paper presented at the ACS Annual Meeting, Agricultural and Food Chemistry Division, Flavors Subdivision, August, in Philadelphia, PA.

Lee, H., W-H. Hong, J-H. Yoon, K-M. Song, S-S. Kwak, and J-R. Liu. 1992. Extraction of indole alkaloids from *Catharanthus roseus* by using supercritical carbon dioxide. *Biotechnol. Tech.* 6 (2):127–130.

Lee, M. L., and K. E. Markides. 1990. *Analytical supercritical fluid chromatography.* Provo: Chromatography Conferences, Inc.

Leyers, W. E., R. A. Novak, and D. A. Lining. 1991. The economics of supercritical coffee decaffeination. In *Proceedings of the second international symposium on supercritical fluids.* Boston, Massachusetts: Johns Hopkins University.

List, G. R., J. P. Friedrich, and J. Pominski. 1984. Characterization and processing of cottonseed oil obtained by extraction with supercritical carbon dioxide. *J. Am. Oil. Chem. Soc.* 61 (12):1847–1849.

Luganskaya, L. N., E. B. Krasnokutskaya, and L. B. Yasinskaya. 1967. Utilization of extract obtained from tobacco dust for the purpose of aromatizing tobacco. *Tabak.* 28 (1):30–33.

Maffei, R. L. 1971. Extraction and Cleaning Processes. US Patent 4 012 194.

Marti, A., P. Krahenbuhl, R. Brugger, and H. Arm. 1991. Chiral stationary phase with (+)-(R)-N-(1-phenylethyl)-N′-[3-diethoxymethylsilyl)propyl] urea for the direct enantiomeric separation with capillary gas chromatography and supercritical fluid chromatography. *Chimia* 45:13–15.

Maxwell, R. J., J. W. Hampson, and M. L. Cygnarowicz-Provost. 1992. Comparison of the solubility in supercritical fluids of the polycyclic ether antibiotics: lasalocid, monensin, narasin, and salinomycin. *J. Supercrit. Fluids* 5:31–37.

McHugh, M. A., and V. J. Krukonis. 1986. *Supercritical fluid extraction: Principles and practice.* Stoneham: Butterworth.

McHugh, M. A., and V. J. Krukonis. 1994. *Supercritical fluid extraction,* 2nd ed. Boston: Butterworth-Heinemanne.

Moyler, D. 1985. Carbon dioxide: The ideal solvent in essential oil extraction. *Prepared Foods.* 11 (November):100–101.

Mukhopadhyay, M., and G. V. Raghuram Rao. 1993. Thermodynamic modeling for supercritical fluid process design. *Ind. Eng. Chem. Res.* 32:922–930.

Nakamura, K. 1987. Enzymes work in supercritical fluids. *Bioproc. Technol.* 9:April.

Nam, K-S., and J. W. King. 1994. Supercritical fluid extraction and enzyme immunoassay for pesticide detection in meat products. *J. Agric. Food Chem.* 42:1469–1474.

Ndiomu, D. P., and C. F. Simpson. 1988. Some applications of supercritical fluid extraction. *Anal. Chim. Acta.* 213:237–243.

Paulaitis, M. E., V. J. Krukonis, R. T. Kurnik, and R. C. Reid. 1983. Supercritical fluid extraction. *In Reviews in chemical engineering.* 1 (2):179–250.

Penninger, J. M. L., M. Radoz, M. A. McHugh, V. J. Krukonis. 1985. Supercritical fluid technology. In *Process technology proceedings,* Vol. 3. Amsterdam, The Netherlands: Elsevier.

Pictet, G., J-P. Bouldoires, and D. Rovero. 1968. Process for the preparation of extracts from natural products. French Patent 1,512,642.

Process for removing cholesterol from milk seen as commercially viable. 1989. *The Cheese Reporter,* April 7.

Ramchandran, B., J. B. Riggs, H. R. Heichelheim. 1992. Nonlinear plant-wide control: Application to a supercritical fluid extraction process. *Ind. Eng. Chem. Res.* 31:290–300.

Randhava, R., and S. Calderone. 1985. Hazard analysis of supercritical extraction. *Chem. Eng.* 6:59–62.

Randolph, T. W., H. W. Blanch, J. M. Prausnitz, and C. R. Wilke. 1985. Enzymatic catalysis in a supercritical fluid. *Biotechnol. Let.* 7 (5):325–328.

Raynie, D. E. 1993. Warning concerning the use of nitrous oxide in supercrical fluid extractions. *Anal. Chem.* 63:3127–3128.

Rizvi, S. S. H., Z. R. Yu, A. R. Rhaskar, and Chid. 1994. *Supercritical fluid processing of food and biomaterials* (pp. 1–26). London: Blackie Publishers.

Roselius, W., O. Vitzthum, and P. Hubert. 1970. Extraction of nicotine from tobacco. West German Patent 2,043,537.

Russel, A. J., and E. J. Beckman. 1991a. Enzyme activity in supercritical fluids. *Applied Biochem. and Biotech.* 31:197–211.

Russel, A. J., and E. J. Beckman. 1991b. Should the high diffusivity of a supercritical fluid increase the rate of an enzyme catalyzed reaction? *Enzyme Microb. Technol.* 13 (December):1007.

Schaeffer, S. T., L. H. Zalkow, A. S. Teja. 1988. Supercritical extraction of crotalaria spectabilis in the cross-over region. *AIChE J.* 34 (10): 1740–1742.

Schaeffer, S. T., L. H. Zalkow, A. S. Teja. 1989a. Extraction of monocrotaline from crotolaria spectabilis using supercritical carbon dioxide and carbon dioxide-ethanol mixtures. *Biotechnol. Bioeng.* 34: 1357–1365.

Schaeffer, S. T., L. H. Zalkow, A. S. Teja. 1989b. Supercritical fluid isolation of monocrotaline from crotalaria spectabilis using ion-exchange resins. *Ind. Eng. Chem. Res.* 28: 1017–1020.

Sims, M. 1982. Process uses liquid CO_2 for botanical extractions. *Chem. Eng.* 89 (January): 50–51.

Srinivasan, G. and J. R. Elliott, Jr. 1992. Microcellular materials via polymerization in supercritical fluids. *Ind. Eng. Chem. Res.* 31:1414–1417.

Stahl, E., and E. Schultz. 1980. Extraction of labile natural substances by supercritical gases. *Planta Med.* 40:12–21.

Stahl, E., K. W. Quirin, and N. Totani. 1982. Extraction and separation of lipids and other natural products using supercritical fluids. *Rev. Fr. Corps Gras.* 29:259–263.

Taniguchi, M., M. Kamihira, and T. Kobayashi. 1987a. Effect of treatment with supercritical carbon dioxide on enzyme activity. *Agric. Biol. Chem.* 51 (2):593–594.

Taniguchi, M., H. Suzuki, M. Sato, and T. Kobayashi. 1987b. Sterilization of plasma powder by treatment with supercritical carbon dioxide. *Agric. Biol. Chem.* 51 (12):3425–3426.

Temelli, F., C. S. Chen, and R. J. Braddock. 1988. Supercritical fluid extraction in citrus oil processing. *Food Technology* 6:145–150.

Velo, E., J. Pagès, M. A. Larrayoz, F. Recasens, and J. Puiggené. 1994. Modeling and sizing of semi-batch units for supercritical fluid extraction plants. *Computers Chem. Engng.* 18:S325–S329.

Vitzthum, O. G., and P. Hubert. 1971. Supercritical carbon dioxide extraction of hops. West German patent 21,27,618.

Vitzthum, O. G., and P. Hubert. 1975. Process for decaffeination of coffee. German Patent 2,357,590.

Wheldon, A. G. 1982. Method and apparatus for providing a hop extract. European Patent Application 61,877.

Williams, D. F. 1981. Extraction with supercritical gases. *Chem. Eng. Sci.* 36:1769–1788.

Witherup, K. M., S. A. Look, M. W. Stasko, T. J. Ghiorzi, G. M. Muschik, and G. M. Cragg. 1990. Taxus Spp. needles contain amounts of taxol comparable to the bark of *Taxus brevifolia:* Analysis and isolation. *J. Nat. Prod.* 53 (5):1249–1255.

Xie, L. Q., K. E. Markides, and M. L. Lee. 1992. Biomedical Applications of Analytical Supercritical Fluid Separation Techniques. *Anal. Biochem.* 200:7–19.

Zosel, K. 1978. Separation with supercritical gases: Practical applications. *Agnew. Chem. Int. Ed.* 17:702–709.

Zosel, K. 1981. Process for the decaffeination of coffee. US Patent 4,260,639.

Zuer, P. 1988. *Chem. Eng. News.* October 10:22.

5

Preparative Separation of Enantiomers

Charles M. Grill
John R. Kern
Scott R. Perrin

A chemical compound that contains one or more sterogenic centers can exist as a racemate (i.e., containing two or more enantiomers or as a single enantiomer). Enantiomers can be described as chemical compounds that are nonsuperimposable mirror images. They have identical physical and chemical properties with the exception that they rotate the plane of polarized light in the opposite direction. In addition, enantiomers react differently when placed in another "chiral environment." The human body can be considered a chiral environment as many biological processes incorporate stereospecific steps. In the past it was accepted practice to develop pharmaceutical products that contain one or more sterogenic centers as the racemate. This was due, in part, to the difficulty in synthesizing single enantiomers coupled with the inability of analytical methodology specific for the quantitation of enantiomers. Over the past 15 years many significant breakthroughs have occurred in the fields of enantioselective synthesis and chromatographic separations.

These new technologies and the growing concerns among regulatory agencies for data specific to individual enantiomers has driven the pharmaceutical industry to prepare and test a larger number of single enantiomer drugs. A major factor driving the market toward enantiomerically pure drugs is the decision of the U.S. patent office to consider each enantiomer of a chiral drug to be separate chemical species. Specifically, the FDA's 1992 policy statement (*Federal Register*) on stereoisomeric drugs, and the U.S. patent office stance of recognizing a pure enantiomer as a new product unique to the racemate have provided this impetus. The FDA's policy statement allows some flexibility to manufacturer racemates, but the patent office decision strikes at the heart of the pharmaceutical industry—its patents. Already, patents have been issued for enantiomerically pure drugs that have been sold as racemates. Sepracor has developed large-scale methods for resolving enantiomers of some racemic drugs that other pharmaceutical companies have successfully brought to market (Stinson 1994). Thus, they are in a good position to profit if the regulatory environment necessitates the use of enantiomerically pure versions of established racemic drugs. Since Sepracor's strategy has become known, many pharmaceutical companies have moved to patent enantiomerically pure versions of their racemic drugs. In the case of new drugs, pharmaceutical companies will undoubtedly seek patents for the preferred enantiomer.

The pharmaceutical industry has taken the lead in developing enantiomerically pure drugs. This is a multibillion dollar industry. It is estimated that the 1993 market for chiral drugs is 35.6 billion dollars, and will grow by 5 percent per year to 40 billion by 1997 (Stinson 1992, 1994).

It is widely accepted in the pharmaceutical industry that the development of a single enantiomer as a candidate for drug development is the preferred option. Therefore, the need for rapid and inexpensive methodology for the isolation of enantiomerically pure compounds is a major goal. Due to the advances in enantioselective synthesis, many of these isomers will be synthesized by skilled synthetic chemists. However, developing a stereoselective synthetic method requires the investigation of several synthetic pathways and, thus, requires a great deal of time. Another popular method of enantiomer isolation is by preparing a diastereomeric derivative of the racemate. The desired isomer is then preferentially isolated via recrystallization, cleavage of the derivatizing agent, and final purification steps. A major drawback in this method is that the undesired isomer is discarded, meaning that there is a 50 percent waste of material. Another problem is that diastereomeric formation, isolation, and recrystallization is very labor intensive and time consuming. All of these items add greatly to the cost of manufacture.

A faster and cheaper way to obtain moderate amounts of enantiomerically pure species is to synthesize the racemate and then use preparative chiral chromatography to resolve the enantiomers. This approach additionally has the advantage of producing both enantiomers for comparative testing. Due to the large number of compounds that are often screened in early development stages, the use of a simple method, such as preparative high pressure liquid chromatography (HPLC) to isolate a single enantiomer may often be more economically feasible than synthesizing individual enantiomers. As a matter of practical thinking, a pharmaceutical company should always consider related costs for synthesis vs. preparative HPLC in early development. In the past, preparative chromatography has not been seriously considered in the large-scale production of enantiomerically pure drugs. Preparative HPLC generally has very high operating costs due to expensive stationary phases, large amounts of mobile phase, and high labor costs. Also, if the undesired enantiomer cannot be racemized economically, half the costs of production will be wasted since the undesired enantiomer will be discarded.

Since the early 1980s, the use of analytical scale enantiomer separations by HPLC has been an accepted means to assist in the determination of the enantiomeric purity of a drug substance. The majority of references on this topic deal mainly with analytical-scale separations (micrograms or less). This research on the separation of enantiomers has mostly been concerned with raw drug material, formulated product and the metabolic fate of enantiomers. However, in the past five years, the scaleup to 1–200 g amounts of a single enantiomer by preparative HPLC is becoming more commonplace among major pharmaceutical companies (Francotte and Junker-Buchheit 1992). The use of preparative HPLC allows for faster screening of individual isomers, and gives synthetic chemists additional time to spend preparing elegant enantiospecific synthetic procedures on compounds that are further along in development.

The choice to use preparative chromatography for isolation purposes infers that there is a mutually respectful working relationship between the analytical chemist, the synthetic chemist, and project management. This type of cooperative project team development means that all parties are involved in the synthetic/isolation strategies from the start. In many cases it may be better to isolate an isomer(s) of an intermediate or desired product by preparative HPLC than by performing an enantiospecific synthesis. Project management needs to assess the use of preparative HPLC at all pertinent steps of the synthesis. If a synthetic chemist is spending a week developing the enantiospecific synthesis, and the isomers of the racemate can be separated in one day by preparative HPLC, then it is obvious that more samples can be screened and evaluated at a lower cost by isolating desired isomers by preparative HPLC.

However, there is opposition from synthetic chemists in that the early work they perform on enantioselective synthesis is essential to the basic chemical understanding of the synthetic process. This is true in many instances however, in the fast-paced environment of today's pharmaceutical industry where the mode of operation is to screen as many candidates as possible in the shortest amount of time, the need for rapid isolation procedures for individual enantiomers is a major challenge. With many pharmaceutical companies pushing to file investigational new drug applications (INDs) in less than one year from discovery, preparative HPLC should be considered an important part of this process.

In some cases the only method that will work for resolving enantiomers is chromatography. It then becomes imperative to use the most efficient chromatography process available. This chapter will discuss three types of preparative chromatography. First we will examine traditional preparative HPLC and discuss how to optimize its efficiency. Next, we will discuss closed loop recycling HPLC, a technique that is proving to be a very efficient method of resolving enantiomers. Finally, we will discuss simulated moving bed (SMB) chromatography, a technique that has been around for about 30 years and is now being applied to the purification of pharmaceuticals and fine chemicals. Simulated moving bed chromatography promises to be the most efficient method of HPLC and will undoubtedly find applications in large-scale processes.

PRACTICAL ASPECTS

In this section we define preparative HPLC as a technique where amounts (>0.1 g) of individual enantiomers are isolated. Chiral preparative HPLC in theory is similar to other types of preparative HPLC. Many of the barriers to economical preparative chromatography are similar to those with the preparative separation of enantiomers. A unique barrier associated with chiral separation is the use of specialized chiral stationary phases (CSPs). These CSPs present the analyst with several unique challenges. Some of these are as follows:

- *There are limited amounts of specialized stationary phases available for preparative purposes.* There are currently greater than 100 CSPs commercially available on the analytical scale. The number of CSPs available for preparative purposes is limited to about 30. Those types available in bulk quantity sufficient for preparative purposes are the polysaccharides (cellulose and amylose), Pirkle-Concept, polyacrylamide, cyclodextrin, and ligand exchange phases.
- *Due to the limited number of binding sites* (chiral recognition sites) and low coverage, some CSPs give rise to lower sample loading capacity.

- *Many CSPs are expensive.* The high costs of many CSPs makes them unattractive for use in the preparative mode. As an example, the cost of a 50 cm × 2″ Chiralcel OD column is about $24,000 which is comparable to a "Pirkle-Concept" CSP. However, bulk material is less expensive.

- *Many CSPs are thought to be chemically unstable.* Certain CSPs have the reputation for being troublesome and show a lack of ruggedness. These problems stem from the inability of some CSPs to tolerate a large range of solvents to actual chemical degradation of the CSP due to the sample injected. Choosing a CSP known to be stable is imperative in preparative separations.

These barriers can be overcome through experimentation, working from the selection of the CSP, performing an analytical-scale separation, scaling up and, finally, optimization of the preparative separation.

Three important aspects of a chromatographic process (speed, resolution and sample loading or capacity) are interrelated. That is, if one of the goals is to be optimized, the other two are affected, usually in a negative fashion. The resolution between two enantiomers is dependent on the differences in binding (or interaction) energies between the solutes and the stationary phase. Any situation that interferes with these binding interactions will lead to decreased resolution and peak shape problems. In preparative HPLC, increased sample loading and throughput are the objective. In order to achieve this, speed and resolution are usually compromised. In order to obtain increased sample throughput, columns of a large diameter (>10 mm) and increased sample loads (>1 mg/g packing) are utilized. Often in preparative HPLC, column overload conditions are used. For further discussions on the theory and practical aspects of preparative HPLC, see Francotte and Junker-Buchheit (1992); Verzele and Dewaele (1986); Snyder and Kirkland (1979); Snyder and Cox (1989); and Snyder et al. (1995). The reader is further directed to Stinson (1992) and Francotte (1994) for an in-depth discussion of preparative chiral separations. Jacobson et al. (1990) discusses theoretical aspects for several examples of overload conditions in preparative HPLC.

In order to successfully separate enantiomers using a CSP, the analyst should be familiar with chiral recognition theory, column types, and different separation modes used in this field. Initially, this appears to be daunting as there are over 100 CSPs to choose from. However, much is known regarding chiral recognition and separation mechanisms. There are several excellent review articles, and books (Pochapsky 1987; Lough and Wainer 1989; Perrin and Pirkle 1991; Ward and Armstrong 1988, Taylor and Maher 1992; Pirkle and

Pochapsky 1989) that detail the processes used in determining initial column and conditions to attempt the chromtographic separation of enantiomers. In addition, column manufacturers (Wainer 1988 and *Application Guide* 1993) have published method development schemes to help assist in chiral separations.

Chiral Stationary Phases

Wainer (1988) proposed a classification scheme for chiral stationary phases based on the formation of solute CSP complexes that is widely used today. The following section discusses these CSPs and their relative merits for preparative separations.

Type I

The mechanisms for enantioselectivity are attractive interactions such as hydrogen bonding, π-π interactions, and steric interactions. This class of CSPs are generally referred to as "Pirkle-Concept" CSPs named after the inventor, Dr. William H. Pirkle of the University of Illinois. These CSPs are often prepared from a silica backbone bonded with a chiral moiety such as an amino acid. This type of CSP is useful for preparative separations and there are about 20 different CSPs of this type commercially available as preparative columns. For the past 12 years Pirkle and coworkers have been separating gram amounts of enantiomers (mostly as precursors to other CSPs) on these CSPs.

These CSPs are best suited for use under normal phase conditions. The majority of preparative separations reported in the literature use mixtures of a nonpolar solvent, such as hexane, with resolution being modified by the addition of a polar modifier, such as ethanol or isopropanol. These CSPs are considered to be fairly structurally sound and hold up well to preparative conditions. A common limitation for their use is the limited solubility of many analytes in the nonpolar mobile phase.

Type II

Type II CSPs includes the numerous cellulose and amylose derivatives. As with the type I CSPs, the enantioselective mechanisms are a combination of attractive interactions, such as hydrogen bonding interactions. Cellulose triacetate 1 (CTA-1), a polymeric material prepared by the heterogeneous acetylation of cellulose is the most commonly used CSP for preparative separations (Francotte and Junker-Buchheit 1992). This CSP is inexpensive, easy to prepare, and it has higher loading capacities than many of the other CSPs commercially available for preparative purposes. The majority of separations on the CTA-1 CSP are performed with reversed-phase conditions, usually with methanol-ethanol-water mixtures. Separations of up to 200 g of material have been reported for this CSP (Lohmann et al. 1988 and Ariens and Simmons 1964). Daicel Chemical Company (Japan) has a line of CSPs based on derivatives of amylose or cellulose that have had a huge success on the analytical scale. However, there are few preparative applications in the literature. A major disadvantage to the cellulose and amylose based CSPs is that they may dissolve in solvents such as dichloromethane, tetrahydrofuran, dioxane, or acetone. This limits their applicability to solvents such as methanol, ethanol, and water. They can also be used in combination with a nonpolar solvent, such as hexane and isopropanol.

Type III

Type III CSPs are characterized by enantioselective interactions such as inclusion complexing that requires entrance of the solute into a "chiral cavity" and, to a lesser extent, involve attractive interactions. The macrocyclic oligomer cyclodextrin CSPs are a common type III CSP. Cyclodextrins offer an advantage over other CSPs in that they can be used in the reversed-phase mode. There are several cavity sizes available and there are several new derivatized cyclodextrin CSPs commercially available. Ward and Armstrong (1986) have reported preparative separations on cyclodextrins. Other type III CSPs are the polymer based CSPs such as poly(triphenyl methyl methacrylate), and poly *N*-acrylamide bonded to silica gel (ChiraSpher). These polymer CSPs offer higher loading capacities than the cyclodextrins and can be use with both nonpolar and polar mobile phases. Separations of up to 5 g of enantiomers have been reported (Francotte and Junker-Bucheit 1992) on polymer CSPs.

Type IV

Type IV CSPs pioneered by Davankov (1988) involves formation of diastereomeric metal complexes and is known as ligand exchange. These CSPs are polymeric and contain moieties such as L-hydroxyproline bonded to the polymer. Copper sulfate is often used as a mobile phase additive. These CSPs can also be a bonded phase, such as C18 coated with a copper amino acid complex. Separations of up to 1.2 g of amino acid enantiomers have been performed on analytical scale columns of coated CSPs (Kurganov et al. 1993). The most common types of enantiomers resolved with ligand exchange chromatography are underivatized amino acids. Up to 20 g of amino acid enantiomers per run (Francotte and Junker-Buchheit 1992) have been isolated using ligand exchange.

Type V

Type V CSPs are the protein-based types where a protein such as α-1 acid glycoprotein (AGP) is bonded to a silica backbone. These CSPs are characterized by low loading capacities and are currently not being used for preparative purposes. However, they have been shown to be very useful for analytical-scale separations. Jacobson and Guiochon (1992) reported the isolation of 26 micrograms of d,l-mandelic acid on a bovine serum albumin (BSA) column.

Solvent Strength

It is desirable to have the analyte dissolved in the mobile phase for preparative HPLC. Dissolving the analyte in a different solvent may cause the sample to precipitate when introduced to the column. The polarity of the dissolution solvent is important to consider. Generally, if the dissolution solvent is weaker (i.e., less polar) than the mobile phase, more sample (larger volumes) can be loaded. If the dissolution solvent is more polar, less sample (lower volume) should be loaded. In general, the loading of a sample in a solvent stronger than the mobile phase results in broad peaks for early eluting species.

Sample Mass

Many preparative separations are performed in column overload conditions. Column overload results in a loss in resolution and poor peak shape. Overload conditions with a small volume sample can occur if the sample is highly concentrated. Snyder et al. (1995) refers to "lightly overloaded" conditions to classify sample loading of an intermediate level. Using lightly overloaded conditions, retention times and peak widths are not compromised by other peaks in the sample matrix.

OPTIMIZATION STRATEGY

A general optimization strategy for the preparative separation of enantiomers must consider column selection, loading capacities, recovery, throughput, and production rate. Production rate is defined as the amount of product separated per unit time (the throughput) divided by the amount of stationary phase. In our opinion the true meaning of production rate should include the time it takes to resolve a given racemate, isolate, and purify each enantiomer. This dictates the true production rate in preparative chromatography.

Column Selection

Choosing the correct CSP and mobile phase condition that yields high enantioselectivity, good resolution, respectable peak shape, and high loadability are crucial in preparative chromatography. The source of the bonded phase should be available on bulk for packing large columns of different configurations. The supplier must readily produce reproducible packings at relatively low costs. The CSP should be durable with good chemical and mechanical stability. It must be sufficiently robust to withstand packing in columns based on axial, radial, and annular expansion processes. High loading capacity should be an intrinsic property of the CSP, and it should exhibit a broad range of applicability. It is also frequently convenient if the CSP is available in both enantiomeric forms, especially in the separation of enriched samples where it is usually preferable to have the smaller peak elute first.

In addition to choosing CSPs that give good resolution of the enantiomers of interest, it would also be advantageous if the k' of the more highly retained enantiomer were small. In preparative chromatography increasing the retention significantly increases the volume of the mobile phase consumed and results in the elution of a more dilute fraction. This makes isolation of the individual isomer of interest more difficult and costly.

The first step in column selection is to identify candidate CSPs exhibiting good chiral selectivity for the resolution under consideration. Predictions based on structural requirements or empirical rules (from previous experience, a literature search, or by searching a commercially available database such as Chirbase (Roussel and Piras 1994) can help in selecting candidate CSPs.

The candidate CSPs must next be screened. This is usually done with analytical columns, and today the process is much more streamlined than in the past due to the large number of publications on chiral separations. Screening is ideally done using the same particle size of stationary phase that will be packed into the preparative column. However, smaller particles can be used as long as allowances are made comparing results obtained on different-sized particles. Several CSPs and mobile phase conditions should be evaluated before selecting the appropriate CSP. Again, the ideal candidate will have good resolution values and low retention times.

Loading Capacity and Recovery

The final step in CSP selection is ascertaining which CSP will give the highest production rate and recovery. This is done through the use of overloading studies. (Of course, in smaller-scale, semipreparative separations this degree of CSP discrimination may not be necessary. However, even for semipreparative methods, one has the incentive to maximize the production rate for the column and the CSP ultimately chosen.) Loading studies should be performed on analytical columns packed with the same particle size of stationary phase that will be used in the preparative column. In loading studies the sample load is increased, either by increasing the volume injected of a given solution or by increasing the concentration of the sample, until the desired throughput and recovery are attained. Within this context at least three situations, involving different trade-offs often arise:

1. *Crude solution is inexpensive.* Here we can maximize throughput using overload conditions, even though we sacrifice some recovery. Our goals are high throughput and high purity.
2. *Crude solution is expensive and stable.* In this case we often use a two-pass approach. The first step produces a moderate enantiomeric excess (e.e.) of about 90 percent. The second polishing step yields an e.e. > 99 percent. In both steps, unoverloaded conditions are used to give high recoveries.
3. *Crude solution is expensive and unstable.* Again, high recoveries are important. Because the compounds are unstable on the column, a fast one-pass procedure is needed. Typically, baseline resolution is sought so that 100 percent recovery can be accomplished in one pass.

Because the price of a CSP can vary greatly with particle size and type, it is important for large-scale separations to determine the optimum particle size. An easy technique of assessing the relative preparative utility of packings of different particle sizes is the "equal-cut-point" method (Perry and Szczerba 1989). A plot is made of particular particle size vs. the load per gram of packing when the observed peak width is 50 percent greater than that at essentially zero load. Taking into consideration the cost of the packing, Perry and Szczerba observed that a column packed with 20 μm particles is almost 10 times more productive per dollar paid for packing than a column packed with 80 μm particles. Thus, if load is critical, then a particle size of no greater than 20 μm is probably the best choice. If resolution is the most important criterion (which in most cases is true for the separation of enantiomers because of low alpha values), then a small particle packing is the best option.

Isolation

An important point to consider in determining the total production cost is that the resolved compounds should be easily isolated. The problem that most chromatographic separations have in common is removal of the mobile phase. Because the usual method is rotary evaporation, lower boiling solvents are preferred. No matter what the means of solvent stripping, it is advantageous to have the collected fractions be as concentrated as possible. This, of course, is a function of the chromatography itself: less retained, Gaussian shaped peaks give the most concentrated fractions.

The isolation problem is particularly important in cases where additives are used in the mobile phase and have to be removed from the desired product after separation. For example, mobile phase additives such as trifluoroacetic acid, acetic acid, diethylamine, triethylamine, ammonium acetate, and formate are commonly used to prevent peak broadening due to superfluous interaction sites. Removal of these additives plays an important role in determining the feasibility of a given separation. Under normal phase conditions for the resolution of amines, the use of acidic modifiers is preferred over basic modifiers. Acidic modifiers can readily be removed by the azeotrope of toluene or under high vacuum.

Column Longevity

An important factor in determining costs of long-term preparative separations is column longevity. Using extreme column-loading conditions can cause deterioration of column packing due to mass buildup, column voids, and other issues. Throughput will increase under mass-loading conditions. However, in our opinion, loss of performance occurs in a relatively short period of time, and excessive regeneration of column packing is required. We have found that column life can be extended if unoverloaded conditions are used. If the system is automated, large quantities of material can be resolved each day since the system can be operated around the clock with very little supervision.

Scaleup

After the loading studies have been completed, the next step is to transfer the separation to a preparative column that will give the desired daily throughput. The scaleup calculations are straightforward. When scaling up, we wish to keep the linear velocity constant. Assuming the preparative column is packed as well as the analytical-scale method column, keeping the linear velocity constant will assure that the preparative column has the same efficiency as the analytical method column. Under the constraint of constant linear velocity, the volumetric flow rate scales as the square of the column diameter, the sample load scales as the volume of the column, and the run time and back pressure as the length of the column.

PREPARATIVE CHROMATOGRAPHIC RESOLUTION OF THE α-BURKE 1 OLEFIN

As an example of the development of a preparative chromatographic resolution of a racemate, we will discuss the preparative resolution of the α-Burke 1 olefin, used to produce a bonded CSP marketed by Regis Technologies, Inc. This separation is an excellent example of situation 1 discussed above: the crude solution is inexpensive. We developed a preparative method under overload conditions that would maximize throughput at the expense of recovery. Our screening determined that a column packed with (S)-*N*-1-*N*-naphthylleucine gave good resolution and tolerates very high load conditions while maintaining reasonable peak shape.

A loading study was undertaken using analytical-scale HPLC equipment. An analytical column (25 cm × 4.6 mm I.D.) was packed with 10 μm silica bound (S)-*N*-1-*N*-naphthylleucine. Pore diameter was 100 Å. Three different loads, shown in Table 5.1, were injected on the column. The chromatograms resulting from the first two loads, 0.1 mg and 15.2 mg, are shown in Figure 5.1.

For the load of 0.1 mg, the chromatogram is characteristic of unoverloaded analytical HPLC as expected. The main peaks are Gaussian and well resolved.

For the load of 15.25 mg, the second peak has the "right triangle" shape characteristic of an overload state. The first peak still retains its

Table 5.1. Loading Study Results Performed on the Analytical Scale

Volume Injected	Quantity Injected	Load[a]	k'_1	k'_2	Rs	Run Time
10 ml	0.1 mg	0.03 mg/g	2.91	8.30	4.31	29.1[b]
200 ml	15.2 mg	5.08 mg/g	4.90	12.90	2.56	39.0[c]
240 ml	30.0 mg	8.98 mg/g	4.72	11.87	0.63	45.8[d]

[a]Load: mg of solute/gram of stationary phase

[b]Mobile phase 1: 85/15 hexane/IPA (v/v); Sample concentration: 10 mg/ml; Wavelength 1: 254 nm

[c]Mobile phase 2: 90/10 hexane/IPA (v/v); Sample concentration: 76.24 mg/ml; Wavelength 2: 295 nm

[d]Mobile phase 2: 85/15 hexane/IPA(v/v); Sample concentration: 126.0 mg/ml; Wavelength 2: 295 nm

Figure 5.1. A comparison between 0.1 mg (left) and 15.25 mg (right) on an analytical column.

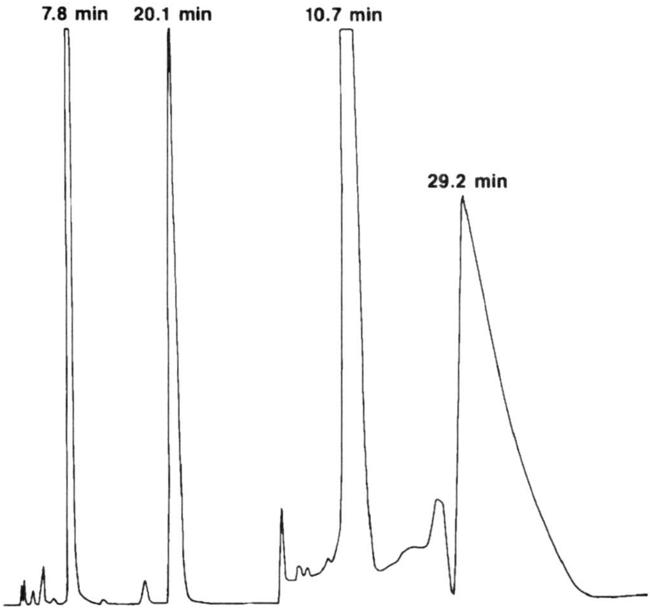

Gaussian shape. Usually the later eluting peaks become overloaded first. Since at equilibrium they are partitioned more to the adsorbed state, they will saturate the chromatographic surface sooner as the injection load is increased. Note that in this run a weaker mobile phase (90/10 hexane/isopropanol) is used. Thus the resolution is still good.

For the highest load, 30.0 mg, no chromatogram is available. The mobile phase was the same used for the load of 15.25 mg. As shown in Table 5.1, the *k*′ for each peak decreased as did the resolution as loading increased. This is common in overload conditions. We chose these conditions to use in our large-scale preparative runs.

Table 5.2 shows the results of scaleup calculations based on the loading studies. Note that for columns 21.1 mm in diameter, the total time to resolve 40 grams of the α-Burke 1 olefin is about 50 hours (2 days). For a column length of 25 cm, a load of 0.63 g is required, and the run time is 45.8 minutes. For a length of 50 cm, the load is 1.26 g, and the run time is 94 minutes. The first option is safer because less material is injected each run. Thus, if a malfunction occurs and sample is lost, less would be lost. Also, less stationary phase would be required. However with the second option, less work is involved since

Table 5.2. Results of the Scaleup Calculations to Resolve 40 grams of the α-Burke 1 Olefin

Column			Quantity[a]	Quantity[b]		Throughput	
Length	I.D.	Factor	Injected	Injected	Injections	Hr.	Days
25 cm	10.0 mm	4.7×	30.0 mg	0.14 g	286	186	7.7
25 cm	21.1 mm	21.0×	30.0 mg	0.63 g	64	49	2.0[c]
50 cm	21.1 mm	42.0×	30.0 mg	1.26 g	32	50	2.1[d]

[a]Analytical: From Table 5.2

[b]Preparative: From Table 5.5

[c]Based on 45.8 minute run time

[d]Based on 94.0 minute run time

half as many injections are required. Therefore, we chose the second option. The entire sets of preparative conditions are summarized in Table 5.3. The actual chromatographic parameters measured during the preparative runs are summarized in Table 5.4.

The preparative chromatogram indicating the points where fractions were collected is shown in Figure 5.2. As shown in Table 5.5 we achieve a recovery of 15.5 g of the S enantiomer and 16.0 g of the R enantiomer for a total yield of 79 percent. The e.e. for each enantiomer was >99 percent.

Overload preparative HPLC is thus an effective means of obtaining high throughputs of purified enantiomers. Although in this instance the middle fraction could be recycled if higher recoveries were desired, a less overloaded procedure would probably have been more economical.

RESOLUTION OF RACEMATES BY CLOSED LOOP RECYCLING CHROMATOGRAPHY

For very difficult separations it may be impossible to resolve the enantiomers by conventional preparative HPLC. In such cases closed loop

Table 5.3. Preparative Conditions for the Separation of the α-Burke 1 Olefin

Compound	α-Burke 1 Olefin
Column	(S)-*N*-1-*N*-naphthylleucine 50 cm × 21.1 mm I.D.
Particle Size	10 μm, 100 Å
Mobile Phase	90/10 hexane/IPA
Flow Rate	21.0 ml/minute
Injection Volume	10 ml
Sample Concentration	0.126 mg/ml
Quantity Injected	1.26 g
Wavelength	295 nm

Table 5.4. Loading Conditions on the Preparative Column

Volume Injected	Quantity Injected	Load[a]	k'_1	k'_2	α	R_s	Run Times
10 ml	1.26 g	12.60 mg/g	4.71	11.35	2.40	0.76	94.0 min[b]

[a]Load: mg of solute/gram of stationary phase

[b]Mobile Phase: 90/10 hexane/IPA (v/v); Sample concentration: 126 mg/ml; Wavelength: 295 nm

Figure 5.2. Preparative chromatogram of 1.26 g of α-Burke 1 olefin. Conditions as described in Table 5.3.

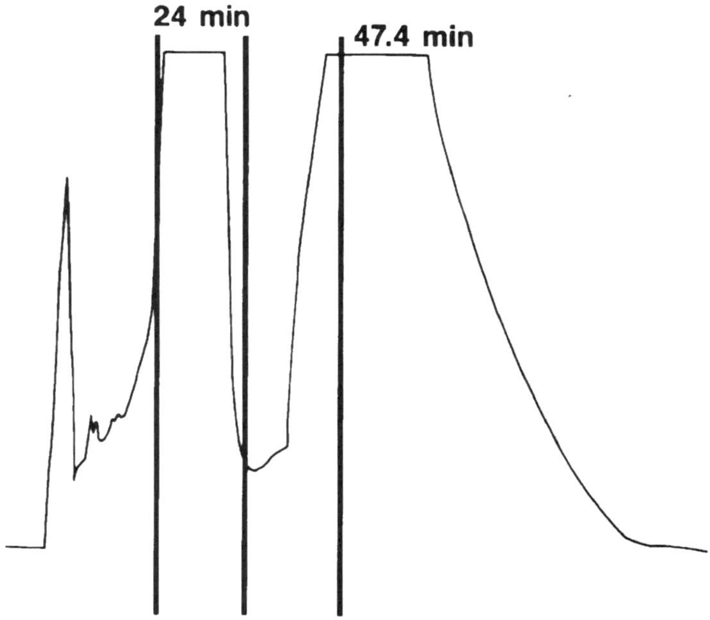

Table 5.5. Summary of Chromatographic Data

	Enantiomeric Excess		
Recovery	S enantiomer (e.e.)	R enantiomer (e.e.)	Yield[a]
15.5 g	99.4%		79%
16.0 g		99.6%	

[a]Based on 40 grams of racemate

recycling chromatography, where the sample is reinjected at the column inlet, may be the answer. Recently Dingenen and Kinkel (1994) discussed in detail the advantages of closed loop recycling chromatography in the preparative resolution of racemates.

Closed loop recycling simulates the use of a longer column, but without the corresponding increase in back pressure that a longer column would cause. A schematic diagram of a commercial preparative recycling system (the ST 140R system, EM Separations Technology) is shown in Figure 5.3, and a photograph of the instrument is shown in Figure 5.4. The sample can either be pumped onto the column or injected through the loop injector. After exiting the column and detector,

Figure 5.3. Schematic diagram of the ST 140R preparative recycling chromatograph.

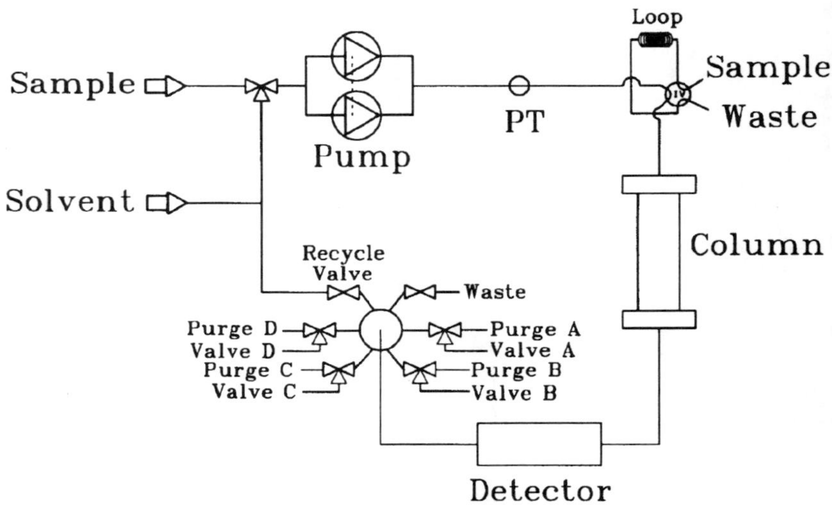

Figure 5.4. The ST 140R recycling chromatograph. Copyright EM Separations Technology (1994). Used with permission.

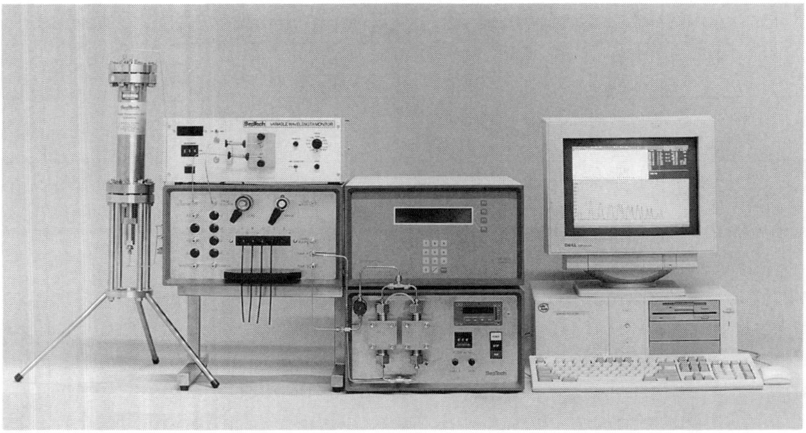

the sample can be collected through one of four collection valves (valves A, B, C, or D), can be sent to waste, or can be directed to the suction side of the pump through the recycle valve. If this last option is chosen, the sample will pass again through the column. Note the absence of a valve between the solvent reservoir and the system. This allows automatic solvent makeup when the collection or waste valves are open. Since there is no solvent valve that could malfunction during fraction or waste collection, the pump cannot become "starved."

For recycling to be effective, the number of theoretical plates must increase each cycle. Each time the sample passes through the column, the separation will progress further, thus increasing the number of plates. However, since mixing occurs in the pump, some of the separation will be degraded. Thus, for recycling to be effective, the increase in efficiency due to repeated passes through the column must be greater than the loss of efficiency each cycle due to mixing in the pump.

Figure 5.5 shows a recycling chromatogram performed on the ST 140R system. The column was an Annular Expansion (A/E) column

Figure 5.5. A recycling experiment used to determine the increase in efficiency with the cycle number. Column: A/E 50 mm I.D. × 250 mm; CSP: ChiraDex 10 μm; sample: 2-nitroaniline, 0.94 mg in 2 ml of mobile phase; mobile phase: methanol/water (0.1 percent triethylamine, pH 4.1) 40/60; flow rate: 100 ml/min. Each peak was used to measure the number of theoretical plates for its respective cycle. These results are plotted in Figure 5.6.

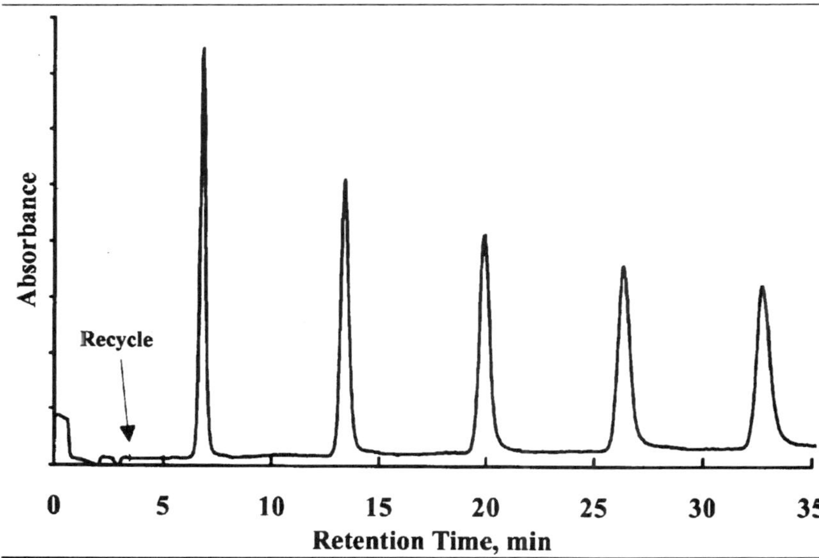

50 mm I.D. × 250 mm (EM Separations Technology, Wakefield, RI). The column was packed with 10 μm ChiraDex (β-cyclodextrin/silica, Merck, Darmstadt, Germany). The mobile phase was methanol/water (0.1 percent triethylamine, pH 4.1) 40/60, and the flow rate was 100 ml/min. The sample was 2-nitroaniline, 0.94 mg dissolved in 2 ml of the mobile phase; this compound is used by Merck to quality control their ChiraDex columns. In Figure 5.3 the recycle valve was opened at 3.4 minutes, after passage to waste of some impurities. Each peak was used to determine the number of theoretical plates as a function of cycle number. This plot is shown in Figure 5.6. As measured for the first cycle, the column efficiency was 3823 plates. The A/E column bed had been compressed to a length of 18.8 cm; thus, the normalized efficiency was 20,300 N/m. The measured efficiency increases with each cycle, indicating that recycling runs are feasible with this column and system. However, as expected, the increase in efficiency each cycle is less than would be predicted from a consideration of the column alone: the mixing in the pump destroyed some of the efficiency.

Two examples of closed loop recycling chromatography to resolve racemates will be presented here. In both cases the ST 140R system and

Figure 5.6. Plate count increases with each cycle, but not as much as would be predicted from a consideration of the column alone. Mixing occurs in the pump, destroying some of the efficiency gained by recycling through the column.

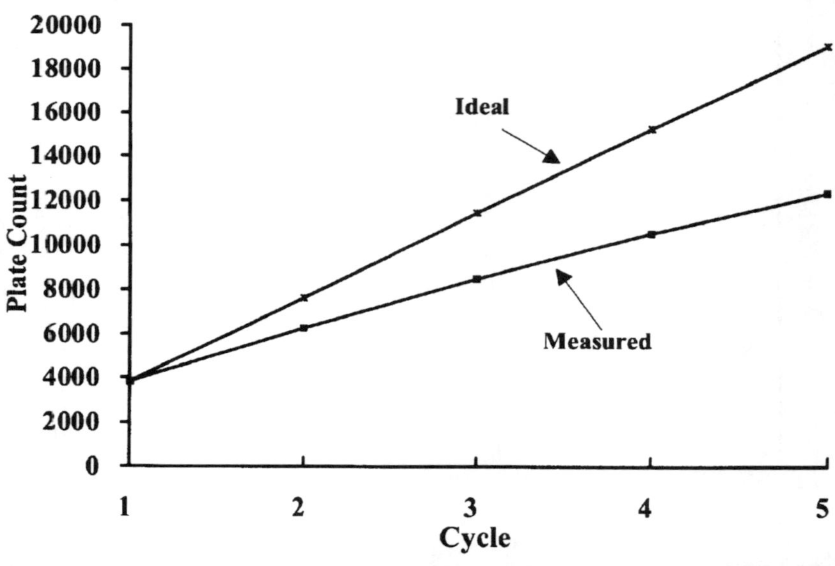

Figure 5.7. Resolution of cromakalim with closed loop recycling chromatography. Column: A/E 50 mm I.D. × 250 mm; CSP: ChiraDex 10 μm; sample load: 29 mg in 100 ml of mobile phase; mobile phase: methanol/water 20/80; flow rate: 100 ml/min.

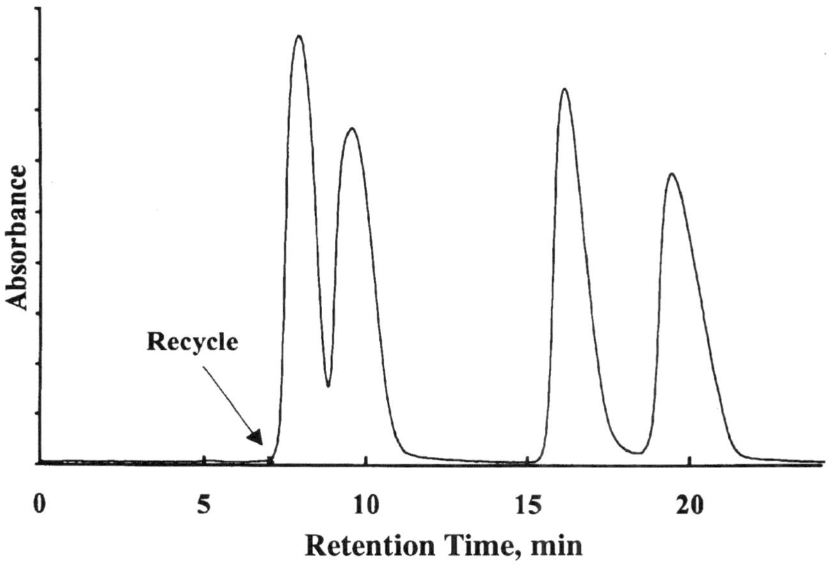

A/E column packed with 10 μm ChiraDex were used. Figure 5.7 shows the resolution of cromakalim, a potassium-channel activator. The mobile phase was methanol/water 20/80, and the flow rate was 100 ml/min. The sample, 29 mg in 100 ml of mobile phase, was injected through the pump. The recycle valve was opened at 7.0 minutes. Essentially, baseline resolution had occurred after the second pass through the column.

The separation of propranolol, a β-blocker, is shown in Figure 5.8. The mobile phase was acetonitrile/methanol/acetic acid/triethylamine 95/5/0.3/0.2, and the flow rate was 100 ml/min. The sample, 10 mg in 1.0 ml of mobile phase, was injected through the loop. The recycle valve was opened at 10.4 minutes. This separation is much more difficult on cyclodextrin than was the previous cromakalim separation. No appreciable separation is seen after the first cycle. Resolution increased during the second through fifth cycles. Thereafter, no further improvement in resolution was noted. In an actual run, therefore, the first half of the first peak and the second half of the second peak would

Figure 5.8. Separation of propranolol with closed loop recycling chromatography. Column: A/E 50 mm I.D. × 250 mm; CSP: ChiraDex 10 μm; sample load; 10 mg in 1 ml of mobile phase; mobile phase: methanol/water (0.1 percent triethylamine, pH 4.1) 40/60; flow rate: 100 ml/min.

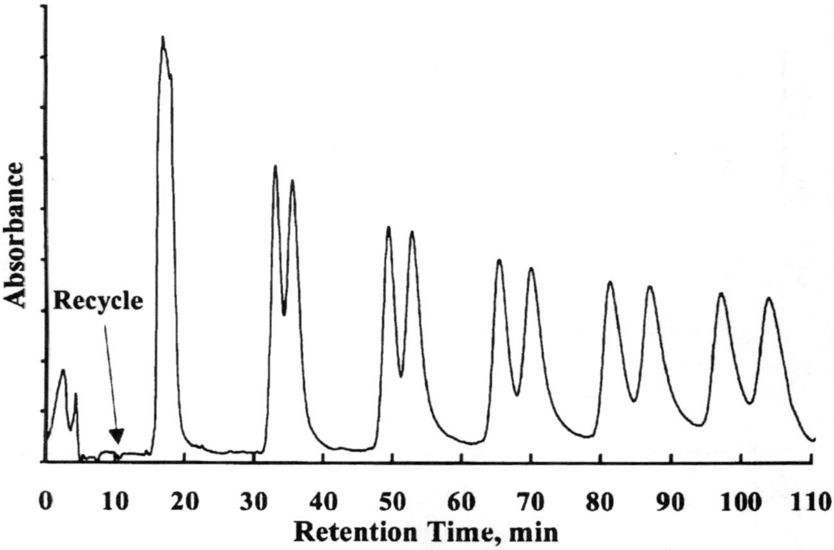

probably be collected during the fourth or fifth cycle. In subsequent cycles essentially all of the enantiomers could be collected.

As with other preparative HPLC techniques, closed loop recycling can be fully automated. Sample can be automatically pumped onto the column at the beginning of each run, and the collection, waste, and recycle valves can be opened by control software at the appropriate times in response to detector output. The reader is referred to the review article by Dingenen and Kinkel (1994) for a thorough discussion of automated closed loop recycling in the preparative resolution of racemic mixtures.

CONTINUOUS PREPARATIVE TECHNIQUES FOR RESOLVING RACEMIC MIXTURES

The best known continuous HPLC technique is simulated moving bed (SMB) chromatography. Crude material is continuously fed into the system and two product streams are continuously collected. Thus, SMB is ideal for the resolution of enantiomers when significant quantities of both enantiomers are desired. Simulated moving bed chromatography gets its name from the rather elaborate system of valves that are switched in such a manner as to simulate the movement of the chromatographic bed in the direction opposite (or countercurrently) to the direction of the liquid phase.

To understand how SMB works and why we should want to simulate the movement of the chromatographic bed, we shall first consider the case of an ordinary, unoverloaded batch chromatography. We will then consider a liquid-liquid countercurrent system that is easy to understand and shares many basic characteristics with SMB. Finally we will look at SMB in some detail.

The chromatographic profiles shown in Figures 5.9 through 5.14 were calculated using a Monte Carlo simulation based on the Craig model (Craig 1944). Note that the units of time and distance along the columns are not specified. Although it would be easy to express the time and distance in familiar units if we were modeling a specific set of laboratory equipment, such specificity is not needed here. The scales of time and distance are consistent from figure to figure, and that is sufficient for our purposes.

Batch Chromatography

Figure 5.9 shows the progress of a chromatographic peak down a column. Two important conclusions can be drawn from Figure 5.9. First, the distance traveled in time 60 is twice that traveled in time 30. This indicates that the peak is traveling at a constant linear velocity. Second, the peak broadens as it travels down the column.

To understand why the peak broadens, we can employ a useful mathematical trick, a transformation of coordinates. Since the peak is traveling at a constant linear velocity, we can choose as our reference frame the coordinate system moving at the same velocity as the peak and whose origin occurs at the peak's maximum. Such a coordinate system is said to be at rest relative to the chromatographic peak. This transformation is similar to the game one could play in a car traveling at constant velocity on the highway. One could imagine that the car is stationary and everything else is moving. For example, if the car were traveling at 100 km/hr, one could imagine that the car was stationary and that the road was moving at a velocity of –100 km/hr.

Figure 5.10 shows the same chromatographic experiment as in Figure 5.9, but plotted in the coordinate system that is at rest relative to the peak. In this frame of reference, the liquid phase appears to move to the right and the solid ("stationary") phase appears to move

Figure 5.9. The peak travels at constant velocity and broadens as it travels down the column.

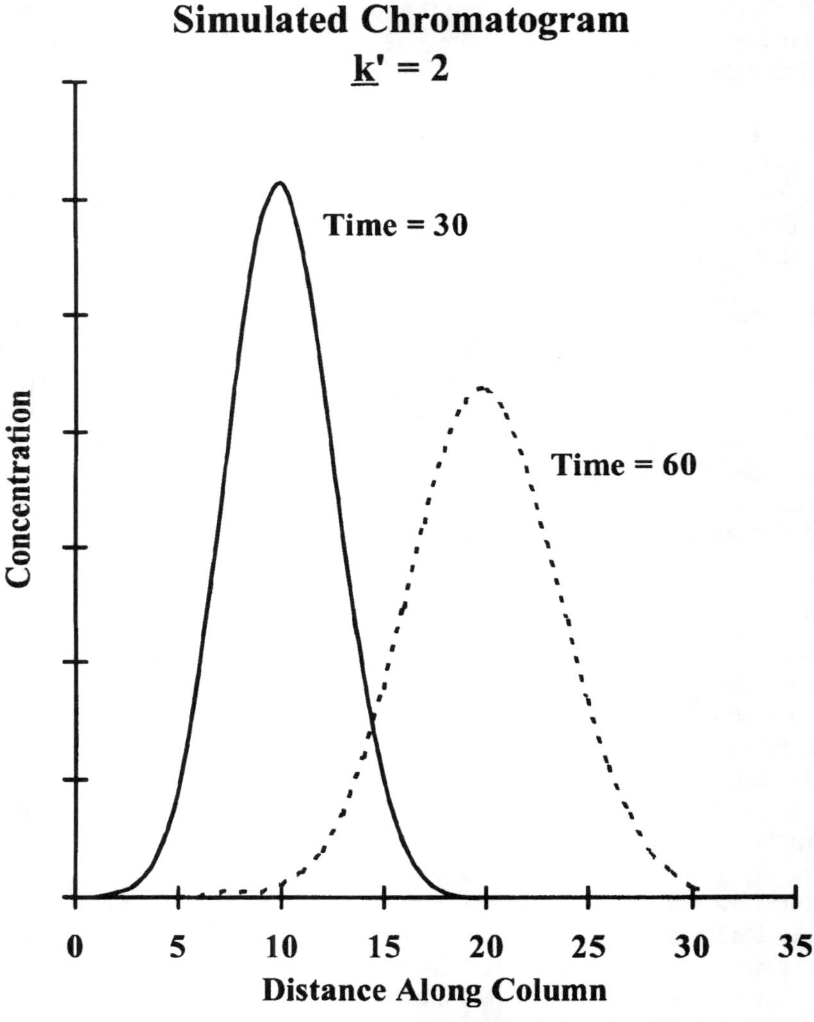

to the left. The reason the peak broadens is now clear. Whenever a molecule is adsorbed onto the solid phase, it is dragged to the left. A molecule dissolved in the mobile phase is transported to the right. This shearing action causes the peaks to broaden. Of course, the molecules are continuously adsorbing and desorbing according to the relevant adsorption isotherm. The result is that, in the unoverloaded case, the broadening peak maintains a Gaussian shape.

Figure 5.10. The same peak shown in Figure 5.9, plotted in the coordinate system at rest relative to the peak.

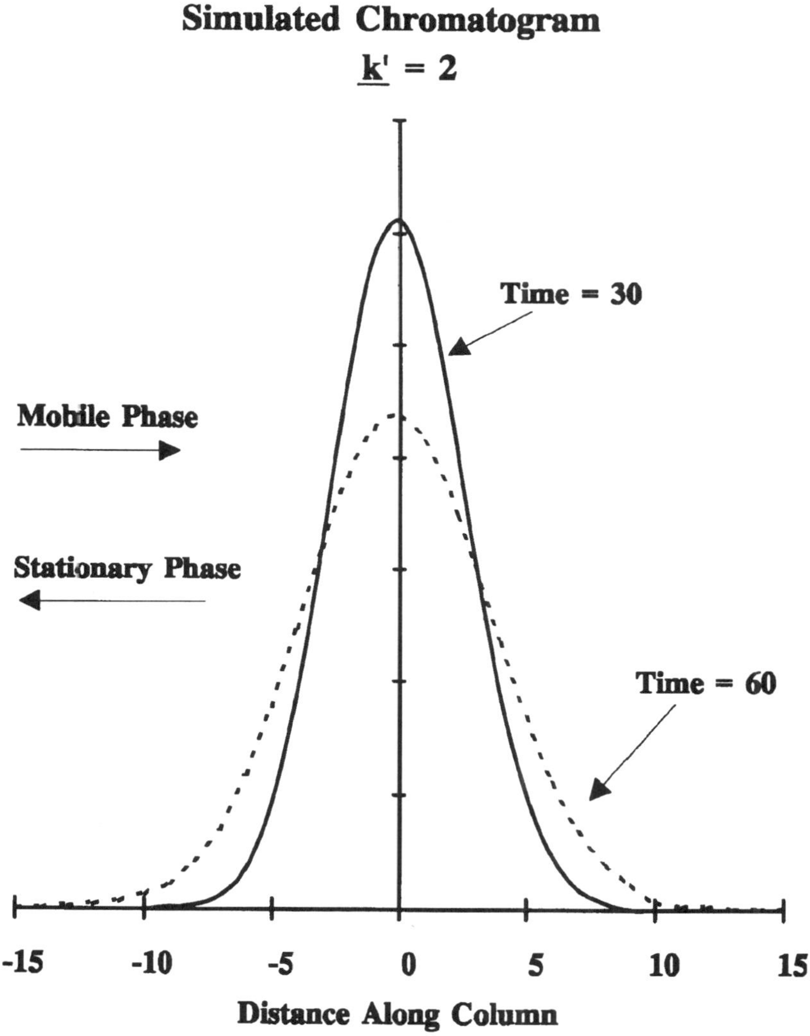

Now consider the case where two different compounds are injected onto a chromatographic column. As shown in Figure 5.11, the two peaks separate because they travel down the column at different velocities. As in the single peak case, each peak travels at a constant velocity. The midpoint between the two peaks also travels at a constant velocity (in fact, the midpoint's velocity is equal to the average

Figure 5.11. Each peak travels at a different constant velocity.

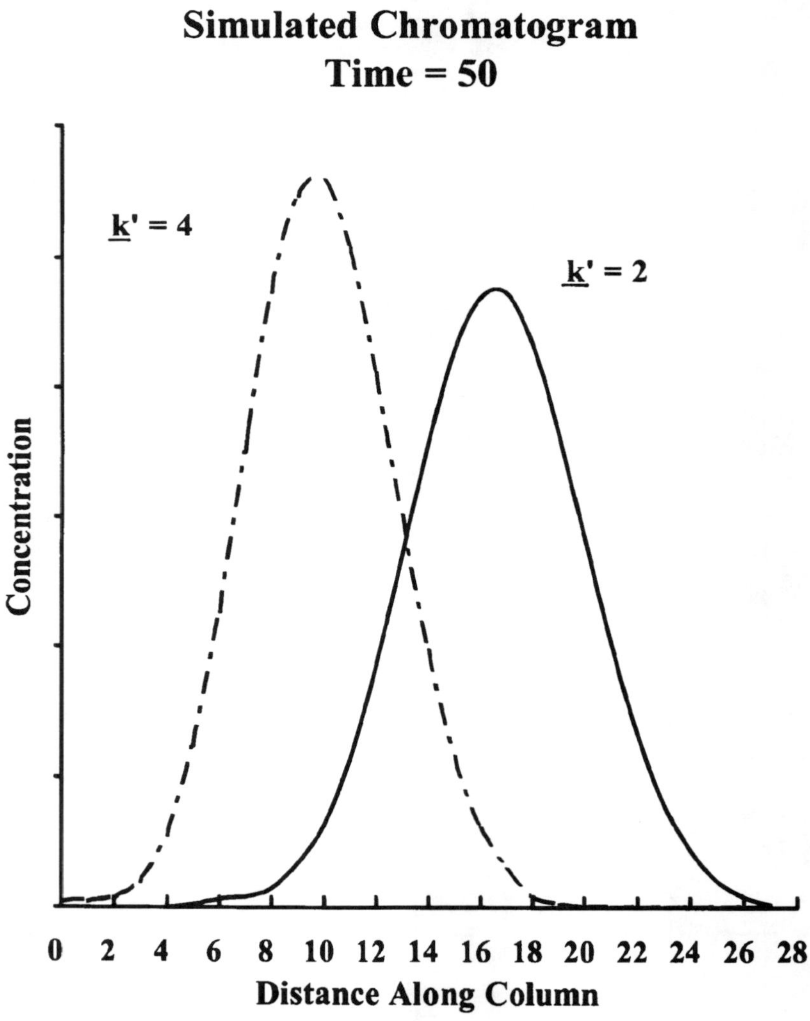

velocity of the two peaks). To use our highway analogy, imagine three cars, one traveling at 110 km/hr, one traveling at 90 km/hr, and the third traveling at 100 km/hr. At time zero all three cars are abreast. From that time on, the car traveling at 100 km/hr will be at the midpoint between the other two cars. Furthermore, if the occupants of the middle car choose to regard themselves as stationary, they will observe that the fastest car is moving in the positive direction at 10 km/hr, the slowest car is moving in the negative direction at 10 km/hr, and the road is moving in the negative direction at 100 km/hr.

We can perform the same type of coordinate transformation on the chromatographic profile shown in Figure 5.11. We will choose as our reference frame the coordinate system at rest relative to the midpoint between the peaks. The result is shown in Figure 5.12. The faster

Figure 5.12. The same chromatogram as Figure 5.11, but plotted in the coordinate system at rest relative to the midpoint between the peaks.

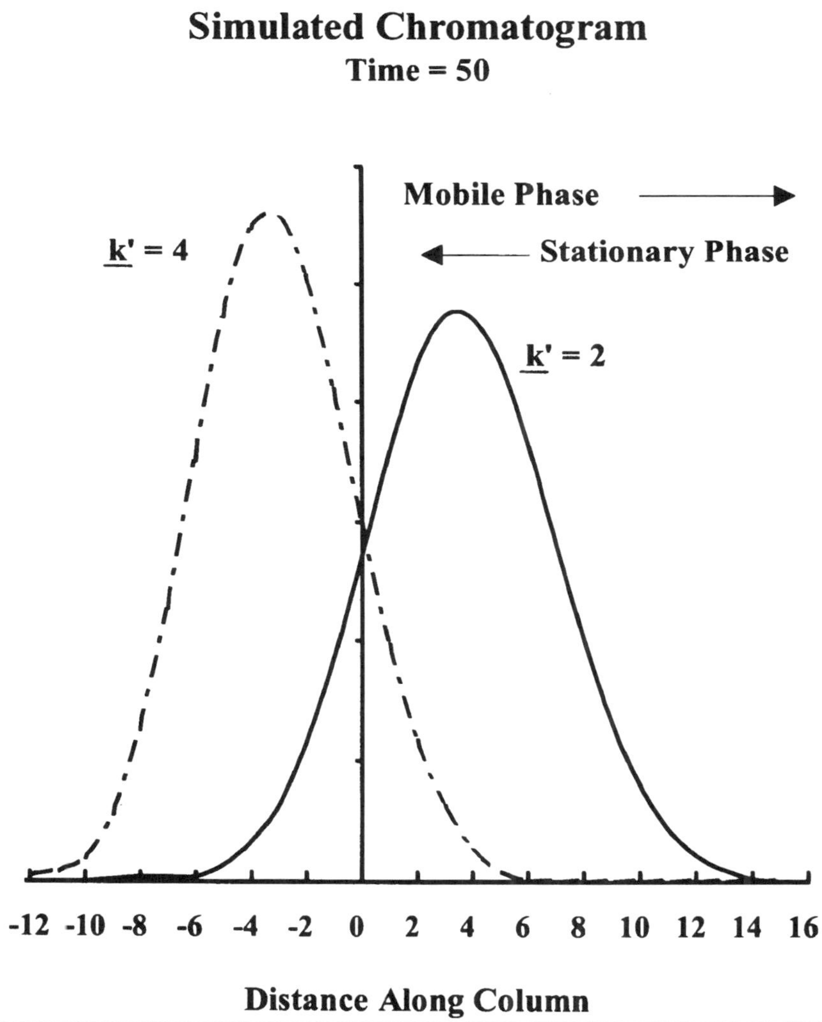

(less-retained) peak appears to move to the right, the slower (more-retained) peak appears to move to the left, the liquid phase appears to move to the right and the solid phase appears to move to the left.

Thus far we have been considering mathematical transformations. Figures 5.11 and 5.12 are in every way equivalent. We are simply using two different inertial reference frames to describe what is happening. However, suppose we could really cause the solid phase to move countercurrently to the liquid phase. We could adjust the velocity at which the chromatographic bed moves to the left and the velocity of the liquid phase to the right such that the midpoint between the peaks would remain stationary relative to an observer outside the system.

Liquid-Liquid Countercurrent Chromatography

Unfortunately, it is not feasible to build a system in which the solid phase is truly moving countercurrently to the liquid phase. However, it is possible to construct a system in which two liquid streams, separated by a membrane, flow countercurrently. Each liquid would have a different affinity for the compounds to be separated. Thus, partitioning between the two liquid streams would occur, and a separation similar to that depicted in Figure 5.12 would develop. This is shown in Figure 5.13 where, compared to Figure 5.12, "Liquid Phase 1" has been substituted for "Mobile Phase" and "Liquid Phase 2" has been substituted for "Stationary Phase." The concept of k' is still meaningful in this two-liquid system, but a degree of arbitrariness has been introduced we could designate either liquid as the "sorptive phase." By analogy with Figure 5.12, we will designate Liquid Phase 2 as the sorptive phase, thus the compound having greater affinity for this phase will be assigned the larger k' (4 in this case). If we had designated Liquid Phase 2 as the sorptive phase, the k's would have been reversed.

In Figure 5.13 note that in the tails of the separation profile (i.e., Distance $\geqslant$ 8 and Distance $\leqslant$ 8), pure fractions of each compound can be collected. This is what allows us to perform a continuous separation. As shown in Figure 5.14, collection valves have been placed at –8 for Liquid Phase 2 and at 8 for Liquid Phase 1. All molecules dissolved in Liquid Phase 2 at –8 will be collected; those molecules in Liquid Phase 1 at –8 will be swept to the right. Similarly, all molecules dissolved in Liquid Phase 1 at position 8 will be collected; those dissolved in Liquid Phase 2 at 8 will be swept to the left. Suppose now that we inject at 0 (i.e., the midpoint), a feed consisting of the total number of moles of each compound collected at 8 and –8. Since we also would be continuously adding fresh Liquid Phase 1 from the left and Liquid Phase 2 from the right, we would be adding to the system

Figure 5.13. Simulated liquid-liquid countercurrent separation.

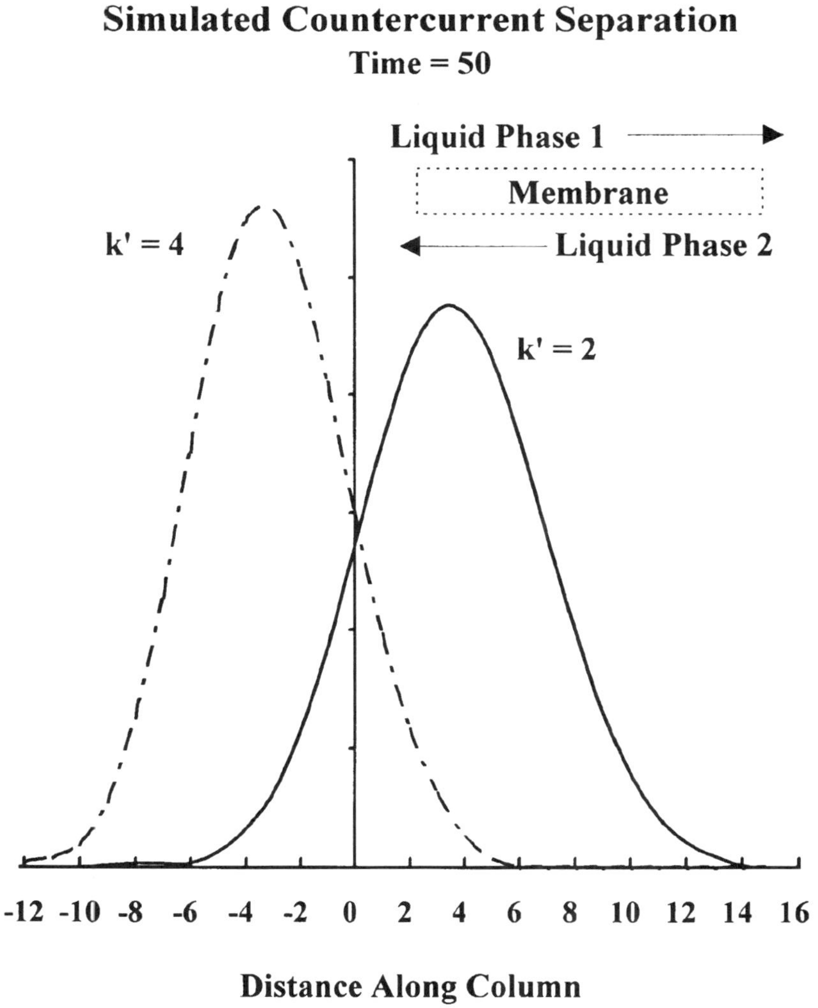

exactly the same components (types and quantities) that we would be collecting. Injecting the feed and collecting the product streams continuously will result in the development of a steady state. In Figure 5.14 the steady state separation profile that has evolved from the two Gaussian peaks in Figure 5.13, will remain stable indefinitely. In the jargon of continuous chromatography, the less-retained material that is

Figure 5.14. A simulated continuous liquid-liquid countercurrent separation. A steady state separation profile has evolved.

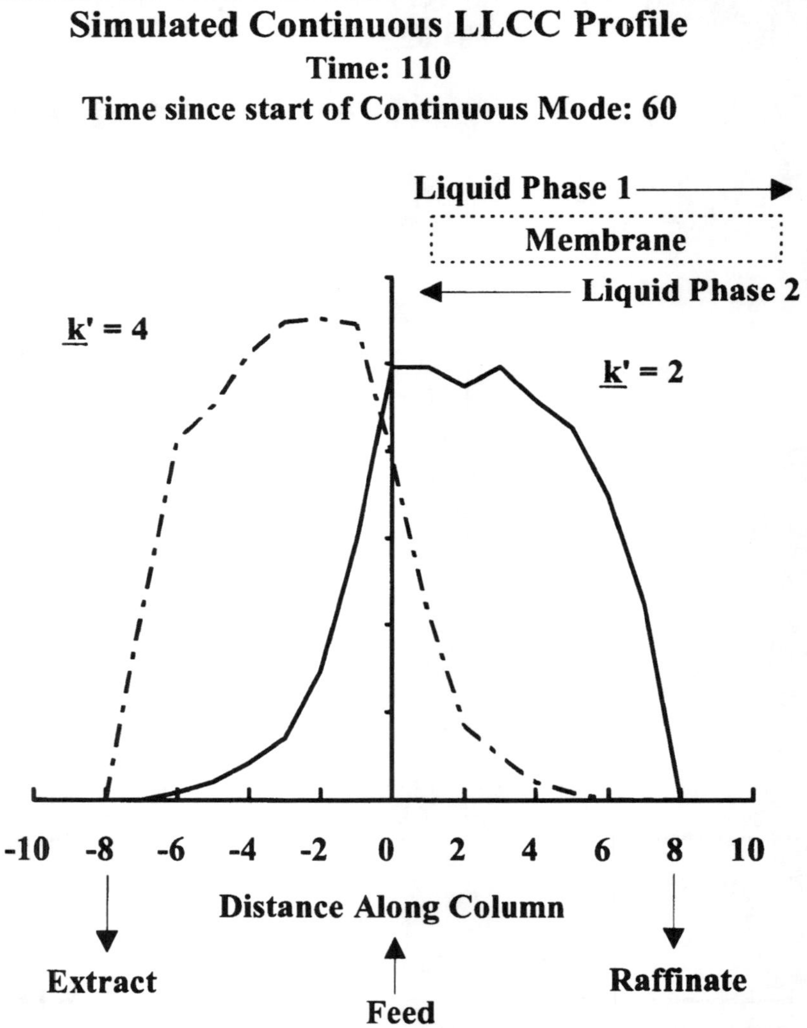

collected is called the raffinate, and the more-retained material collected is called the extract. In Figure 5.14 we have followed this convention. Designation of extract and raffinate is arbitrary in the liquid-liquid countercurrent case, since, as mentioned above, designation of the sorptive phase is arbitrary.

A system based on liquid-liquid countercurrent chromatography (LLCC) has been reported by workers at Akzo International Research (Keurentjes et al. 1994). Fourteen artificial kidney membrane modules were connected in series. One liquid stream, containing S,S-dihexyltartrate as the chiral selector (10 wt percent in heptane) was pumped through the system on one side of the membrane; the other stream (10 wt percent R,R-dihexyltartrate in heptane) was pumped countercurrently on the other side of the membrane. The feed, a solution of racemic norephedrine, was introduced between two of the modules in the center of the system. The product streams were collected at the ends of the system opposite from their respective pumps. During the course of the separation, the enantiomers of the feed are partitioned between the two liquid streams. Each feed enantiomer is preferentially concentrated in different streams, depending on the strength of the complexes each enantiomer forms with the chiral selectors. Thus, a separation profile similar to Figure 5.14 develops, and the enantiomers are continuously separated. A final separation step is needed to separate the products from the chiral selectors. Using the Akzo system, racemic norephedrine was resolved to >99 percent purity for each enantiomer.

Simulated Moving Bed Chromatography

Suppose we could build a system in which the chromatographic bed could move countercurrently to the liquid phase. Such a true moving bed (TMB) system would have a few differences with the liquid-liquid countercurrent system. In the LLCC case both liquid streams can be collected, one as the extract and the other as the raffinate. In TMB one of the moving phases is a solid, but only the liquid phase can be collected. Thus, the extract as well as the raffinate is collected in the liquid phase.

Figure 5.15 shows a schematic representation of a TMB profile. The extract has a greater affinity for the solid phase and is thus dragged to the left by the solid phase (see Figure 5.12). The raffinate, having a greater affinity for the liquid phase, is pulled to the right. Each zone in Figure 5.15 has a different flow rate. The highest flow rate occurs in zone 3 and is equal to the pumping rate of the mobile phase that enters the system at –8. Extract is drawn off continuously at –4; therefore, the flow in zone 2 is lower than in zone 3. Feed is injected at 0; thus the flow rate in zone 1 is greater than in zone 2. Finally, raffinate is continuously drawn off at 4; therefore, the flow rate in zone 4 is lower than in zone 1. The concentration of the less-retained component goes to zero between 7 and 8, and pure mobile phase occurs thereafter. This pure mobile phase is available for recycling and could be

Figure 5.15. A schematic representation of a true moving bed profile. Each zone has a different flow rate.

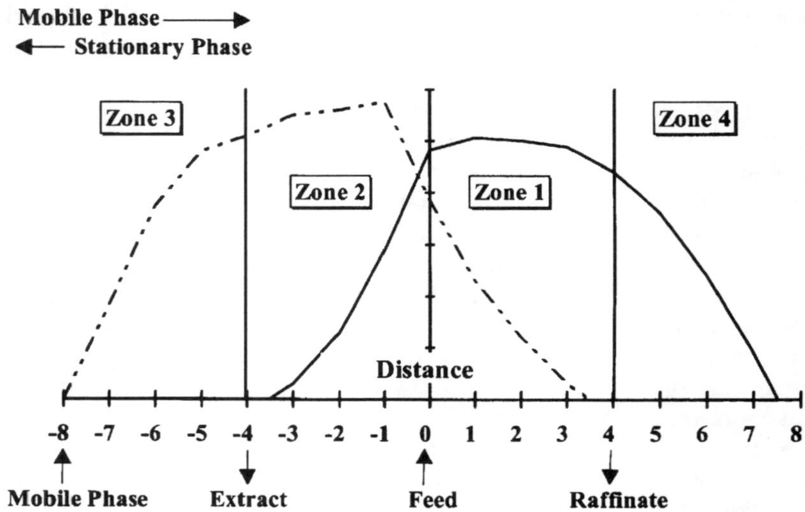

directed (not shown) by several methods to the beginning of the profile at -8.

Not all of the more-retained component is drawn off in the extract stream. Some is dragged further to the left than the extract collection point. What prevents the solid phase from dragging the more retained component indefinitely to the left? The zone 3 flow rate, which is the highest flow rate, must be increased to the point where transport of the more retained component to the right by the liquid phase equals its transport to the left by the solid phase. Similarly, not all of the less-retained component is drawn off in the raffinate stream. The zone 4 flow rate, which is the lowest flow rate, must be low enough so that transport of the less retained component in the liquid phase can be nullified by its transport in the solid phase.

As mentioned previously, TMB chromatography is not feasible. Chromatography columns are often difficult to pack. It is unlikely that the integrity of the chromatographic bed could be maintained if it were somehow caused to move inside a tube. If we are to perform continuous chromatography using these principles, we must find a way to simulate the movement of the chromatographic bed.

Figure 5.16 shows a computer simulation of an SMB profile that has developed in a 16-column system. This profile is similar to Figure 5.15. In Figure 5.16, however, the profile is stretched so that its length

Figure 5.16. A schematic representation of an SMB profile. The movement of the bed is simulated by the rate at which the injection and collection valves are switched.

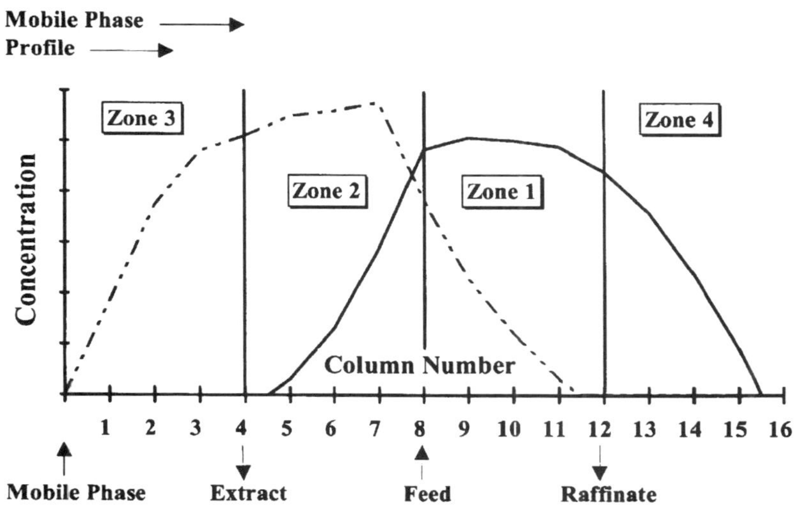

is nearly equal to the length of the 16-column system. Injection of feed and mobile phase, and collection of extract and raffinate are carried out between the columns. Because the bed is not moving, the entire profile moves to the right in the direction of the mobile phase. Therefore, the injection and collection points must be moved periodically to keep up with the moving profile. The movement of the bed is simulated by the rate at which these injection and collection points are switched.

The columns are connected in a circular array so that the outlet of column 16 is connected to the inlet of column 1. The profile, therefore, travels continuously around the circuit, the collection and injection valves switching so as to simulate the movement of the bed around the circuit in the sense opposite to that of the mobile phase and profile. The total composition of the injected materials (feed + mobile phase) is equal to the total composition of the collected materials (extract + raffinate). Thus, the profile is at steady state and will be stable indefinitely.

Overt means must be taken to ensure that the flow of mobile phase in each zone is in the same direction. There are two ways to do this. In all large commercial systems a pump is connected in series with the beds. Because this pump, called the pump-around pump, does not move relative to the beds, each zone will pass through it once each cycle. Thus, the pump-around pump must change its flow rate four times each cycle. Pump-around pumps are used in most smaller systems also. The exception is the UOP (Des Plaines, IL) design, The Sorbex Prep™ SMB System, marketed by EM Separations Technology (Broughton 1961). This design uses a check valve system to ensure correct flow direction in each zone. The extra dead volume of the pump-around pump and its associated flowmeter, which can be significant at the preparative scale, is thus avoided. The Sorbex Prep™ SMB System is shown in Figure 5.17.

Advantages of Simulated Moving Bed Chromatography Relative to Batch High Pressure Liquid Chromatography

As shown in Figure 5.16, SMB utilizes the entire stationary phase in a manner not possible with batch modes of HPLC. This more efficient use of the stationary phase coupled with the continuous nature of the

Figure 5.17. The Sorbex Prep™ SMB System. The system is designed and built by UOP, Des Plaines, IL, and marketed by EM Separations Technology, Wakefield, RI. Copyright EM Separations Technology (1994). Used with permission.

process means that the production rate (amount of product collected per amount of stationary phase per unit time) is usually higher with SMB than with batch HPLC. For much the same reason, the utilization of mobile phase is also more efficient with SMB.

A further savings in mobile phase is also possible. Often the optimized batch process will involve gradient elution, which requires extra solvent at the end of a run to regenerate the column. Simulated moving bed chromatography by its continuous nature is isocratic; thus, regeneration of the column system is unnecessary.

Simulated moving bed chromatography works best when it is allowed to operate around the clock. Because SMB is continuous, it can be automated to a very high degree. In some systems the ideal of unattended operation is approached. With these systems the operator measures the purity of the extract and raffinate streams once or twice daily, monitors the computerized log periodically, and performs maintenance as needed. Thus, labor costs are minimized with highly automated SMB systems.

The major advantage of SMB compared to batch HPLC can be summed up as lower operating costs. The amount of stationary phase and mobile phase for a given throughput is lower with SMB. Labor costs can also be much lower.

Examples of the Use of Simulated Moving Bed Chromatography to Resolve Racemic Mixtures

Nicoud et al. (1993) used a 12-column SMB system to separate the enantiomers of 1a,2,7, 7a,-tetrahydro-3-methoxynaphth-(2,3b)-oxirene, an epoxide used in the synthesis of enantiomerically pure compounds. The glass Superformance® columns 2.6 cm ID × 11 cm (E. Merck, Darmstadt, Germany) were packed with CTA-1 25–40 mm (E. Merck). The results of an optimized batch preparative HPLC technique under overloaded conditions were compared to those of a nonoptimized SMB run. For each technique the optical purity of each isomer was >99 percent. The recovery was 70 percent for the batch HPLC technique and 100 percent for SMB. The mobile phase usage (volume of mobile phase per weight of separated enantiomer) was 8 times greater with the batch HPLC technique than with SMB. The production rate was significantly lower for SMB than for batch HPLC. Under conditions of 100 percent recovery (i.e., baseline separation), the production rate of the HPLC technique would probably have been lower than that of SMB. The authors experimented with further optimization of the SMB method by increasing the flow rates. This allowed them to obtain the same production rate as that of the batch HPLC method, but the purity of the collected products decreased to

92 percent. The authors speculated that the purity could have been maintained at 100 percent with more precise control of the pumping rates.

Barder (1994) reported the separation of the enantiomers of propranolol, a β-blocker, using a 16-column SMB system. The columns were packed with 50 μm Chiralcel OD (Daicel). Propranolol and other β-blockers are used in the treatment of angina, cardiac arrhythmias, high blood pressure, and relief of intraocular pressure. The S enantiomer of propranolol is 100 times more active as a β-blocker than the R isomer (Davies 1990). Thus, the SMB separation was developed with the aim of optimizing the production of the S enantiomer. The mobile phase was a mixture of hexane and isopropanol. Several mobile phase compositions were modeled. The highest production rate (4.2 g racemate/ℓ CSP/hr) was predicted to occur with a composition of 80/20 hexane/isopropanol; however, the mobile phase usage was predicted to be quite high (0.52 ℓ/g racemate). The minimum in mobile phase usage (0.32 ℓ/g racemate) was predicted to occur with a composition of hexane/isopropanol 50/50; the production rate was predicted to decrease slightly to 3.8 g racemate/ℓ CSP/hr. This latter set of conditions was chosen for the actual run. The actual production rate was found to be 3.3 g racemate/ℓ CSP/hr, and the actual mobile phase usage was 0.30 ℓ/g racemate. Compared to an optimized preparative HPLC method, the SMB results represented a 4-fold increase in production rate and a 20-fold reduction in mobile phase usage. The preparative HPLC work was performed by workers at the Daicel Corporation in Japan.

Negawa and Shoji (1992) performed a study comparing conventional batch chiral chromatography (50 cm × 1 cm column) with a SMB system that contained eight 15 × 2 cm Chiralcel OD columns. For a sample of 1-phenylethanol, the SMB system was found to be superior to the batch mode in the following ways:

- Productivity (g/hr/ℓ bed) was 66 times greater.
- Solvent consumption was 1/87th of the batch mode.
- The concentration of the raffinate was 200 times higher.
- The overall yield was 4 times higher.
- Optical purities were equivalent.

Ikeda and Murata (1993) compared SMB with conventional batch chiral chromatography for three β-blockers (propanolol, pindolol, and altenol). Their findings showed that the productivity was 2 to 3 times greater and solvent usage 2 to 10 times less with SMB.

NEW TECHNIQUES

Hollow-Fiber Membrane Reactors

This is a flow technique where an enzyme is immobilized in a hollow-fiber membrane module. The racemate (derivatized with a nonpolar functionality) flows though the hollow-fiber membrane. An aqueous solvent is then introduced to the system and the enzyme in the fibers enantiospecifically hydrolyzes the derivative. The hydrolyzed enantiomer is carried away in the aqueous phase whereas the nonhydrolyzed enantiomer is collected at the other end of the reactor. A single reactor containing one membrane module has the capacity to resolve 165 kg per year for low molecular weight compounds (Stinson 1994). Estimates of greater than 100,000 kg per year have been made for reactors containing 20 production-scale membranes.

REFERENCES

Ariens, E. J., and A. M. Simmons. 1964. *Molecular pharmacology,* vol. 1. New York: Academic Press.

Application guide for chiral column selection. 1993. Exton, PA. Chiral Technologies, Inc.

Barder, T. J. 1994. A chiral separation using SMB. Paper presented at PrepTech 94. Secaucus, NJ.

Broughton, D. B. 1961. U.S. Patent 2 985 589 for UOP, Des Plaines, IL.

Craig, L. C. 1944. *J. Biol. Chem.* 155:519.

Davankov, V. A., J. D. Navratil, and H. F. Walton. 1988. *Ligand exchange chromatography.* Boca Raton, FL: CRC Press.

Davies, C. L. 1990. *J. Chromatogr.* 531:131.

Dingenen, J., and J. N. Kinkel. 1994. *J. Chromatogr.* 666:627.

Federal Register 57:22249 (May 27, 1992); also published in *Chirality* 4:338–340 (1992).

Francotte, E. 1994. *J. Chromatogr. A.* 666:565–601.

Francotte, E., and A. Junker-Buchheit. 1992. *J. Chromatogr.* 576 (1):1–46.

Ikeda, H., and K. Murata. 1993. Paper presented at the International Symposium on Chiral Discrimination, Montreal, Canada.

Jacobson, S. C., S. Golshan-Shirazi, and G. Guiochon. 1990. *J. Am. Chem. Soc.* 112:6493.

Jacobson, S. C., and G. Guiochon. 1992. *J. Chromatogr.* 590:119–126.

Keurentjes, J. T. F., L. J. W. M. Nabuurs, and E. A. Vegter. 1994. In *Proceedings of the Chiral '94 USA Symposium.* 39.

Kurganov, A. A., T. M. Ponomaryeva, and V. A. Davankov. 1993. Paper presented at the International Symposium on Chiral Discrimination, Montreal, Canada.

Lohmann, D., K. Auer, and E. Francotte. 1988. Paper presented at the International Symposium on Chromatography at the Weizman Institute, Israel.

Lough, W. J., and I. W. Wainer. 1989. In *Chiral liquid chromatography,* edited by W. J. Lough. New York: Chapman and Hall.

Negawa, M., and F. Shoji. 1992. *J. Chromatogr.* 590:113.

Nicoud, R. M., G. Fuchs, P. Adam, M. Bailly, E. Küsters, F. D. Antia, R. Reuille, and E. Schmid. 1993. *Chirality* 5:267.

Perrin, S. R., and W. H. Pirkle. 1991. In *Chiral separations by liquid chromatography,* edited by S. Ahuja. Washington, DC: American Chemical Society.

Perry, J. A., and T. J. Szczerba. 1989. *J. Chromatogr.* 484:267.

Pirkle, W. H., and T. C. Pochapsky. 1989. *Chem. Rev.* 347:362.

Pochapsky, T. C. 1987. *BioChromatogr.* 2 (1):28–36.

Roussel, C., and P. Piras. 1994. In *Procedings of the Chiral 94 USA Symp.*

Snyder, L. R., and G. B. Cox. 1989. *J. Chromatogr.* 483:85.

Snyder, L. R., J. L. Glajch, and J. J. Kirkland. 1995. *Practical HPLC method development.* New York: Wiley Interscience. (in press).

Snyder, L. R., and J. J. Kirkland. 1979. *Introduction to modern liquid chromatography.* New York: Wiley Interscience.

Sofer, G. K., and L. E. Nystrom. 1984. *Process chromatography, a practical guide.* New York: Wiley.

Stinson, S. C. 1994. *Chem and Eng. News* 72:38–72.

Stinson, S. C. 1992. *Chem and Eng. News* 70:46–79.

Taylor, D. R., and K. Maher. 1992. *J. Chrom Sci.* 30:67–85.

Verzele, M., and C. Dewaele. 1986. *Preparative high performance chromatography.* Ghent: TEC.

Wainer, I. W. 1988. *A practical guide to the selection and use of HPLC chiral stationary phases.* Phillipsburg, NJ: J.T. Baker.

Ward, T. J., and D. W. Armstrong. 1986. *J. Liq. Chromatogr.* 9:407.

Ward, T. J., and D. W. Armstrong. 1988. In *Chiral liquid chromatography,* edited by M. Zeif and L. J. Crane. New York: Marcel Dekker.

6

Chromatographic Purification of Recombinant Protein Products

Eugene P. Kroeff

The practical application of recombinant DNA (rDNA) technology has resulted in the availability of more than a dozen therapeutically important protein products since the early 1980s (Klausner 1993). Critical to the successful commercialization of these products is the development and implementation of a suitable downstream recovery process capable of providing a high purity, active protein product, in good yield. Chromatography plays a key role in the recovery process because it is a highly versatile and selective technique that can be incorporated into a production facility.

PURIFICATION CHALLENGES

High purity is a universal requirement for therapeutic products in order to eliminate or minimize the potential for undesirable side effects. Because rDNA products are derived from microbial fermentations and have complex primary, secondary, and tertiary structure, they pose a unique set of purification challenges. The purification process must be capable of isolating the product from a wide range of potential impurities. These impurities may be placed into two major categories:

1. Macroheterogeneity, which represents components that are totally unrelated to the product of interest
2. Microheterogeneity, which results from structural modification of the product

Examples of macroheterogeneity include the following: proteins, lipids, and nucleic acids derived from the host cell expression system; fermentation additives and nutrients; and biochemical or chemical reagents added to facilitate the processing of the product. Depending upon the expression system, macroheterogeneity may be initially present at a 10–50-fold weight excess relative to the product (Marston 1986). By contrast, the final purified product typically contains part per million (ppm) levels or less of these components (Ross et al. 1982; Kroeff et al. 1989; Garnick et al. 1991). This requires a process that can provide a purification factor of 100,000-fold or greater!

Microheterogeneity results from modification of the protein's structure and can occur during the purification and processing of the product. Deamidated or carbamoylated product forms as well as improperly formed disulfide bonds are examples of potential microheterogeneity (Kroeff and Chance 1982; Kumagaye et al. 1985; Furman et al. 1987; Smith et al. 1989). Depending upon the nature and the biological characteristics of these related components, their level in the final product may range from 1 percent to <10 ppm. The purification process, therefore, must deal effectively with the large excess of unrelated components that result from the fermentation process as well as reduce significantly the level of very closely related components that arise during processing.

The purification process must also reproducibly and efficiently produce the product at a rate and scale that is capable of meeting the demands of the marketplace. Depending upon the nature of the product, annual bulk requirements may range from 100s of grams for a product like factor VIII, to kilograms for human growth hormone, to 1000s of kilograms for human insulin (Datar et al. 1990; Ladisch and Kohlmann 1992). This requires that the process be transferred from the lab bench, to a versatile pilot plant, and ultimately to a production facility. It is imperative that the purification process deliver a consistent quality product as the product moves through clinical trials and eventually into the marketplace. The process must, therefore, be developed with ruggedness and reproducibility in mind. Finally, the purification process must be efficient and cost-effective.

A purification process based on chromatography is capable of reproducibly generating a high purity product in good yield and at a scale sufficient to meet market demands. This chapter will discuss the philosophy of developing such a process and will examine several practical examples.

SYSTEM SELECTIVITY

The ability of the process to deliver a product in both high purity and good yield is directly related to the *selectivity* of the process. A highly selective process is one that effectively exploits the physicochemical differences that exist between the product of interest and the impurities that must be removed. The selectivity of the process can be manipulated by varying both the modes of chromatography used for purification and the form of the product that is purified.

Chromatographic Mode

Table 6.1 lists the four major modes of chromatography that are typically used for the purification of rDNA products. Each resolves components on the basis of different types of molecular characteristics. Ion-exchange chromatography (IEXC) separates solutes on the basis of ionic charge (Karlson et al. 1989; Henry 1990; Wang 1990). Hydrophobic chromatography (HC) exploits differences in the polarity of molecules (Eriksson 1989; Hearn 1989; Rassi et al. 1990; Roettger et al. 1990). Affinity chromatography (AC) is a highly selective technique that separates on the basis of unique biological interactions (Goding 1986; Carlsson et al. 1989; Kagedal 1989; Bailon and Roy 1990; Clonis 1990; Nau 1990). Size-exclusion chromatography (SEC) separates compounds on the basis of apparent molecular size and shape (Kelley et al. 1986; Hagel 1989). Because of the complex nature of rDNA products, no single mode of chromatography is able to produce a high purity product in good yield. Thus, a multimodal approach, which employs combinations of the different types of chromatography, is used. Typically, the different types of chromatography are run sequentially. In reality, however, separations may be based upon mixed mode chromatography; that is, two or more modes of separation functioning during a single operation (Kennedy et al. 1986; Samanta et al. 1988). While mixed mode interactions can be useful, it is important that one recognize and understand the factors that influence such separations in order to insure that the chromatographic process behaves in a consistent and reproducible manner.

The IEXC, HC, and AC modes are based upon adsorption processes. Components in the process stream are retained on the stationary phase and are then selectively desorbed. Separations for these three modes of chromatography are greatly influenced by a wide range of chromatographic variables (see below). By contrast, ideal SEC separations involve no adsorption processes since the stationary phase matrix serves simply to filter the solute molecules. While the SEC technique can be very useful, it is more limited in its capacity and resolving capabilities than the adsorptive modes (Hagel 1989; Prouty 1991).

Table 6.1. Modes of Chromatography for rDNA Product Purification

Ion-Exchange Chromatography (IEXC) • Cation • Anion
Hydrophobic Chromatography (HC) • Hydrophobic Interaction (HIC) • Reversed-Phase (RPC)
Affinity Chromatography (AC) • Immobilized Metal Affinity • Antibody-Antigen Complex • Dye Binding
Size-Exclusion Chromatography (SEC)

For each mode of chromatography, there are a variety of chromatographic variables that can be manipulated in order to maximize the resolution. These are classified in Table 6.2 into three major categories: stationary phase, mobile phase, and operational variables.

Stationary Phase Variables

A significant amount of research has gone into the development of stationary phase matrices for protein separations. The ideal matrix is one that has no inherent affinity for a protein until it is derivatized with the interactive ligand of choice. This eliminates the potential for secondary interactions that can result in a stationary phase that is difficult to regenerate. The matrix should be stable throughout the entire pH range, permitting the use of acidic and basic mobile phases during regeneration. Additionally, the ideal support should be rigid, to allow for high mobile phase flow rates and it should be available in a range of particle parameters: pore and particle size, and surface area. Organic polymers of agarose and dextran (Andersson et al. 1985; Hagel 1989), polystyrene divinylbenzene (Lloyd and Warner 1990; Afeyan et al. 1990), and hydrophilic vinyl copolymers (Kusano et al. 1990), as well as inorganic matrices such as silica and zirconia (Stout and DeStefano 1985; Chicz and Regnier 1986), have all been successfully used for peptide and protein purification.

Table 6.2. Chromatographic Variables

Stationary Phase

- Matrix characteristics
  - organic polymers: agarose, dextran, acrylic, polystyrene divinylbenzene
  - inorganic: silica, zirconia
- Ligand
  - ionic: cation/anion; weak/strong
  - hydrophobic: C_1–C_{18}, phenyl
  - affinity: antibody, metal, sugar, dye
  - ligand density
- Particle characteristics
  - particle size: ⩾100 μm–10 μm
  - pore size
  - surface area

Mobile Phase

- pH and buffer type
- Additives
  - organic modifiers
  - denaturants: urea, SDS, chaotropic agents
  - ion pair reagents
  - salts

Operational Variables

- Temperature
- Gradient shape
- Flow rate
- Loading
- Column dimensions

In the broadest sense it is the nature of the stationary phase ligand (or lack thereof in the case of SEC) that dictates both the mode and the type of interactions that will be used to accomplish the separation. The three interactive modes of chromatography can be further subdivided according to the type of stationary phase ligand. A matrix derivatized with either acidic or basic groups yields cation or anion exchangers, respectively. Further, the pK_a of these sites and the makeup and length of the spacer that links the ion-exchange groups to the matrix will effect the selectivity of the chromatographic process (Karlson et al. 1989; Henry 1990). For HC there are a number of commonly used nonpolar ligands. These include straight chain alkyl groups ranging in length from C_1 to C_{18}, as well as phenyl ligands (Eriksson 1989). In general, it is the density of the ligands on the matrix that dictate if the stationary phase will be useful for hydrophobic interaction chromatography (HIC) or reversed-phase chromatography (RPC) (Eriksson 1989; Barford 1990; Rassi et al. 1990). Supports having ligand densities of $\leqslant$500 μM/g are useful as HIC supports, while supports having ligand densities of $\geqslant$2 mM/g are used for RPC separations. The lower ligand density of the HIC stationary phase produces relatively mild hydrophobic interactions between solute molecules and the matrix. These interactions are diminished by lowering the salt content of the mobile phase. By contrast, the higher ligand density of a RPC matrix yields strong interactions that can only be reduced by the addition of organic modifiers. (Reversed-phased separations are also conducted using underivatized organic polymer matrices. The high surface area of these media produce interactions that require the addition of an organic modifier for elution.) Ligands for AC include immobilized antibodies, metals, sugars, nucleic acids, dyes, and enzymes (Carlsson 1989; Kagedal 1989; Bailon and Roy 1990; Clonis 1990; Nau 1990). As with the other adsorptive methods, the density of the affinity ligand and the length of the spacer arm will effect chromatographic selectivity.

For all of the chromatographic modes, the particle size of the stationary matrix is selected to be compatible with both the nature of the process stream and the efficiency required to obtain the separation. Particle sizes ranging from $\geqslant$100 μm to $\leqslant$10 um play an important role in the purification of an rDNA product. A large particle ($\geqslant$100 μm) is especially useful for early purification steps where large volumes of a crude, dilute product are processed. The product of interest can be adsorbed onto the stationary phase while the insoluble particulates, such as cell debris, flow through the column. The large particles allow for relatively high flow rates (100–1000 cm/hr) with minimal back pressure since the large interstitial space of the packed column allows the particulate matter to flow through the packed bed. Large particles have also been successfully used in fluidized, expanded, and magnetically stabilized bed chromatography (Chetty and Burns 1991; Chase and Draeger 1992; Dasari et al. 1993). In all of these applications, the appropriate mode of chromatography in conjunction with large particles can produce a partially purified and concentrated product in a clarified stream.

After the initial purification step, chromatography using smaller, intermediate size particles (30–100 μm) will provide for better resolution of both micro- and macroheterogeneity. This particle size range is used to further process the product derived from the initial purification step and/or to generate a product of appropriate purity for subsequent chemistry steps. Flow rates of 20–100 cm/hr are typical.

Final "polishing" of the product (that is, reduction of microheterogeneity) may require the use of small, high performance particles (≤30 μm). Small particles provide for excellent resolution of structurally similar components (Kroeff et al. 1989). Additionally, the chromatography can be carried out at high speed (>100 cm/hr), thereby minimizing the amount of time the product is in solution and prone to degradation.

In addition to particle size, other important particle parameters to optimize include pore size and surface area (Henry 1990). The pore size should be large enough to provide for good mass transfer of the product into and out of the stationary phase. High surface area media combined with high ligand densities are especially useful for preparative applications, since they will have high capacity and perform more efficiently at the high loads.

Mobile Phase Variables

The selection of the mobile phase provides the development scientist with the greatest opportunity for optimizing the selectivity of a given mode of chromatography. This is especially true of the adsorptive forms of chromatography where solute proteins are retained by the stationary phase matrix. Mobile phase composition significantly impacts retention and selectivity because it directly effects the molecular characteristics of the protein during the purification process. The pH of the mobile phase, for example, will determine the ionization state of the protein. Likewise, the presence of a protein denaturant such as urea, or an organic modifier such as acetonitrile, will effect the protein's conformation and its interactive surface area, commonly referred to as the chromatographic footprint (Kopaciewiez et al. 1983; Regnier 1987). The larger this footprint, the greater the interaction with the chromatographic stationary phase.

The mobile phase is also critical in determining selectivity because unlike the stationary phase, the mobile phase can be varied throughout the operation. Preparative separations are commonly divided into four discrete segments: loading, washing, elution, and regeneration/equilibration. For each of these segments, the composition of the mobile phase is varied in order to accomplish the desired result. During the loading segment of an ion-exchange separation, for example, the pH and conductivity of the charge solution are adjusted to insure that the product is retained and concentrated on the stationary phase. (An exception to this is passive chromatography, in which the charge conditions are adjusted to allow the product to flow through the column, while impurities are selectively retained. While this can be a very useful technique, it does not allow the product to be concentrated.) The loaded column is then washed. During this segment of an ion-exchange separation, the mobile phase pH and/or salt concentration are adjusted to provide for selective desorption of impurities while allowing the product to be retained. During elution, the mobile phase composition is manipulated to bring about selective product desorption. Salt or pH gradients are typically used for ion-exchange separations. Once the product is eluted, the column is regenerated in order to remove highly retained components from the stationary phase matrix. The regeneration step for an ion-exchange operation usually involves acidic or basic pHs in combination with high salt. Finally, the column is reequilibrated to conditions conducive for loading the column.

Operational Variables

Operational variables include temperature, gradient shape, flow rate, loadings, and column dimensions. Adjustments in gradient shape (step, linear, or curved) and temperature can bring about substantial changes in the selectivity of the system. These selectivity changes can be especially useful in the separation of closely related components. For example, using analytical loads, it has been demonstrated that the separation between insulin and proinsulin could be significantly enhanced by increasing the operational temperature from 25°C to 55°C (Dimarchi et al. 1990). Likewise, the retention order of two protein components was inverted by changing the gradient slope for a reversed-phase separation (Ghrist and Snyder 1988).

The interrelationship between flow rate, loading, and resolution has been discussed in a number of papers (Stadalius et al. 1985; Snyder et al. 1988; Cox et al. 1988). For preparative separations, throughput can be optimized by increasing loadings and/or flow rates up to the point when the product and the impurities begin to overlap (Prouty 1991).

Product Form

The downstream processing of an rDNA product often requires that posttranslational modifications be made to the amino acid sequence (primary structure) and/or the conformation (secondary and tertiary structure) of the expressed protein. These controlled modifications or conversions are accomplished using specific enzymatic or chemical procedures. Such conversions will change the basic molecular characteristics of the product and hence its chromatographic properties. If the chromatographic properties of the product are changed relative to those of the unwanted contaminants, then these conversions provide a means for modifying the selectivity of the purification system. By strategically combining these conversions with the appropriate modes of chromatography, a highly selective purification process can be developed.

There are a number of posttranslational modifications that are typically carried out on rDNA products. One example is the expression of the product in the form of a fusion protein and its subsequent conversion to the product of interest (Ross 1981; Carter 1990). This approach is especially useful for products that are expressed as intracellular proteins in such organisms as *E. coli.* The fusion protein is composed of two discrete segments: the first being an amino acid leader sequence that is not part of the active form of the molecule and the second is the product of interest. The leader sequence may represent a portion of a protein that is native to the host cell organism or it may represent a naturally occurring "pro" form of the active molecule. After the fusion protein is expressed and isolated, a chemical or enzymatic process is used to liberate the product from the fusion. Fusion protein technology offers a number of processing advantages. Relatively small protein products, which might otherwise be degraded by intracellular host cell enzymes, can accumulate within the cell. The removal of the fusion leader sequence provides a convenient means for removing the N-terminal methionine residue that codes for the initiation codon, AUG, in all *E. coli* proteins. A fusion protein also provides the opportunity to incorporate a purification handle. Leader sequences containing amino acid residues that selectively bind to immobilized metal ions have been successfully used for the purification of IGF-II expressed in *E. coli* (Smith et al. 1990). Highly cationic amino acid sequences have been incorporated into the leader sequences of rDNA products to aid in their recovery using cation exchange chromatography (Sassenfeld and Brewer 1984). In each of these examples, the fusion protein is purified with the aid of the purification handle, the leader sequence containing the purification handle is cleaved, and the liberated product is then further purified.

Modification of cysteine residues in rDNA products provides another opportunity to selectively manipulate the product form. Products produced in *E. coli* typically form inclusion bodies or granules (Marston 1986; Krueger et al. 1989; Mitraki and King 1989). In this granular state, proteins containing cysteine residues can become highly polymerized as a result of intermolecular disulfide bond formation. These products must be depolymerized and converted to a monomeric species prior to chromatographic purification. Reduction with dithiothreitol to yield cysteinyl residues, reaction with excess cysteine to yield cysteinyl *S*-cysteine residues, or oxidative sulfitolysis with sodium sulfite to yield cysteinyl *S*-sulfonate residues, are three chemical procedures for accomplishing this process. After purification of one of these product forms, the protein is renatured and folded. The folding process allows the protein to assume its correct secondary and tertiary structure and to form the appropriate intramolecular disulfide bonds. The properly folded protein molecule will have a very different chromatographic footprint than its denatured, unfolded form. Thus, chromatography before and after the folding process can significantly increase the purity of the product.

Certain rDNA products are chemically modified in order to generate the desired product. Immunoconjugates, for example, are prepared by chemically modifying antibodies with either reporter groups for diagnostic applications (Nau 1990) or cytotoxic agents, such as vinca alkaloids, for therapeutic applications (Bumol et al. 1988). The preparation of these compounds involves producing the desired antibody in a cell culture fermentation, followed by chromatographic purification. The purified antibody is then conjugated with the reporter protein or drug of interest. This derivitization process typically results in a product that is more hydrophobic due to the addition of the drug or reporter protein to the antibody. Also, many of the conjugation chemistries involve derivitization of lysine residues, thereby altering the ionic properties of the antibody. In these cases the chromatographic properties of the immunoconjugate will be very different from that of the native antibody.

Chromatographic purification post conjugation, therefore, has a threefold effect:

1. It allows for the removal of excess unreacted drug.
2. It provides a means for selecting the immunoconjugate population that has the desired conjugation ratio.
3. Because the chromatographic characteristics of the product have been altered, it provides another opportunity to reduce heterogeneity.

System Selectivity: Practical Considerations

Given the large number of variables that can be used to maximize the selectivity of a chromatographic purification process, it is virtually impossible to evaluate each individual variable and all of the resulting combinations. In order to establish a blueprint for the process, general purification strategies for rDNA products have been reviewed and discussed (Belter et al. 1988; Asenjo 1990; Ho 1990; Lo 1990). There are also a number of reports on the use of expert systems for designing a purification process (Asenjo et al. 1989; Kenney 1990; Prokopakis and Asenjo 1990). These systems attempt to predict the ionic, hydrophobic, and size characteristics of the target protein on the basis of its primary structure. By comparing these properties to a list of common host cell proteins (i.e., *E. coli*) and their respective properties, the program predicts a potential purification sequence for the target protein.

With a basic purification sequence in place, chromatographic variables for the individual steps can then be evaluated and optimized. Again, there has been some progress in constructing models that allow for the prediction of optimal operational parameters, such as gradient slope, flow rate, and protein loading (Stadalius et al. 1985; Snyder et al. 1988). Using a minimal amount of empirical data, these programs attempt to predict the maximum quantity of product that can be purified, while maintaining a given level of product purity and yield.

While the literature and expert systems can provide an initial framework from which to start, the results *must* be coupled with empirical data in order to confirm and further optimize the process. This is a consequence of the complex nature of both the starting material and the protein product. The expert systems have only limited data regarding host cell contaminants and their respective chromatographic properties. The situation is further complicated when posttranslational modifications are conducted, since they can also alter the structure and chromatographic characteristics of the impurities. Even the behavior of the target protein can be difficult to predict since its secondary and tertiary structures, which determine the chromatographic footprint, will vary depending upon the composition of the mobile phase. It is clear then, that a substantial amount of experimentation is necessary during the development process. Screening studies that evaluate various chromatographic sequences, as well as the chromatographic variables of individual steps, can be very useful. Automation of the screening process allows for a substantial number of parameters to be evaluated in a relatively short period of time (Kenney 1990; Warren et al. 1990). The use of Perfusion Chromatography™ (Afeyan et al. 1990; Salzman et al. 1993), when coupled with an automated system for gradient generation, column switching, and data handling, provides the potential for very rapidly scouting and optimizing a wide range of chromatographic variables. In addition, extensive analytical characterization of intermediate products provides valuable insight into the nature and levels of impurities. This analytical information is very useful when evaluating subsequent chromatography steps.

PROCESS REPRODUCIBILITY

Equally as critical as product purity is product consistency. In order to insure product consistency, the purification process must be reproducible. The reproducibility of the chromatographic process is maximized by maintaining the integrity of the stationary phase media and reproducing the composition of the mobile phase.

Stationary Phase Integrity

In order to maintain the selectivity of the stationary phase and to insure a reasonable stationary phase lifetime, it is necessary that the interactive surface of the media remain constant. Changes in the selectivity of the surface result from chemical degradation and/or chemical fouling. Chemical degradation results from an irreversible, chemical modification of the stationary phase. The acid hydrolysis of the alkyl ligand from reversed-phase silica matrix or the dissolution of a silica matrix at alkaline pH (>8) are examples of such degradation (Henry 1990; Rassi et al. 1990). Chemical degradation, in general, can be minimized by using the stationary phase within the operational limits established by the manufacturer.

Chemical fouling occurs when components in the process stream (either product or impurities) become irreversibly bound to the chromatographic surface and thereby alter its selectivity. In many instances this also results in a reduction of the capacity of the stationary phase. The elution and regeneration steps should, therefore, be optimized in order to insure complete recovery of *both* the product and the impurities. A balance must be reached in finding elution and regeneration conditions that are strong enough to elute all adsorbed components, while remaining within the operating limits of the media.

As with particle size, the choice of the stationary phase matrix will often be dictated by the nature of the charge material. Early steps that must handle crude, heterogeneous process streams should employ robust media capable of withstanding rigorous regeneration procedures, such as extremes of pH and strong protein denaturants. Porous polymer- and carbohydrate-based stationary phases are especially useful. As the process progresses and the product becomes more homogeneous, media with more limited operational ranges, such as silica, may be employed. In order to insure a reasonable lifetime for such a matrix, it may be necessary to establish a minimum purity requirement for the charge material at that step.

By following the points discussed above, cycle-to-cycle reproducibility and good stationary phase lifetime can be achieved. However, even the most robust media exhibit finite lifetimes and it will be necessary to replace the stationary phase. It is important to insure that the new lot of media be *chromatographically* equivalent to the previous batch. The number of media lot changes can be minimized by obtaining a large quantity of the desired batch of packing. When changes in lots are made, it is important that the properties of the packing meet predetermined specifications. The specifications should cover not only physical parameters (i.e., particle and pore size, surface area, ligand density) but also chromatographic parameters for the product of interest, such as selectivity, recovery, and accountability (Kroeff et al. 1989). In many instances it will be necessary to work closely with the manufacturer in order to guarantee good reproducibility. Identifying an alternative supplier of a chromatographically equivalent packing is also beneficial.

The ability to consistently produce and maintain an efficiently packed column is of the utmost importance. A well-packed column will allow one to take full advantage of the efficiency that can be obtained with intermediate and high performance particles, thereby maximizing yield and throughput. Columns will need to be repacked when the useful lifetime of the stationary phase media is exceeded. Likewise, during the normal lifetime of a media, it may become necessary to repack the chromatographic bed if column efficiency becomes compromised. There are a variety of packing procedures that can be used to obtain high efficiency columns. These include pressure packing, radial compression, and axial compression techniques (Godbille and Devaux 1974; Colin 1987; Verzele et al. 1988). The procedure used will depend on the size of the column, the nature and particle size of the media, and the anticipated frequency of column repacking. Once packed, the integrity of the column bed can be maintained by eliminating particulates that can plug frits and disrupt flow. Additionally, it is imperative to use conditions that minimize the potential for precipitation that can lead to significant back pressures and perturb the flow characteristic of the packed bed.

Mobile Phase

Reproducibility of the mobile phase is also a necessary requirement for consistent operation. Gradient elution techniques are commonly used, and the ability to reproduce both the composition of the elution solvents and the gradient profile is critical. Specifications for the composition of the elution solvents should be established. For example, gas chromatographic analysis can be used to determine the organic modifier concentration of eluents used for RPC. Likewise, specifications can be established for the conductivity and pH of buffers used for IEXC, HC, and SEC.

Once the composition of the individual buffers has been insured, it is necessary to generate the appropriate gradient in a reproducible manner. Step gradients, which are produced by changing from a weak to a strong elution strength mobile phase, are the simplest and easiest to generate. Generally, however, step gradients do not provide the selectivity that is required to accomplish the desired separation. Continuous gradients are, therefore, often used. As mentioned previously, the shape of the continuous gradient can be either linear or curved. Both low and high pressure gradient systems may be used in production settings. In either case it is necessary to calibrate flow rate, gradient shape, and the absolute mobile phase composition generated by the system. Routine maintenance of the equipment and periodic calibrations of the gradient generating equipment are also necessary. A closed loop system, which continuously monitors the gradient profile using pH, conductivity, or spectroscopic measurements, is very useful. Based on feedback from the monitor, the system adjusts the gradient composition so that it is equivalent to the preprogrammed profile.

The temperature at which a separation is carried out will also effect the chromatographic reproducibility. Temperature extremes can result in product degradation and alter the selectivity of the chromatographic system. By operating in a controlled temperature environment and/or using thermostatted solutions and columns, temperature variations can be minimized.

PROCESS SCALEUP CONSIDERATIONS

Scaleup involves the transfer of the process from the lab bench to the manufacturing facility. The scaleup procedure typically involves three phases: lab, pilot plant, and production.

Lab Scale Studies and Process Design

The initial lab phase allows the development scientist to evaluate and assess which modes of chromatography are best suited for purifying the product. It is at this stage that any posttranslational processing steps are coordinated with the chromatographic purification. Various combinations of steps may be evaluated in order to determine the optimal sequence. For each purification step, the chromatographic variables are optimized to provide for maximum purity while maintaining good yield and throughput. Typically, milligram to gram quantities of final product are generated in order to determine the quality of the final bulk product and to estimate process yield.

It is critical during this phase that the development scientist remain aware that the ultimate destination of the process is the production facility. In the ideal scaleup situation, the only changes that should occur result from changing the size of the operation and the equipment. The sequence of the steps, the chromatographic variables for each of the steps, and the quality of the reagents used should all remain constant. Thus, the chromatographic media and other critical reagents used for the lab studies should be available in quantities that can support a production process. In many instances it may be necessary to work closely with the suppliers of these raw materials in order to establish tests and specifications that will insure consistent quality.

Designing a process that provides smooth product flow is also important. The sequence of chromatographic steps should be chosen so that at the completion of one step, the product is in a matrix that is compatible with the next step. This helps to eliminate ancillary processing such as solvent exchanges and serves to speed the processing of the product. In certain instances, it is important to build into the process a convenient product holding point, such as a crystallization or precipitation step. In this state the product will not degrade and it can be conveniently accumulated before further processing. Such a step is essential when the process stream "narrows" from a step that generates a relatively large quantity of material to one whose capacity is more restricted. The transfer of product from a step employing an intermediate performance particle to one using a high performance particle is one such example. Typically, smaller columns are used for the high performance steps, requiring that they be cycled numerous times relative to the previous step.

Worker safety and environmental issues must also be considered in the design of the process. The equipment used in the pilot and production facilities must be explosion proof when solvents are used. Further, these facilities must be capable of handling waste solvent streams and designed to minimize solvent release to the environment.

Pilot Plant Studies

From the lab the process next moves to a versatile pilot plant facility. The pilot plant offers the opportunity to evaluate how the process operates with larger equipment. Larger diameter columns are used at this phase in order to increase throughput, while maintaining proportional column loads. The wider diameter columns may require that new or modified column packing methods be developed in order to generate columns that exhibit efficiencies similar to those used in the lab. The column packing methods should be simple, with minimal emphasis on technique. This will insure a high degree of reproducibility at the pilot and production scales.

During the pilot phase cycle times and yields can be carefully assessed for both individual steps and the process as a whole. Process pinch points, that is, those steps that limit the overall rate at which the process can operate, are determined. At the pilot scale 100s of grams to kilograms of bulk material are prepared for toxicology and clinical trial studies. Based on the quality of these materials, specifications for the bulk drug are determined. From this point forward it is imperative that the process operate in a reproducible fashion in order to insure that future lots will meet these specifications. To further insure the quality of the final bulk product, specifications are often established for intermediates in the process. Pilot plant operations also provide an opportunity to determine equipment reliability and to assess the effective lifetime of both the chromatographic media and the packed columns. (Automated lab studies can also prove useful for these lifetime studies. However, due to differences in column geometry at larger scales, packed bed lifetime may be difficult to predict with lab studies.) Process automation and on-line analytical monitoring of the process may be developed and evaluated at this scale.

Production Operations

From the pilot plant the process moves to a production facility that is designed to deliver the appropriate quantities of bulk drug for the marketplace. Unlike the pilot plant that is designed to be highly flexible in order to accommodate a variety of processes and products, the production facility is typically a dedicated facility. Automation often plays a significant role in the operation of the equipment at the production scale (Kenney 1990). Computer monitoring and control of such parameters as flow rates and gradient profiles helps to insure consistent, reproducible chromatographic operations. On-line monitoring of product purity using spectroscopic and/or fast HPLC techniques, further serves to insure a smooth flow of a product through the process (Prouty 1991). Automation also minimizes the need for operator monitoring and intervention.

Throughout the course of production, complete documentation of each process step is necessary. The operational details for each process step are recorded on a manufacturing ticket. The analytical information required to determine the quality and/or quantity of the product entering and leaving each step in the process is also obtained and recorded. Thus, a substantial amount of information must be accumulated for each unit operation. By limiting the number of cycles that are required for a given column step, one can significantly limit the amount of information that must be collected and maintained. All other things being equal, in the production environment it is advantageous to use large columns with limited cycles, as opposed to running many cycles on a small column.

EXAMPLES OF CHROMATOGRAPHIC PURIFICATION OF rDNA PRODUCTS

As a means of demonstrating the versatility and usefulness of chromatography for the purification of rDNA products, several examples from the literature will be briefly reviewed. In addition, chromatographic strategies for the purification of human insulin and chimeric human proinsulin will be examined. All of these examples make use of the multimodal and/or the multiproduct form approach in order to optimize the selectivity of the purification system.

Literature

Most of the processes reported in the literature deal with laboratory scale operations, though many of these processes are potentially amenable to larger scale operations. Furman et al. (1987) discuss the expression of IGF-II in *E. coli* in the form of a fusion protein. The fusion protein is isolated, cleaved with CNBr, and the resulting IGF-II is purified by anion exchange chromatography in the form of a cysteinyl *S*-sulfonate derivative. The purified product is folded and subjected to multimodal purification using high performance RPC and SEC. The production of rDNA–derived human calcitonin has been described by Gilligan et al. (1987). Calcitonin, expressed in *E. coli* in the form of a fusion protein, is purified using both anion exchange and SEC. The calcitonin sequence is cleaved from the fusion protein with trypsin and purified by RP–HPLC. The calcitonin is then enzymatically α-amidated and purified by a final RP–HPLC step. Thus, three discrete forms of the molecule are chromatographically purified in this process. A multimodal purification process for producing gram quantities of rDNA hirudin expressed in yeast was reported by Bischoff et al. (1989). A 50 μm, intermediate performance C_8 matrix was used to initially purify the product. After precipitation and concentration, hirudin is further purified using a high performance, 15–20 μm, anion exchanger followed by RPC using a high performance, 10 μm, C_8 matrix. Hochuli (1988) has reported the use of multimodal chromatography for the purification of a recombinant interferon expressed in *E. coli.* The process uses immunoaffinity, metal affinity, and cation exchange steps. The affinity purification strategy provides for a high purity product with a minimal number of process steps. Though not a therapeutic product, recombinant HIV-1 proteinase has been produced as a direct expression product in *E. coli* (Singh et al. 1991). The isolation of milligram to gram quantities of this enzyme was deemed necessary for rational drug design studies. Purification of the product was accomplished using a combination of cation exchange and SEC steps.

The use of affinity purification handles has also been reported for a number of rDNA products. Insulin-like growth factors (IGF) I and II have been expressed in *E. coli* in the form of fusion proteins that are excreted into the growth medium (Nilsson et al. 1991). The IGF products are fused to an amino acid sequence of staphylococcal protein A, which is recognized by an immunoglobulin G (IgG) antibody. The fusion protein is affinity purified using an IgG immobilized on a Sepharose matrix. After purification, the leader sequence is removed using a variety of chemical and enzymatic cleavage protocols. Smith et al. (1990) have developed a series of metal-chelating peptide sequences that can be added to a recombinant protein of choice. After expression as a fusion protein, the product is selectively isolated by immobilized metal ion affinity chromatography. The chelating peptide is then removed, and the protein of interest is further purified. Fusion proteins containing a polyarginine leader sequence have been selectively purified by cation exchange chromatography (Sassenfeld and Brewer 1984).

Production and Purification of Biosynthetic Human Insulin

Since 1982 Eli Lilly and Co. (Indianapolis, IN) has produced biosynthetic human insulin (BHI) by means of rDNA technology. Two independent routes, the chain process (Chance et al. 1981; Burnett 1983) and the proinsulin process (Frank et al. 1981; Burnett 1983), have been successfully used for the production of BHI. The proinsulin process, which is now the sole method of manufacturing BHI, is outlined in Figure 6.1. In this process, human proinsulin (HPI), the precursor to BHI, is expressed in the form of a fusion protein (see below). Treatment of the fusion protein with CNBr yields the HPI sequence. The HPI molecule is initially purified by chromatography as the cysteinyl *S*-sulfonate derivative. The purified cysteinyl *S*-sulfonate is then subjected to a renaturation-folding process in which the three intramolecular disulfide bonds are formed and the properly folded HPI is chromatographically purified. Treatment of the purified HPI with trypsin and

Figure 6.1. Production of biosynthetic human insulin: proinsulin process.

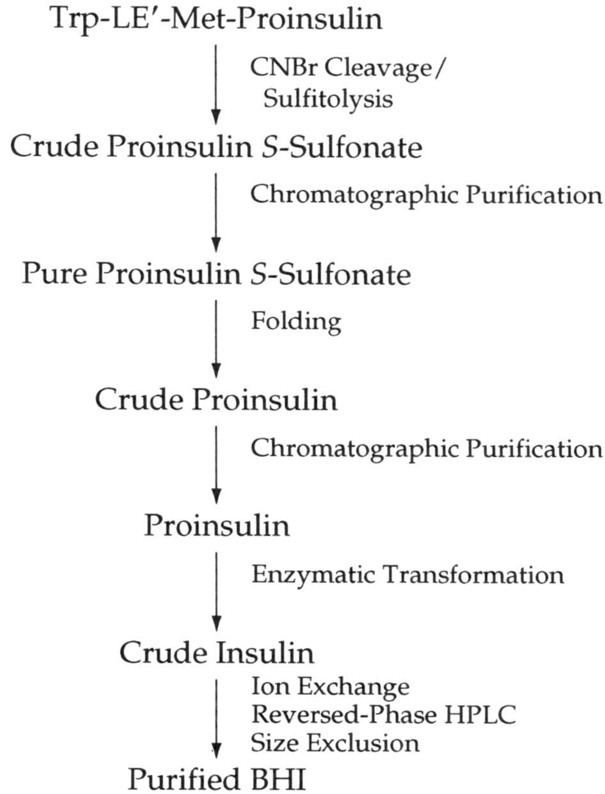

carboxypeptidase yields the BHI molecule. The resulting insulin is then subjected to multimodal chromatographic purification using IEXC, RPC, and SEC techniques. While ion exchange and size separations have traditionally been used for the purification of the bovine and porcine pancreatic insulins (Lilly Research Laboratories, Indianapolis, IN 1980), the incorporation of the reversed-phase technique further extends the multimodal approach to include a separation mechanism based upon the hydrophobic characteristics of the molecule. The reversed-phase purification of BHI has been discussed in depth by Kroeff et al. (1989).

Much of the general philosophy discussed earlier has been incorporated into the BHI production process. Selectivity is maximized by the fact that three distinct product forms are chromatographically purified: HPI cysteinyl *S*-sulfonate, HPI, and BHI. Furthermore, three discrete modes of chromatography are used for the purification of the BHI molecule. The high performance RPC step is incorporated toward the end of the process where it can serve as a polishing step, thereby reducing the level of insulin-like components. The placement of the RPC step late in the process insures that the 10 μm C_8 silica-based media is not exposed to high levels of components that might physically or chemically foul the packing. As a result, mild conditions (60 percent acetonitrile, 40 percent buffer, pH = 7.4) are effective for regenerating the C_8 packing material. By contrast, more robust matrices, capable of being regenerated with very acidic (pH $\leq$ 3) or basic (pH $\geq$ 11) mobile phases, are used during the early processing steps. The operational conditions of the RPC step provide for quantitative recovery ($\geq$97 percent) of the protein. Zinc insulin crystals are prepared prior to and immediately after the RPC step. The crystals provide both a simple means for isolating the insulin from the mobile phase and a convenient, stable storage form for the insulin until further processing occurs.

Scaleup of the reversed-phase step of the insulin process posed some unique development challenges due to the complications of efficiently packing 10 μm particles into large diameter columns. While conventional slurry packing techniques could be used to pack 10–100 ml columns for lab studies, an alternative approach was necessary for the pilot and production operations. This problem was effectively addressed by using axial compression column technology (Godbille and Devaux 1974; Colin 1987; Kroeff et al. 1989). This column packing technology yielded pilot scale columns (15 cm I.D.) as well as production columns (30 cm and 45 cm I.D.) that exhibited efficiencies exceeding 40,000 plates/meter. Product quality and yield were maintained at ≥98.5 percent and 80 percent, respectively, as the process was transferred from the lab to the pilot plant and finally to the production facility (Kroeff et al. 1989). In the production environment cycle-to-cycle reproducibility of the RP–HPLC step was demonstrated by the fact that the mainstream purity for 25 consecutive runs was 98.8 percent ±0.3 percent.

Preparation of Chimeric Human Proinsulin Reference Standard

As mentioned above, HPI, the biosynthetic precursor to BHI, is expressed in *E. coli* in the form of a fusion protein. The general structure of the fusion protein, known as chimeric human proinsulin (CHPI), is shown in Figure 6.2. The CHPI molecule is composed of two discrete

Figure 6.2. General structure of chimeric human proinsulin.

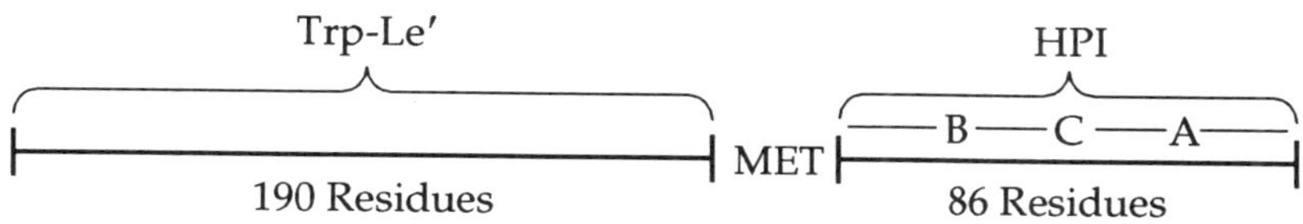

amino acid segments joined by a methionine residue. The Trp-Le′ leader sequence is made up of 190 amino acids found in the enzyme tryptophan synthetase. The product of interest, HPI, is 86 amino acids and is composed of the A and B chain of insulin joined by the C-peptide. The HPI molecule is liberated from the fusion protein by treating the CHPI with CNBr. In order to carefully monitor the early processing steps—fermentation, granule isolation, and cleavage—it is necessary to accurately quantitate CHPI. A purified reference standard of CHPI was, therefore, required in order to develop the appropriate assay methodology. Because the CNBr cleavage is conducted very early during processing, the proinsulin process outlined above does not provide for extensive purification of the CHPI molecule. It was, therefore, necessary to develop a process that could generate 10s to 100s of grams of pure CHPI in a pilot plant facility.

The purification process developed for CHPI is shown in Figure 6.3. Granules containing CHPI are isolated and dissolved in 7 *M* urea, in the presence of 5 m*M* cysteine. Except for the final solvent exchange step, 7 *M* urea is used throughout the process, to insure good solubility of the CHPI molecule and to dissociate potential impurities from the product. The solution of solubilized HPI is clarified by a combination of centrifugation and filtration and loaded onto a column packed with Fast Flow Q™ (FFQ™) anion exchange media (Pharmacia). After loading the column is washed with a 7 *M* urea/50 m*M* tris solution (pH = 8.0) to remove impurities. The CHPI is then eluted with a linear salt gradient ranging from 0 to 0.3 *M* sodium chloride. The column effluent is fractionated and the desired fractions for the mainstream are chosen on the basis of protein assays and SDS gel electrophoresis. The mainstream pool from the column is then subjected to oxidative sulfitolysis, concentrated using a 10 kD molecular weight cutoff membrane, and dialyzed against a 7 *M* urea/50 m*M* tris solution (pH = 8.0) to remove the sulfitolysis reagents. The CHPI cysteinyl *S*-sulfonate is then further purified using a second FFQ™ column that is run virtually identical to the first. Again the effluent from the column is fractionated and characterized. The desired fractions are combined,

Figure 6.3. Purification of chimeric human proinsulin.

Dissolve CHPI Granules
(7 *M* Urea, 5 m*M* Cysteine)

↓

Clarification

↓

FFQ™ Anion Exchange Chromatography #1

↓

Sulfitolysis, Concentration, Dialysis

↓

FFQ™ Anion Exchange Chromatography #2

↓

Solvent Exchange, Lyophilize

solvent exchanged into 10 m*M* ammonium bicarbonate to remove the urea, and the protein containing effluent is lyophilized. In the pilot plant two 20 ℓ FFQ™ columns and a 100 ℓ G-15 column were used to produce >100 g of purified CHPI.

The process purifies two discrete forms of CHPI by anion exchange chromatography. The two forms result from changing the chemistry of the nine cysteine residues of the CHPI molecule (Figure 6.4). When CHPI is initially solubilized, the nine cysteinyl residues of the CHPI molecule form disulfides with the excess cysteine in the dissolution buffer to form cysteinyl *S*-cysteine residues. At pH = 8 the free carboxyl and amine groups of these residues will both be charged, yielding neutral functionalities. The nine cysteinyl *S*-cysteine residues will, therefore, contribute very little to the overall retention of the CHPI during anion exchange chromatography. After purification, the product is subjected to oxidative sulfitolysis and the nine cysteine residues are converted into cysteinyl *S*-sulfonate residues. At pH = 8.0 the acidic *S*-sulfonate groups will be fully charged and the cysteinyl *S*-sulfonate groups will increase the retention of CHPI during the second anion exchange chromatography step.

Figure 6.4. Chemistry of CHPI cysteine residues during purification. The CHPI molecule contains nine cysteine residues: three in the Trp-LE′ leader and six in the HPI sequence of the molecule.

Cysteinyl *S*-Cysteine Form: FFQ #1 Purification

$$CHPI(S-S-CH_2-\overset{\displaystyle COO^-}{\overset{|}{C}}-NH_3^+)_9$$

Neutral Functionality

Cysteinyl *S*-Sulfonate Form: FFQ #2 Purification

$$CHPI(S-SO_3^-)_9$$

Anionic Functionality

The effectiveness of purifying two discrete forms of the product is demonstrated in the chromatograms shown in Figure 6.5. These analytical anion exchange chromatograms represent samples of the CHPI process stream taken from various points during purification. Studies by Lo (1990) have demonstrated that the analytical Mono Q™ matrix provides selectivity that is comparable to that obtained with the larger FFQ™ matrix used for the preparative operations. The analytical chromatographic profiles of Figure 6.5, therefore, mimic the separations that were obtained using the FFQ™ matrix in the pilot plant. Chromatogram A demonstates the high degree of heterogeneity that is associated with the starting CHPI granules prior to the first anion exchange step. By contrast, chromatogram B represents the mainstream from the first FFQ™ column. Note that a significant amount of the early eluting peaks (<250 seconds) as well as the large envelope of later eluting peaks (≈1600–2100 sec) have been removed by the chromatography. Chromtaogram C represents the same FFQ™ mainstream after sulfitolysis. Note that after the reaction a peak representing CHPI cysteinyl *S*-sulfonate appears at ≈1100 seconds. Thus, the chromatographic mobility of the product has been selectively modified and it

Figure 6.5. Mono Q™ analysis of CHPI. Chromatography was conducted with a Mono Q HR5/5 (Pharmacia) and detection at 280 nm, 0.05 AUFS. Gradient elution was accomplished with the following: A buffer = 7 *M* urea/tris, pH = 8.0; B = A buffer + 1.0 *M* NaCl. Gradient profile shown in the chromatogram. Chromatogram A = solubilized CHPI granules; chromatogram B = mainstream from FFQ™ #1 column; chromatogram C = sulfitolyzed mainstream from FFQ™ #1 column; chromatogram D = mainstream from FFQ™ #2 column.

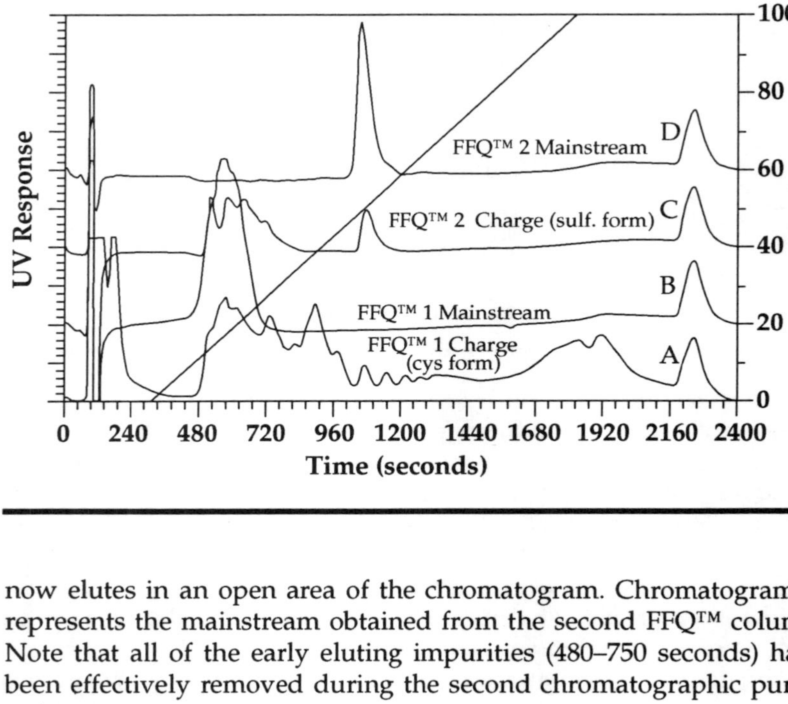

now elutes in an open area of the chromatogram. Chromatogram D represents the mainstream obtained from the second FFQ™ column. Note that all of the early eluting impurities (480–750 seconds) have been effectively removed during the second chromatographic purification step.

The high purity of the CHPI cysteinyl *S*-sulfonate preparation is also demonstrated by the additional analytical information shown in the SDS–PAGE profile (Figure 6.6) and the amino acid analysis (Table 6.3). The SDS–PAGE profile demonstrates a progressive improvement in the purity of the CHPI preparation when comparing the starting material (lane 2) to the first intermediate (lane 3), and to the final material (lane 6). Analytical high performance SEC confirmed that the CHPI preparation was ≈95 percent pure. Note also that the amino acid composition (Table 6.3) is in good agreement with the theoretical

Figure 6.6. SDS–PAGE of CHPI during purification. Conditions: 15 percent cross-linked gels, commassie blue stain for detection. Lanes 1, 7 = molecular weight markers; lane 2 = solubilized granules (starting material for FFQ™ #1 CHPI-Cys); lane 3 = FFQ™ #1 mainstream; lane 4 = sultitolyzed, dialyzed, concentrated FFQ™ #1 mainstream (starting material for FFQ™ #2 CHPI-SSO_3); lane 5 = mainstream from FFQ™ #2 mainstream; lane 6 = lyophilized final material.

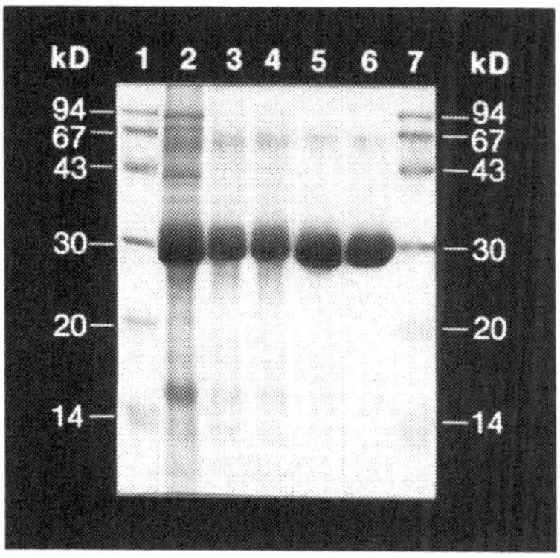

amino acid content. Thus, selective modification of the product form produced a highly purified preparation of CHPI using only two adsorptive chromatographic steps.

SUMMARY

The development and implementation of an efficient chromatographic purification process for an rDNA product poses a variety of challenges and opportunities. While there are a wide range of variables that determine the overall effectiveness of the process, it is the sheer number and combination of these variables that makes a chromatography-based purification process highly selective. Posttranslational modifications of the protein using chemical or enzymatic procedures should be integrated with multimodal chromatography in order to maximize the selectivity of the process. Chromatographic variables for individual steps should be carefully evaluated and optimized in order to maximize the purity, consistency, and throughput of the product. Finally,

Table 6.3. Analytical Characterization of CHPI

- **Amino acid ratios**

Asp*	22.1 (22)**	Thr	11.8 (12)
Ile	8.7 (10)	Leu	32.5 (32)
Ser	13.7 (16)	Tyr	9.8 (10)
Glu	30.4 (30)	Phe	5.8 (6)
Pro	7.1 (6)	His	10.3 (10)
Gly*	25.1 (25)	Lys*	7.9 (8)
Ala*	28.0 (28)	Arg	22.0 (22)
Cys	5.9 (9)	Met	5.1 (8)
Val	21.8 (23)	Met-O	0.7

- **% CHPI by amino acid analysis—86%**

* 24-hour acid hydrolysis (asp, gly, ala, lys residues used to calculate molar unity)

** Values in ( ) represent theoretical ratios

the procedures must be designed and developed with the overall goal of implementing the process in a production environment.

The development of such a process is best accomplished by drawing upon the expertise of individuals from a variety of disciplines—molecular biology, biochemistry, analytical chemistry, organic chemistry, and chemical and biochemical engineering. This group should work closely with suppliers of equipment and media as well as remain abreast of the regulatory issues concerning rDNA products. The combined talents of the group can fully exploit the power of modern chromatographic purification technology.

REFERENCES

Afeyan, N. B., N. F. Gordon, I. Mazsaroff, L. Varady, S. P. Fulton, Y. B. Yang, and F. E. Regnier. 1990. *J. Chromatog.* 519:1–29.

Andersson, T., M. Carlsson, L. Hagel, P. A. Pernemalm, and J. A. Janson. 1985. *J. Chromatog.* 326:33–42.

Asenjo, J. A. 1990. In *Separation processes in biotechnology,* edited by J. A. Asenjo. New York: Marcel Dekker.

Asenjo, J. A., L. Herrera, and B. Byrne. 1989. *J. Biotechnol.* 11:275–298.

Bailon, P., and S. K. Roy. 1990. In *Protein purification,* edited by M. R. Ladisch, R. C. Willson, C. C. Painton, and S. E. Builder. Washington, DC: American Chemical Society.

Barford, R. A. 1990. In *High performance liquid chromatography in biotechnology,* edited by W. S. Hancock. New York: J. Wiley and Sons, Inc.

Belter, P. A., E. L. Cussler, and W. S. Hu. 1988. *Bioseparations.* New York: John Wiley and Sons.

Bischoff, R., D. Clesse, O. Whitechurch, P. Lepage, and C. Roitsch. 1989. *J. Chromatog.* 476:245–255.

Bumol, T. F., A. L. Baker, E. L. Andrews, S. V. DeHerdt, S. L. Briggs, M. E. Spearman, and L. D. Apelgren. 1988. In *Antibody mediated delivery systems,* edited by J. D. Rodwell. New York: Marcel Dekker, Inc.

Burnett, J. P. 1983. In *Experimental manipulation of gene expression,* edited by M. Inouye. New York: Academic Press.

Carlsson, J., J. Janson, and M. Sparrman. 1989. In *Protein purification,* edited by J. Rydén. New York: VCH Publishers, Inc.

Carter, P. 1990. In *Protein purification,* edited by M. R. Ladisch, R. C. Willson, C. C. Painton, and S. E. Builder. Washington, DC: American Chemical Society.

Chance, R. E., J. A. Hoffmann, E. P. Kroeff, M. G. Johnson, E. W. Schirmer, and W. W. Bromer. 1981. In *Peptides: Synthesis-structure-function; Proceedings of the seventh American peptide symposium,* edited by D. H. Rich, and E. Gross. Rockford, IL: Pierce Chemical Co.

Chase, H. A., and N. M. Draeger. 1992. *Sep. Sci. and Technol.* 27:2021–2039.

Chetty, A. S., and M. A. Burns. 1991. *Biotechnol. Bioeng.* 38:963–971.

Chicz, R. M., Z. Shi, and F. E. Regnier. 1986. *J. Chromatog.* 359:121–130.

Clonis, Y. D. 1990. In *Separation processes in biotechnology,* edited by J. A. Asenjo. New York: Marcel Dekker.

Colin, H. 1987. *Sep. Sci. Technol.* 22:1851–1856.

Cox, G. B., P. E. Antle, and L. R. Snyder. 1988. *J. Chromatog.* 444:325–344.

Dasari, G., I. Prince, and M. T. W. Hearn. 1993. *J. Chromatog.* 631:115–124.

Datar, R., C. Datar, and R. Rosen. 1990. In *Separation processes in biotechnology,* edited by J. A. Asenjo. New York: Marcel Dekker.

Dimarchi, R. D., H. B. Long, E. P. Kroeff, and R. E. Chance. 1990. In *High performance liquid chromatography in biotechnology,* edited by W. S. Hancock. New York: J. Wiley and Sons, Inc.

Eriksson, K. 1989. In *Protein purification,* edited by J. Rydén. New York: VCH Publishers, Inc.

Frank, B. H., J. M. Pettee, R. E. Zimmerman, and P. J. Burck. 1981. In *Peptides: Synthesis-structure-function; Proceedings of the seventh American peptide symposium,* edited by D. H. Rich, and E. Gross. Rockford, IL: Pierce Chemical Co.

Furman, T. C., J. Epp, H. M. Hsiung, J. Hoskins, G. L. Long, L. G. Mendelsohn, B. Schoner, D. P. Smith, and M. C. Smith. 1987. *Bio/Technology* 5:1047–1051.

Garnick, R. L., M. J. Ross, and R. A. Baffi. 1991. In *Drug biotechnology regulation,* edited by Y. Y. Chu, and J. L. Gueriguian. New York: Marcel Dekker.

Ghrist, B. F. D., and L. R. Snyder. 1988. *J. Chromatog.* 459:43–63.

Gilligan, J. P., T. G. Warren, J. A. Koehn, S. D. Young, A. H. Bertelsen, and B. N. Jones. 1987. *BioChromatog.* 2:20–27.

Godbille, E., and P. Devaux. 1974. *J. Chromatog.* 2:564–569.

Goding, J. W. 1986. *Monoclonal antibodies: Principles and practices.* London: Academic Press, Inc.

Hagel, L. 1989. In *Protein purification,* edited by J. Rydén. New York: VCH Publishers, Inc.

Hearn, M. T. W. 1989. In *Protein purification,* edited by J. Rydén. New York: VCH Publishers, Inc.

Henry, M. P. 1990. In *High performance liquid chromatography in biotechnology,* edited by W. S. Hancock. New York: J. Wiley and Sons, Inc.

Ho, S. V. 1990. In *Protein purification,* edited by M. R. Ladisch, R. C. Willson, C. C. Painton, and S. E. Builder. Washington, DC: American Chemical Society.

Hochuli, E. 1988. *J. Chromatog.* 444:293–302.

Kagedal, L. 1989. In *Protein purification,* edited by J. Rydén. New York: VCH Publishers, Inc.

Karlson, E., L. Ryden, and J. Brewer. 1989. In *Protein purification,* edited by J. Rydén. New York: VCH Publishers, Inc.

Kelley, J. K., G. Y. Wang, and H. Y. Wang. 1986. In *Separation recovery and purification in biotechnology: Recent advances and mathematical modeling,* edited by J. A. Asenjo, and J. Hong. Chicago, IL: American Chemical Society Symposium Series 314.

Kennedy, L. A., W. Kopaciewicz, and F. E. Regnier. 1986. *J. Chromatog.* 359:73–84.

Kenney, A. C. 1990. In *Separation processes in biotechnology,* edited by J. A. Asenjo. New York: Marcel Decker.

Klausner, A. 1993. *Bio/Technology* 11:S35–37.

Kopaciewicz, W., M. A. Rounds, J. Faushaugh, and F. E. Regnier. 1983. *J. Chromatog.* 266:3–21.

Kroeff, E. P., and R. E. Chance. 1982. In *Hormone drugs; Proceeding of the FDA–USP workshop on drug and reference standards for insulins, somatropins, and thyroid-axis hormones,* edited by J. L. Gueriguian, D. D. Bransome, and A. S. Outschoorn. Rockville, MD: United States Pharmacopeial Convention, Inc.

Kroeff, E. P., R. A. Owens, E. L. Campbell, R. D. Johnson, and H. I. Marks. 1989. *J. Chromatog.* 461:45–61.

Krueger, J. K., M. H. Kulke, C. Schutt, and J. Stock. 1989. *BioPharm.* 2:40–45.

Kumagaye, K. Y., M. Takai, N. Chino, T. Kimura, and S. Skakibara. 1985. *J. Chromatog.* 327:327–332.

Kusano, H., E. Miyata, H. Takayanagi, and T. Itagaki. 1990. In *High performance liquid chromatography in biotechnology,* edited by W. S. Hancock. New York: J. Wiley and Sons, Inc.

Ladisch, M. R., and K. L. Kohlmann. 1992. *Biotechnol. Progress* 8:469–478.

Lilly Research Laboratories. 1980. In *Diabetes mellitus,* 8th ed. (1st rev.), edited by S. O. Waife. Indianapolis: Eli Lilly and Co.

Lloyd, L. L., and F. P. Warner. 1990. *J. Chromatog.* 512:365–376.

Lo, D. 1990. In *High performance liquid chromatography in biotechnology,* edited by W. S. Hancock. New York: J. Wiley and Sons, Inc.

Marston, F. A. O. 1986. *Biochem J.* 240:1–12.

Mitraki, A., and J. King. 1989. *Bio/Technology* 7:690–697.

Nau, D. R. 1990. In *High performance liquid chromatography in biotechnology,* edited by W. S. Hancock. New York: J. Wiley and Sons, Inc.

Nilsson, B., G. Forsberg, and M. Hartmanis. 1991. *Methods in enzymology,* Vol. 198. London: Academic Press, Inc.

Prokopakis, G. J., and J. A. Asenjo. 1990. In *Separation processes in biotechnology,* edited by J. A. Asenjo. New York: Marcel Dekker.

Prouty, W. F. 1991. In *Drug biotechnology regulation,* edited by Y. Y. Chu, and J. L. Gueriguian. New York: Marcel Dekker.

Rassi, Z. E., A. L. Lee, and C. Horvath. 1990. In *Separation processes in biotechnology,* edited by J. A. Asenjo. New York: Marcel Dekker.

Regnier, F. E. 1987. *Science* 238:319–323.

Roettger, B. F., J. A. Myers, M. R. Ladisch, and F. E. Regnier. 1990. In *Protein purification,* edited by M. R. Ladisch, R. C. Willson, C. C. Painton, and S. E. Builder. Washington, DC: American Chemical Society.

Ross, M. J. 1981. In *Insulins, growth hormone, and recombinant DNA technology,* edited by J. L. Gueriguian. New York: Raven Press.

Ross, J. W., R. S. Baker, C. S. Hooker, I. S. Johnson, J. R. Schmidtke, and W. C. Smith. 1982. In *Hormone drugs; Proceeding of the FDA–USP workshop on drug and reference standards for insulins, somatropins, and thyroid-axis hormones,* edited by J. L. Gueriguian, D. D. Bransome, and A. S. Outschoorn. Rockville, MD: United States Pharmacopeial Convention, Inc.

Salzman, R. T., S. Fulton, N. Gordon, M. Meys, and R. Carberry. 1993. *Amer. Biotechnol. Lab.* 11:40–44.

Samanta, A., G. A. Cordis, M. R. Prasad, and D. K. Das. 1988. *J. Chromatog.* 459:221–228.

Sassenfeld, H. M., and S. J. Brewer. 1984. *Bio/Technology* 2:76–81.

Singh, O. M. P., D. S. Baines, R. M. Hall, N. M. Gray, and W. P. Weir. 1991. *J. Biotechnol.* 21:127–136.

Smith, M. C., J. A. Cook, T. C. Furman, and J. L. Occolowitz. 1989. *J. Biol. Chem.* 264:9314–9321.

Smith, M. C., J. A. Cook, T. C. Furman, P. D. Gesellchen, D. P. Smith, and H. Hsiung. 1990. In *Protein purification*, edited by M. R. Ladisch, R. C. Willson, C. C. Painton, and S. E. Builder. Washington, DC: American Chemical Society.

Snyder, L. R., G. B. Cox, and P. E. Antle. 1988. *J. Chromatog.* 444:303–324.

Stadalius, M. A., M. A. Quarry, and L. R. Snyder. *J. Chromatog.* 327:93–113.

Stout, R. W., and J. J. DeStefano. 1985. *J. Chromatog.* 326:63–78.

Verzele, M., M. DeConinck, J. Vindevogel, and C. Dewaele. 1988. *J. Chromatog.* 450:47–69.

Wang, N. L. 1990. In *Separation processes in biotechnology*, edited by J. A. Asenjo. New York: Marcel Dekker.

Warren, W., N. E. Astephen, and T. E. Wheat. 1990. *J. Chromatog.* 512:13–22.

7

Affinity Chromatography

Wayne P. Olson

Markets for biotech pharmaceuticals likely will continue to grow at 15 percent per annum over the next decade; the 1993 gross for these products was about $3.7 billion. In 1993 erythropoietin (approved in 1989) grossed $825 million, primarily for Amgen, and insulin (approved in 1982) grossed $350 million, primarily for Lilly. Baxter-Hyland's recombinant human clotting factor VIII (F-VIII) received regulatory approval in 1992 and, at this writing, sales figures are unavailable. Of these, only the F-VIII is purified by affinity methods, specifically immunoaffinity. Affinity methods are preferred for process because they reduce the number of steps in product purification, often reduce costs, and almost always improve yield.

DEFINITIONS

Affinity chromatography usually is defined as the purification of a molecule based on its specific retention by a second, immobilized molecule. However, there are semantic problems with this definition. Affinity refers to the tendency of two things to associate; specificity is implied, meaning that one type of entity can associate only with a severely limited class of other entities. Specificity is not an absolute. For example, an antibody (Ab) often appears to be specific for a particular protein in a mixture of proteins but may cross-react with a protein from the same or an entirely different source. Protein A of some strains of *Staphylococcus aureus* binds some immune globulins but not others and is nonspecific as to the mammalian source of the immune globulins. Particular Abs, usually monoclonal Abs (MAbs), and Protein A, when immobilized, are considered sufficiently specific sorbents to qualify as matrices for affinity chromatography. Some affine ligands and their specificities are detailed in Table 7.1.

Specificity usually is defined in terms of the dissociation constant, K_d, in this way (Scatchard 1949):

$$\underset{\text{(bound ligand)}}{\mathrm{L}} \quad + \quad \underset{\text{(free protein)}}{\mathrm{P}} \quad \underset{k_{-1}}{\overset{k_1}{\leftrightarrow}} \quad \underset{\text{(ligand-bound protein)}}{\mathrm{L{-}P}}$$

and at equilibrium,

$$K_d = \frac{k_{-1}}{k_1} = \frac{[\mathrm{L}][\mathrm{P}]}{[\mathrm{L-P}]}$$

The K_d is in units of m*M* for nonspecific interactions and in units of μ*M* and even p*M* for specific interactions. Therefore, as a rule of thumb, interactions with K_ds of the order of 10^{-3} or 10^{-4} are considered nonspecific, whereas interactions with K_ds of the order of 10^{-6} and lower are considered specific. The K_d is of practical consequence; if the value is not sufficiently low, batch affinity adsorption-elution methods cannot be used because too much of the desired protein may be lost during rinsing (Graves and Wu 1974). However, even a low-affinity system can be useful (e.g., recycling filtration, wherein the peak of interest is recycled and all other eluate is run to drain).

Many proteins are capable of binding to a variety of ligands. Therefore, several affinity approaches to purification may be successful (Table 7.2).

Usually, whether an interaction is considered specific or not is determined somewhat arbitrarily by publishing investigators. For example, early investigators referred to hydrophobic interaction chromatography as a type of affinity chromatography; semantically, that is true. In an aqueous system hydrophobic moieties *do* have an affinity for one another, but aside from their hydrophobic nature may have no specificity for one another (i.e., the K_d is too high). Currently, hydrophobic interaction is placed in a class by itself, like ion exchange or size exclusion.

Affinity separations may be based on the interactions of peptides with proteins, saccharides with proteins, proteins with proteins, and

Table 7.1. Affinity Ligands[a]

Ligand	Specificity and Applications
Acriflavin	Plasmids, oligonucleotides
Dyes	
Cibacron Blue F3G-A	Albumin, lipoproteins, kinases, dehydrogenases
Procion Red HE-3B	Dehydrogenases, reductases
Heparin	Polyanion that reacts with coagulation proteins in the clotting cascade; lipoproteins; lipases; steroid receptors; restriction endonucleases
Imidodiacetic acid	Metal chelator; the proteins and nucleotides that are bound are characteristic of the chelated metal (e.g., Ca^{2+}, Fe^{3+}, Ni^{2+}, Zn^{2+})
Lectins	
Concanavalin A	Polysaccharides, glycoproteins, and glycolipids with alpha-D-mannosyl or alpha-D-glucosyl or similar residues
Lentil lectin	Glycoproteins with alpha-D-glucosyl or alpha-D-mannosyl residues
Wheat germ lectin	Polysaccharides and glycoproteins with *N*-acetyl-beta-D-glucosaminyl residues
Helix pomatia lectin	Glycoproteins and lymphocytes with *N*-acetyl-alpha-D-glalactosaminyl residues
Protein A	Immunoglobulins
Protein G	Immunoglobulins

[a]A few of the affine ligands that are available commercially, arranged by application.

nucleic acids with proteins or other nucleic acids. Metals also show a degree of affinity and specificity for some proteins (e.g., Ni^{2+} for proteins with histidine), and some organic species of relatively high

Table 7.2. Affinity Methods for the Purification of Fibroblast and Leukocyte Interferons

Immobilized Ligand	Reference
Albumin	Huang et al. (1974)
Dye (Cibacron Blue F3G-A)	Jankowski et al. (1976)
Imidodiacetic acid + Zn^{2+}	Edy et al. (1977)
Immunosorbent (PAb)[a]	Berg et al. (1978)
Immunosorbent (MAb)[b]	Secher and Burke (1980)
Lectin (concanavalin A)	Davey et al. (1978)
Poly(U)	DeMaeyer-Guignard et al. (1978)

[a]Polyclonal antibodies
[b]Monoclonal antibodies

molecular weight, other than proteins or peptides, may have a degree of fit for one another. In the main, affinity interactions are restricted to macromolecules of biologic origin.

ORIGINS OF AFFINITY SEPARATIONS

In 1936 Landsteiner (a name known well to immunologists) coupled diazotized haptens to blood cell stroma (a hapten is that portion of an antibody that elicits an immune response). The stroma-hapten-antibody complex was particulate and readily removed from suspension. The complex served as a source of antibody. In 1949 Michael and Evers covalently bound physiologically active protein to carboxymethyl cellulose. Campbell et al. (1951) immobilized ovalbumin to diazotized 4-aminobenzyl cellulose, equilibrated the substituted cellulose with rabbit antiserum, washed the matrix to remove unreacted proteins, and eluted specific polyclonal antibodies with a glycine-HCl buffer. Since that time, many other types of interactions have been exploited. However, the Landsteiner and the Campbell works were largely ignored, probably because most immunologists at that time were microbiologists and somewhat unsophisticated chemically.

Furthermore, Campbell's work had been published in a journal that, although prestigious, was not read by the average biologist. Consequently, affinity separations did not catch the interest of the biological community until (1) the need for pure proteins became compelling, and (2) a very simple, cookbook method had become available with commercially prepared matrices. That occurred when Jerker Porath and colleagues invented cyanogen bromide activation of polysaccharides (Axen et al. 1967). The invention was licensed to Pharmacia AB (Uppsala, Sweden); Pharmacia popularized the activation procedure as well as the beaded agarose matrix used by most investigators and most industrial-scale users. Affinity methods became widespread, even among the chemically inept, with the publication of a useful review (Cuatrecasas and Anfinsen 1971) and the manufacture and distribution of a relatively stable, dried, CNBr-activated beaded agarose matrix by Pharmacia.

AFFINITY PAIRS: EXAMPLES

Antigen-Antibody

The use of immobilized antibodies (Abs) as immunosorbents is reviewed elsewhere (chapter 9). As a rule-of-thumb, Ag-antibody (Ab) interactions are of high specificity.

Protein A/Protein G-Antibody

Protein A-Sepharose commonly is used in the preparation of monoclonal Abs. Andrew and Titus (1992) have published basic protocols for the purification of murine MAbs from mouse ascites. Francis et al. (1990) automated such a column and found that, over 25 cycles, there was no significant deterioration in column efficiency. However, a considerable bleed—approximately first order—was shown, with a minimum of 3–4 ppm Protein A leaching from the column over protracted use. Such residues are intolerable in an injectable product and must be removed. A small amount of Protein A might be tolerable in a diagnostic.

Protein G binds more IgGs than Protein A (Schneider et al. 1982), hence is in widespread use as a specific sorbent for the purification of IgGs. However, Protein G also might immobilize a MAb by the Fc fragment and form the basis of an immunosorbent. Pierce Chemical Co. (Rockford, IL) proposes the covalent cross-linking of Protein G to the IgG of interest so as to provide a more stable immunosorbent. The proposed cross-linker is dimethyl pimelimidate-2 HCl (DMP).

Lectin-Glycoproteins and Saccharides

Lectins are proteins, from either plants or animals, that react with particular sugars in oligosaccharides or polysaccharides or glycoproteins. Seven of the lectins in use are concanavalin A, erythroagglutinating phytohemagglutinin (E_4-PHA), lentil lectin, pea lectin, leukoagglutinating phytohemagglutinin (L_4-PHA), *Datura stramonium* agglutinin (DSA) and *Ricinus communis* agglutinin I (RCA I). These have been used serially in columns for the fractionation of *N*-linked oligosaccharides (Merkle and Cummings 1987). Specificities for some of the lectins are detailed in Table 7.3.

Animal cell glycoproteins may have asparagine-linked (*N*-linked) or serine/threonine-linked (*O*-linked) oligosaccharides or a mixture of both. *N*-linked oligosaccharides are termed *complex*, and those of higher molecular weight usually are branched.

Many of the common allergens are glycoproteins and lectins have been used in their purification [e.g., pollen (Corbi and Carreira 1985; Haavik et al. 1985; Karlstam and Nilsson 1982)], cow's hair and dander (Bog-Hansen et al. 1978), and mugwort pollen (Nilsen et al. 1990). Purified allergens are seldom used in the practice of allergy; there are virtually no standards for allergens and, for practical purposes, the production of allergens as injections for human use is barely regulated in the U.S. However, some large practices of allergy purify their own materials (e.g., Isadore Pitesky, M.D., in Long Beach, CA).

The lectin most commonly used is concanavalin A, immobilized on beaded agarose (e.g., Con A Sepharose). When Nilsen et al. (1990) loaded Con A Sepharose with mugwort extract and washed with tris-buffer, tris-NaCl, then tris-buffer-ethylene glycol, elution with alpha-methyl glucoside followed by alpha-methyl mannoside resulted in the elution of mugwort Ag7 allergen with each of the sugars. The allergen in each of the two fractions was contaminated with one other major glycoprotein. Yields for the single-step affinity purification were slightly better than 50 percent.

Recombinant human clotting factor VIII (rF-VIII) has been recovered from the spent tissue culture fluid of F-VIII-recombinant mammalian cells (Tsay 1991). Immobilized lentil lectin sorbs rF-VIII and, after washing of the resin, the F-VIII can be recovered in 0.25 *M* alpha-methyl mannoside. The concentrate of F-VIII, contaminated with some other glycosylated proteins, then was further purified on a column containing immobilized anti-F-VIII MAbs.

Lentil lectin affinity chromatography, followed by MAb immunosorbent chromatography, has been used in the "large-scale" production of Vaccinia recombinant-derived HIV-1 gp160 (Barrett et al. 1989). The immunosorbent procedure was followed with a

Table 7.3. Applications of Lectins in the Separation of Glycoproteins and Cells

Glycoprotein or Cell Type[a]	Immobilized Lectin	Eluting Sugar	Reference
Glycophorin (erythrocytes)	WGA[b]	100 m*M* *N*-acetylglucosamine	Adair and Kornfeld (1974)
Thy-1-antigen (rat thymocytes)	LCA[c]	500 m*M* methyl-alpha-D-glucopyranoside	Letarte-Muirhead et al. (1975)
Rhodopsin (bovine retina)	Con A[d]	100 m*M* glucose	Steinemann and Stryer (1973)
Membrane glycoprotein (platelets)	LCA	2% methyl-alpha-mannopyranoside	Clemetson et al. (1977)
Bone marrow cells (stem cells)	WGA	*N*-acetyl-D-glucosamine	Nicola et al. (1978)
T cells (2 subpopulations)	WGA	*N*-acetyl-D-glucosamine	Hellstrom et al. (1976a)
T cells (neuraminidase-treated)	HPA[e]	*N*-acetyl-D-galactosamine	Hellstrom et al. (1976b)

[a]Unless otherwise noted, tissue is of human origin
[b]Wheat germ agglutinin
[c]*Lens culinarus* agglutinin
[d]Concanavalin A
[e]*Helix pomatia* agglutinin

second lentil lectin column step and the overall yield was 11 percent (50 mg).

Wheat germ lectin is specific for *N*-acetylglucosamines, including *N,N′,N″*-triacetyl-chitotriose, and likely can be used in the preparation of some bacterial cell-wall polymers (Lotan et al. 1975). Considering the utility of bacterial cell wall products as immunostimulants (Olson 1992), there is cause to consider immobilized lectins for preparative-scale procedures.

Mannan-binding protein (MBP) is a mannose and *N*-acetylglucosamine-specific lectin of mammalian plasma and, structurally, is homologous to C1q (Ohta et al. 1990). Although immobilized C1q binds IgM (Nethery et al. 1990), it also binds some IgG, whereas MBP binds IgM only, and not by the Fab and $F(ab')_2$ fragments, but by the Fc only. The 1993 immunoreagents catalogue of Pierce Chemical Co. (Rockford, IL) presents a nine-step method for the purification of IgM on immobilized MBP in 88-90+ percent purity in a single step, depending on the source.

An invaluable resource on lectins is Ginsburg (1987). A partial listing of the applications of lectins in the separation of glycoproteins and cells is given in Table 7.3.

Enzyme-Inhibitor

Inhibitors of enzymes usually are considered as competitive (reacting with the active site directly) or noncompetitive (reacting with a domain of the enzyme other than the active site [e.g., allosteric inhibitors]). The two types can be differentiated on the basis of the kinetics of the inhibition (White et al. 1973). An interesting example is that of diisopropylfluorophosphate (DFP) and related species, which are powerful competitive inhibitors of acetylcholinesterase (AChE) and other esterases with a serine hydroxyl in the active site. These organophosphorus AChE inhibitors react covalently with that serine hydroxyl (O'Brien 1960). Let us consider paraoxon and the reaction of it with AChE. If one of the ethoxy groups is substituted so as to act as a coupler to an insoluble matrix, then AChE (or another esterase with a serine hydroxyl in the active site) is covalently bound to the matrix when the enzyme reacts with the immobilized paraoxon analogue. One would recover the matrix and, with it, the inhibited enzyme. This is an extreme example in that the enzyme is bound covalently to the inhibitor. More commonly, a competitive inhibitor will bind reversibly to the active site, in which instance a high concentration of substrate (or of the inhibitor in solution) may push the reaction to the left (see, for example, Flanagan and Barondes 1975). Whereas I see no therapeutic utility to AChE at this time, a more practical extension would be the removal of proteases from a solution containing a desired protein (e.g., human clotting factor VIII) onto ligand-immobilized benzamidine. The advantages of this approach would be (1) fewer hydrolytic products of the F-VIII in the purified product and (2) a modest, but likely significant, reduction in protein contaminants.

Cytokine-Receptor

A form of purification leapfrog has been practiced from time to time, whereby a partial purification of one component of a biological pairing leads to immobilization of it, and a high degree of purification of the second. For example, interleukin-2 (IL-2) activates macrophages and stimulates tumor-infiltrating leucocytes (TILs) (Baars et al. 1992). Interleukin-2 has therapeutic value (Allison et al. 1989; Parkinson et al. 1990; Rosenberg et al. 1987; West 1989; Whitehead et al. 1991; Whiteside et al. 1993) and has been cloned (Cetus-Chiron Corp., Emeryville, CA) as an IL-2 receptor (Cosman et al. 1984; Leonard et al. 1984; Nikaido et al. 1984). Presumably, a crude preparation of the receptor could be coupled to a solid-phase support and used in a single-step purification of the cytokine. If the human IL-2 receptor were sufficiently pure, there should be modest concern for the inclusion of small amounts of receptor or a portion of the receptor in the product. However, this strategy for preparing IL-2 (or any other cytokine [e.g., TNF]) for injections requires some additional thought. The IL-2 receptor requires cloning, purification, and immobilization strategies that may prove more involved, time-consuming, and expensive than a MAb-based immunosorbent purification. The receptor-based approach is commercially unlikely, in that MAb methods are fairly well known and well characterized, and high-visibility MAb-purified products are in the marketplace (e.g., human F-VIII), hence a similar approach is most likely of regulatory and market acceptance.

An *enriched* cytokine *receptor*, produced by partial purification (for example, to 90 percent purity) is intolerable as an injectable, but may be adequate for use as a diagnostic (i.e., horseradish peroxidase-coupled IL-2 receptor for an enzyme-linked receptor assay [ELRA], comparable to enzyme-linked immunosorbent assays [ELISAs]).

Generally, for cytokines that are overexpressed in, for example, *E. coli*, there is no advantage to purification on the corresponding receptor. Downstream of the recovery of inclusion bodies by centrifugation, ion-exchange methods (using at least one strong exchanger) usually are suitable for producing a product that is homogeneous and immunologically pure.

Heparin Sulfate-Clotting Factors and HS-Antithrombin

Heparin (= heparin sulfate or HS) is a highly sulfated glycosaminoglycan and is an anticoagulant that acts by selectively binding to the plasma protease inhibitor, antithrombin III (AT-III). Heparin binding increases the normal rate at which AT-III inhibits thrombin. The binding sites on AT-III for clotting enzymes (apart from thrombin) and for heparin are separate. Soluble heparin is available as an anticoagulant as is Sepharose-immobilized heparin (Pharmacia) and Heparin Hyper D™ (Sepracor). When plasma is passed through a bed of immobilized heparin, AT-III is retained and, following a saline rinse of the column, AT-III can be eluted at low ionic strength with soluble heparin (Jordan et al. 1982). Hoffman (1989) developed large-scale methods.

There is a market for AT-III for replacement therapy in patients with a congenital deficiency (Menache et al. 1992; Vinazzer 1987), probably an orphan drug-size market (less than 200,000 U.S. patients), and renewed interest in the product and the process (Lebing et al. 1994). The Lebing group from Miles, Inc. (nee Cutter Laboratories, Clayton, NC) found that the Heparin Hyper D™ resin (Sepracor) has higher capacity, higher flow rates, and lower operating pressures than heparin-agarose. As a consequence, they were able to process 1,800 ℓ of Cohn Fraction IV-1 paste (21 g/ℓ) through a 40 ℓ column in 5 hours. Methods were as follows:

> *The paste was in 0.005* M *Tris buffer and 0.11* M *NaCl. The solution was applied at 20°C and a linear flow rate of 225 cm/hour into a 45 cm diameter × 25 cm high column. The Heparin Hyper D™ resin had been equilibrated with 0.01* M *Tris, pH 7.8 at 5°C. Loosely bound protein was eluted with 0.02 M Tris, 0.30* M *NaCl, pH 7.5 at 20°C and a linear flow rate of 225 cm/hour. Elution of the AT-III was with 0.02* M *Tris containing 2.0* M *NaCl (pH 7.5) at a linear flow rate of 75 cm/hour.*

Elution with 2.0 *M* NaCl was a significant improvement over the soluble heparin used by Jordan et al. (1982), and the more rapid processing made possible by the new resin should speed the process and cut costs further. The authors observed that (based on bench experiments) column life should exceed 80 cycles without loss of capacity or resolution.

Clotting factors that have been purified on immobilized heparin include AT-III (Miller-Andersson et al. 1974), factor VII (Kisiel and Davie 1975), factor IX (Andersson et al. 1975), factor XI (Kurachi and Davie 1977), factors XII and XIIa (Fujikawa et al. 1977), and thrombin. However, HS is a polyanion and consequently interacts with many cationic biologics or with cationic residues on macromolecules. For example, immobilized HS also has been used in the preparation of low-density and very-low-density lipoproteins from human plasma (Iverius 1972), as well as lipoprotein lipase (Greten and Walter 1973) and many other proteins.

Because HS is an anticoagulant, it is appropriate that any HS leaching into the product be removed or neutralized. Therapeutically, protamine is administered, following surgery, as an HS antagonist. Therefore, immobilized protamine, downstream of immobilized HS, should act as a scavenger for any HS eluted with product (see Hou et al. 1990).

Hen's egg lysozyme is useful in the lysing of native or recombinant bacteria; it often is useful in the recovery of bacterial products. (The preference of biochemical engineers is to lyse cells by homogenization in a Manton-Gaulin or in a French Press; however, some cells [e.g., *Micrococcus lysodeikticus*], are difficult to lyse at moderate pressures and so are lysed enzymically or with an homogenizer following enzyme treatment.) Lysozyme is competitively inhibited by heparin (Kerby and Eadie 1953), which likely interacts with the active site of the enzyme. However, the interaction is primarily ionic as has been shown in solution and on a column of heparin immobilized on Ultrogel A4R (Boschetti et al. 1981). The NaCl concentration required to elute the enzyme from the column was inversely proportional to the pH (i.e., 0.6 *M* at pH 4 and 0.3 *M* at pH 7) and with a linear relationship between. However, heparin is anionic; hence it is a most effective ion exchanger at alkaline pHs. For proteins in the egg white other than lysozyme, the ion-exchange properties of heparin are important. Consequently, the most efficient method for purification of the enzyme is as follows:

> *A salt gradient is applied for elution in 0.01* M *phosphate buffer, pH 7.4 at 22°C; proteinaceous contaminants elute in the void volume and 84.0 percent pure lysozyme elutes at 0.25* M *NaCl.*

The purification factor was 20-fold and the specific activity essentially identical with that of the crystalline enzyme. It should be possible to prepare lysozyme in this single-step manner from human urine, saliva, placenta, maternal milk and leukocytes, but I see no particular market or use for them other than for research or in the preparation of recombinant products. A possible exception is the use of lysozyme in toothpastes if the enzyme is stable within such emulsions and is highly active against the organisms responsible for dental caries. Incidentally, because lysozyme reacts with microbial cell walls, Yoshimoto and Tsuru (1974) purified the enzyme on an agarose-supported cell-wall hydrolysate of *Micrococcus luteus*.

Gelatin-Fibronectin

Human plasma fibronectin (pFN) is found in human plasma at a concentration of about 300 mg/ℓ and 20 to 100 variants are present in tissues as cellular FNs (cFNs) (Goodheart and Silverman 1991). The most critical function of the cFNs is to act as part of the pericellular glue that holds cells in position on the collagen matrix (Hedman and Vaheri 1989). The spreading of mammalian cells seems to depend on interactions between cytoskeletal elements and the FN of the pericellular matrix (Singer 1989). Natural killer (NK) cells spread on cFN or pFN and fragments of these. Platelets contain FN that, together with fibrinogen, thrombospondin, and von Willibrand's factor (vWF), hold in place the platelet-cellular-thrombin plug that seals wounds in vessel walls (Ginsberg and Plow 1989). Despite these functions, pFN has not found use as a therapeutic entity, although it is readily prepared in volume. Because FN binds to collagens, it also binds to partially denatured collagens (i.e., gelatins). The preparation of pFN from plasma is straightforward. The following is a practical protocol:

> *Add 500 mg benzamidine/liter plasma and add this to a column of gelatin-agarose in the proportion of 1.25 ℓ plasma/liter affinity gel. The addition rate is 10 to 15 cm/hr. The gel is washed with 0.05* M *tris-HCl buffer, pH 7.5, followed by the same buffer containing 4* M *NaCl, followed by 0.1 M urea in the buffer. The FN then is eluted with the same buffer containing 4* M *urea, but at a linear flow rate of 2.5 to 4 cm/hr.*

Janson (1982) has provided some details of scaleup.

Chelated Metal-Protein

Immobilized metal affinity chromatography (IMAC) was introduced by Porath et al. (1975). Chelation usually is by iminodiacetic acid coupled to polysaccharide or silica supports (Porath et al. 1975; El-Rassi and Horvath 1986). A method for the preparation of F-VIII on a metal affinity column is as follows:

> *Biscarboxymethylamino (= iminodiacetic acid) Sepharose 4B, bearing Zn^{2+} and equilibrated in 20 m*M *Mg^{2+}, 20 m*M *Tris, 2.5 m*M *$CaCl_2$ (pH 7.4), is loaded with sample. The column is developed with an imidazole gradient from 0 to 50 m*M*. Factor VIII and vWF are eluted at about 15 m*M *imidazole.*

An attempt at an IMAC purification of F-VIII on a Ca^{2+} IMAC, achieved 85-fold purification in a single step from tissue culture medium containing F-VIII (Mantovaara et al. 1991). This is inadequate for the purification of rhF-VIII but may prove satisfactory as a purification method for single donor or solvent-detergent treated plasma-sourced F-VIII. The chelating resin for this work was carboxymethylated aspartic acid agarose (Mantovaara-Jonsson et al. 1989). This mechanism should be compared with the Scripps Clinic immunoaffinity method for the purification of F-VIII, wherein vWF is trapped on a column by anti-vWF Abs, F-VIII sorbs noncovalently to the vWF and is desorbed with Ca^{2+}.

Although the cost of IMAC is low and selectivity is "low-medium," ligand stability and protein loading are high and elution conditions mild (Arnold 1991). Side chains of histidine dominate protein binding to Ni^{2+}, Zn^{2+}, Co^{2+}, amd Cu^{2+} (Sulkowski 1989). Ferric ion reacts strongly with phosphoproteins and phosphoserine (Chaga et al. 1989). The most powerful potential use of the technique may be with recombinant proteins using histidine peptide fusions (Ljungquist et al. 1989; Smith et al. 1987). Histidine is relatively rare (about 2 percent of the amino acids in globular proteins) and a 100-residue protein may have only one histidine on its surface (Klapper 1977), therefore a histidine-focused separation is relatively selective. Reduced cysteine binds equally as strongly as histidine (Arnold 1991), but reducing conditions may induce denaturation of the protein.

Histidine fusions may not interfere with protein function but the immunogenicity, although low, may be sufficient to cause problems. It would be imperative that the hexahistidine be removed from an injectable product. Quite the opposite might obtain for a diagnostic. A Zn^{2+}-chelate support (e.g., polystyrene latex or other beads) would immobilize the enzyme or the antibody for ELISA reagents. The purified fusion protein could be recovered directly from the column and, following a buffer change, reacted with the chelate-modified beads.

Iminodiacetate-substituted resins are available as off-the-shelf items from Pharmacia-LKB (Bromma, Sweden), Pierce Chemical (Rockford, IL), Sigma Chemical (St. Louis, MO), and Behringer-Mannheim (Mannheim, Germany). Other chelating groups, with references, are cited in the review by Yip and Hutchins (1992).

Dye-Protein with a Complementary Cleft

Cibacron Blue F3G-A (Figure 7.1) and analogues are capable of acting as biomimetic ligands (Lowe et al. 1992). Downstream of a DEAE-Trisacryl column for the removal of immune globulins (Igs) by ion-exchange chromatography, Cibacron Blue F3G-A-substituted trisacryl sorbs human serum albumin with a high degree of specificity (Saint-Blancard et al. 1982). A system somewhat like this is in use in France for the preparation of human albumin from thawed and macerated frozen placentas (France does not permit the selling of blood; and professional donors constitute a significant proportion of the plasma source in the U.S. and elsewhere). Since all HSA is pasteurized at 60°C for 10 hr so as to destroy any and all viruses or vegetative bacterial cells that may be present in the final container (Olson 1987b), this material does not constitute a viral hazard to the recipient.

In a nonkinetic study of immobilized Cibacron Blue as a biospecific sorbent, Easterday and Easterday (1974) found that kinases and

Figure 7.1. Cibacron Blue F3G-A.

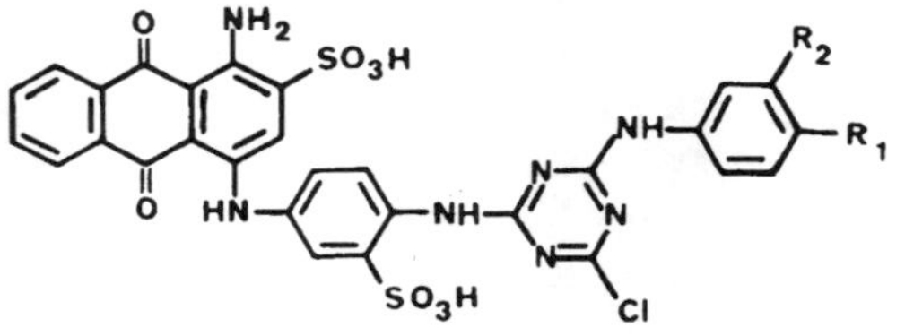

CIBACRON BLUE F3GA
R_1 - H R_2 - SO_3H [m IIIA] R_1 · SO_3H R_2 · H [p IIIA]

dehydrogenases could be purified on such matrices. Studies of the binding of a variety of derivatives of Cibacron Blue with horse liver alcohol dehydrogenase (ADH), reviewed in Lowe et al. (1992), indicated that the *m*-COO^- analogue had the highest affinity for ADH of the analogues tested, with a K_d of 60 n*M*. The authors indicated that the dye binds in the NAD^+-binding site of the enzyme.

Cibacron Blue is one of the Procion dyes that contain an epoxy group for coupling to a matrix bearing hydroxyl groups. Such coupling is very stable; ligand leaching from the column is very small and renders the matrix amenable to cleaning with a wide variety of agents, and steam sterilizable. Columns of such material often can be validated for use over a large number of cycles.

Such dye-ligand supports often are far cheaper than immunosorbent columns, can be synthesized and also cleaned under rigorous conditions, and, in some instances, can replace two or three ion-exchange steps in the purification of peptides and proteins (Hodgson 1992). Amicon (Lexington, MA) offers a variety of dye matrices.

Polymyxin B-Endotoxin for Depyrogenation

Ideally, endotoxin (lipopolysaccharide or LPS) from Gram-negative organisms is not present in an injectable drug during process. In practice, some level of LPS *is* present and prophylactic removal of LPS from simple solutions (e.g., antibiotics) can be done by ultrafiltration with a 10 kDa cutoff membrane. Water and small molecules pass through the ultrafilter but endotoxin is retained. Small peptides (e.g., human chorionic gonadotropin) can be depyrogenated in the same way but larger proteins cannot because both the LPS and the proteins are retained by the membrane.

One method of depyrogenation of proteins in solution is by exposing the solution to immobilized polymyxin B (PMB). Zhang et al.

(1992) show that a 5 ml column of PMB-Sepharose 4B can bind 450 μg of LPS, a huge amount! But most interesting is the finding that blood serum and ascitic fluid can be depyrogenated without the loss onto the PMB of important proteins. The implication is that PMB agarose or a PMB filter might be useful prophylactically in manufacturing any solution, a component of which does not sorb to the immobilized PMB. The Chinese workers also found that PMB-Sepharose could be regenerated to 85 percent of the initial activity by extensive washing with deoxycholate and SDS. Where LPS is a potential problem in manufacturing, immobilized PMB may be of interest.

Covalent Disulfide Chromatography with SH-Proteins

Many peptides and proteins contain thiol groups (e.g., the disulfide bridges in Abs). These can be reduced, and disulfide bonds re-formed with an immobilized thiol ligand. Such re-formed disulfide bonds can be broken after nonbinding proteins are washed from the matrix (Brocklehurst et al. 1973). Half-antibodies can be immobilized in this way.

Thiophilic Adsorption of Immunoglobulins

Another of Jerker Porath's methods is thiophilic adsorption (Porath et al. 1985), wherein a ligand containing both a thioether (R-S-R′) and a sulfone (R-SO_2-R′) is the matrix to which particular proteins sorb at elevated salt levels. This is a special case of hydrophobic interaction chromatography (HIC). In practice, this has become a method for the depletion of immunoglobulins from calf serum (Hutchins et al. 1990; Hutchins and Porath 1987) and for the recovery of immunoglobulins. The method for the recovery of monoclonal Abs, our interest here, is rapid, gives an excellent yield, and is mild (Nopper et al. 1989).

Immunoglobulin recovery is as follows:

To the sample (ascites fluid, human serum, hybridoma cell-culture medium) add $(NH_4)_2SO_4$ solid to 10–12 percent (w/v) and adjust the pH to about 7.4 with dropwise 0.1 M *NaOH or HCl and constant stirring. Remove precipitated material by centrifugation. Load the sample at a linear velocity of 20 cm/hr, loading 20-25 ml sample/5 ml of gel in the column. Wash the column with 20 to 50 column volumes of buffer (10 percent ammonium sulfate, 0.01* M *HEPES, pH 7.4) until A_{280} of the column effluent is essentially zero. Elute the adsorbed Abs with 20 m*M *HEPES buffer, pH 7.4 containing 0.5* M *NaCl. Elute also with 50 percent ethylene glycol in*

20 mM HEPES, pH 7.4 to recover remaining Abs and recondition the column.

This is a simple, relatively rapid, relatively inexpensive method for recovering Igs.

Fusion Proteins with Metals, Glutathion, or Maltose

Purification fusions are sequences or regions of other proteins added to the protein of interest. Fusions may confer immunoaffinity, substrate affinity, metal chelation, or high charge density (Ford et al. 1991).

An hexahistidine-substituted protein is alluded to above (Affinity Pairs: Examples—Chelated Metal-Protein). The hexahistidine might be fused or covalently coupled to a protein. Fusion proteins are of considerable interest but have not as yet found their way into the biotech community as therapeutic products. An overview of some possibilities is in order. Over 150 proteins have been fused with hexahistidine for purification purposes (Hochuli and Piesecki 1992).

One of the issues can be the masking of the fusion site by the tertiary structure of the proteins, creating difficulties in cleaving off the fused protein. Hakes and Dixon (1992) may have found a way around this. They fused glutathione S-transferase (GST) to a model protein, with a glycine-rich "kinker" immediately after the thrombin cleavage site. Therefore, the protein of interest, fused to the GST, was retained on a glutathion-agarose column. After elution, thrombin *apparently* removed all amino acids *N*-terminal to the glutathion cleavage site.

Let us say that a recombinant mouse protein is fused to GST. Let us say that the GST is excised from the recombinant protein with thrombin and the *apparently* GST-free recombinant mouse protein is repeatedly injected into mice. If the recombinant protein is consistently nonimmunogenic in the mouse, GST fusion may merit consideration in the purification of some injectables. The GST approach might be cheaper than immunoaffinity purification, but requires a most rigorous risk assessment. Regardless, the method should be readily applicable to diagnostic reagents.

A maltose binding protein (MBP) has been fused to poliovirus 2C protein (Rodriguez and Carrasco 1995). MBP incorporates a clotting factor Xa cleavage site at the carboxy terminus. The site that Xa recognizes is Ile-Glu-Arg (Nagai and Thogersen 1987). The addition of 5 glycine residues adjacent to the cleavage site facilitates correct proteolysis and release of the target protein.

A clever approach that I believe to be unsuitable for the production of injectables, but suitable for diagnostics, is the use of DNA–binding proteins. My example is the *lac* repressor fused to, for example, an enzyme. If the *lac* operator of *Escherichia coli* is coupled to a solid support, it will bind a protein fused to the *lac* repressor. [The *lac* repressor is a protein that binds specifically and with high affinity and avidity to that part of the *E. coli* DNA that expresses the proteins required to take up and metabolize lactose. When lactose is present and binds to the *lac* repressor protein, the protein changes tertiary structure and leaves the *lac* operator. In the *E. coli* genome, when the operator is free, expression occurs and the proteins required for lactose utilization are expressed.] The *lac* repressor is released from the immobilized operator DNA by lactose, or a lactose analogue (Ljungquist et al. 1993).

Proof of the suitability of fusion systems for diagnostic use is given in Robben et al. (1993). The authors developed a plasmid for construction and high-level production of C-terminal chloramphenicol acetyltransferase (CAT) fusion proteins in *E. coli*. CAT was fused to a multiple-epitope fragment of the rubella virus E1 membrane protein. The expressed protein worked well in an IgG enzyme-linked immunosorbent assay. Other papers on fused proteins used in several ways are Beer et al. (1992), Ginjaar et al. (1992), Kiefer et al. (1992), Mohapatra (1992) and Veldhoven et al. (1992).

Other Affinity Pairs

An interesting possibility that seems not to have not been exploited is immobilized vWF as an affine ligand for the capture and purification of recombinant human clotting factor VIII. The advantage is that the inevitable bleed of vWF from the matrix into product would be far more innocuous than the bleed of a mouse monoclonal antibody. None of the manufacturers who currently purify F-VIII on immunosorbents are likely to consider the vWF approach since they have in use a purification system, and a product, that work.

MATRICES OF CHOICE

Agarose

The preponderance of industrial-scale affinity separations is with beaded agarose matrices. Beaded agaroses were pioneered by Pharmacia, which continues to be a primary source. However, beaded agarose now is supplied by French, Japanese, and other American firms (e.g., Bio-Rad, Richmond, CA). Sepracor (Marlborough, MA) specializes in the tailoring of systems for particular applications. Agarose, when isolated from agar, contains the uncharged polymer mixed with a highly sulfated agaropectin; the agaropectin must be removed prior to use. Residual sulfate groups usually are reduced with $NaBH_4$. The agarose beads maintain structure by H-bonding between adjacent triple strands. Because H-bonds are interdicted by urea, guanidine-HCl, and many detergents, and because the H-bonded beads are notoriously unresistant to compression, all of the beaded agaroses in industrial use for biotechnology are cross-linked.

The advantages of agarose are low, nonspecific adsorption to the matrix; high binding capacity throughout the matrix; uniformity; ready availability from several reliable sources; and ready acceptance by U.S. FDA and other regulatory bodies since many other firms use agarose-based systems.

Cellulose

Cellulose is relatively inexpensive and available in many grades and forms. Various derivatives are readily available or prepared (e.g., carboxycellulose, which is suitable for carbodiimide activation). For example, carboxymethyl cellulose (*not* the ester) can be washed in dilute acid and then reacted with a suitable water-soluble carbodiimide for subsequent coupling to a ligand bearing a primary amine. Unaltered cellulose has hydroxyl groups to which ligands can be coupled, like agarose, and is more tolerant of handling and pressure than agarose, but does not have the high porosity of agarose. Some beaded forms of cellulose are available, but most cellulose-based chromatographic matrices consist of irregular particles that offer more resistance to flow than beaded materials. All coupling is via hydroxyl groups.

Inogamov et al. (1991) provide detailed information on the preparation of a regenerated cellulose matrix.

> *The recovered cellulose is treated with 1.6 percent* $NaBH_4$ *in pyridine, followed by cross-linking with 0.01–0.05 percent (w/v) epichlorohydrin followed by 0.01–0.04 percent (w/v)* $NaBH_4$ *to improve mechanical strength of the matrix. Activation is with* $NaIO_4$ *prior to coupling to the protein for immobilization.*

The Inogamov paper is recommended to anyone planning to prepare matrices from scratch.

A particularly interesting alternative to packed columns of cellulose or cellulose derivatives are the radial-flow systems of Ken Hou and others and described in Huang et al. (1988). A comparison between axial and radial chromatography is shown in Figure 7.2, the structure of a radial-flow cartridge in Figure 7.3, and the schematic of a system in Figure 7.4. Cuno Inc. (Meriden, CT), the producer of these systems, opted out of this growth market several years ago. However, their new, *market*-oriented management, may renew the commitment

Figure 7.2. Comparison of axial- and radial-flow chromatography.

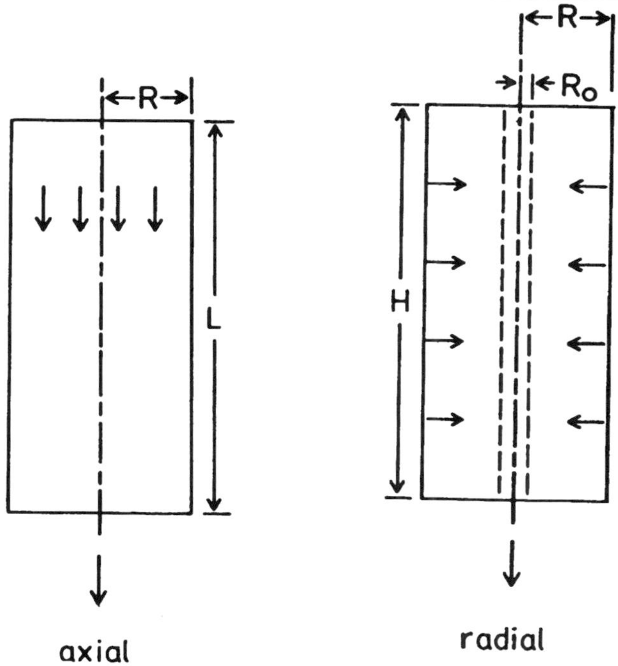

to product development and the support of products like Zeta-Prep. The issue is: If a company resumes a product line that has been dropped, will customers have the confidence to integrate these into process, knowing that they have once been eliminated? There is a lesson in this for every manufacturer of goods used in industry.

Acrylamide and Related Materials

Acrylate and methacrylate offer the possibilities of chemical or photochemical vinyl polymerization or copolymerization to produce cross-linked beads or graft copolymers. These beaded supports are resistant to protein and cell adhesion and upon breakdown yield nontoxic degradation products (Sawhney et al. 1993). Beaded polyacrylamide (Bio-Rad, Richmond, CA) tends to be more resistant to deformation than uncross-linked agarose. It also offers diversity in coupling (e.g., to the carboxyamide side chains that can be generated by alkaline hydrolysis, and to which amines [as on lysine] can be coupled using

Figure 7.3. Structure of a radial-flow cartridge.

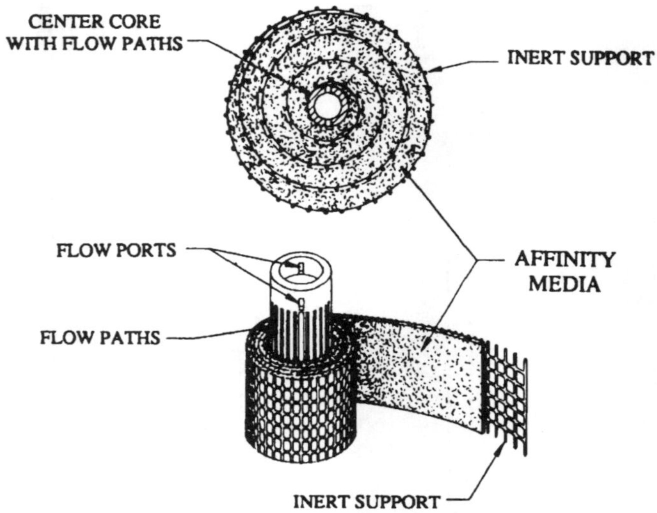

Figure 7.4. Schematic of a radial-flow purification system.

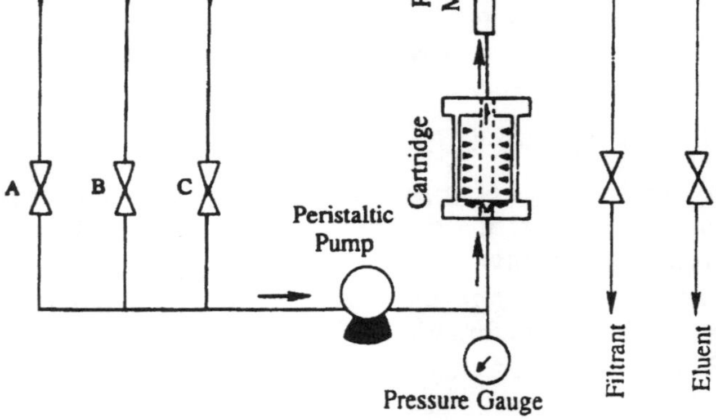

carbodimmides [Bauminger and Wilchek 1980]). However, acrylamide is more adsorptive of aromatics than agaroses.

A copolymer of hydroxyethyl methacrylate (HEMA) and ethylene glycol dimethacrylate (cross-linker) has good stability and pore-size range (Turkova et al. 1981). It is called poly(HEMA) and is relatively rigid; coupling is via hydroxyl groups, using the same chemistries as with agaroses.

Microporous Glass

The emergence of microporous glass as an affinity matrix coincided with the recruitment of Howard Weetall by the Corning Glass Works. Weetall had studied affinity systems at CalTech with Dan Campbell's group and was a collaborator of Ralph Messing at NASA (Pasadena, CA). Weetall was the founding editor of *Solid-Phase Biochemistry*. The advantage of the glass is its incompressiblility. However, glass tends to dissolve at pH > 8. On occasion, this has posed problems. If an alkaline solution containing soluble glass is taken to neutral pH and filled, over time the dissolved glass precipitates as fine particles. Also, glass often is used as a sorbent for proteins (e.g., fibronectin; Olson 1992); hence, particular care is required in the masking of adsorptive sites before or after the ligand has been coupled. I am unaware of any industrial-scale systems using microporous glass as a support matrix. However, a variety of chemistries are possible with glass (e.g., diazotization, carbodiimide activation, the formation of imidocarbonates, etc.). See, for example, Mattiasson et al. (1978) and Hatchikian and Monsan (1980).

Cellulose-Based Nonwoven Filters

Ken Hou and collaborators at Cuno, Inc. (Meriden, CT) developed a series of nonwoven cellulosic filter supports for affinity and ion-exchange chromatography (Zeta-Prep series). In the main, they were radial flow devices. A sheet of nonwoven cellulose and appropriate supports were wound around a porous central core and the device had the appearance of a conventional cartridge filter for particle removal; liquid flow was from the entire exterior to the interior. Coupling chemistries were conventional. There are many strengths to this system, not least of which is that handling of the matrix, plumbing of the filter system and the like, is simplified. Flow rates can be much higher than in column systems because the matrix does not compress. Resistance to flow is far less than in columns. However, although the diffusion distance was shorter than in most beaded matrices (since the ligand is on the surface of the fibers), some care had to be taken in determining the flow rate of the feedstream as a function of the efficiency of the removal of the desired protein from the feedstream. There are fewer theoretical plates in almost all radially operated nonwoven filter systems than in conventional beaded gel columns with a high length-to-diameter ratio. In part, this is because much of the mass of the cellulosic fibers is impermeable (which also is why the material is largely incompressible). The other issue is that product eluted from nonwoven cartridge filter units tends to be much more dilute than product eluted from columns. The mixing volume on the downstream side, and the relative proportion of the bed volume that is void, are far greater in the nonwoven-cartridge system than in a beaded gel column. Unfortunately, this line was dropped when Cuno became consumer product-oriented.

Cellulose-Based Membrane Filters

Millipore (Bedford, MA) has married several types of membrane filters to plastic support screens to produce an unique disposable or semi-disposable filtration system with no upstream supports. The affinity-use matrices are based on regenerated cellulose. Regenerated cellulose, as contrasted with conventional phase-inversion membrane filters, tends to be isotropic (uniform in cell size throughout), is less hydrophobic than the conventional cellulose-based membrane filters (exhibits less nonspecific binding of proteins), and has larger pores than phase-inversion membranes. These systems have not been introduced to the biotech community as a production device competitive with beaded agarose.

Regenerated cellulose is insoluble in all common solvents (as contrasted with many of the phase-inversion membranes that may be soluble in acetone, dimethyl sulfoxide, or dimethyl formamide; and the agaroses that may shed matrix in, for example, 6 *M* urea). Cartridge filter systems bearing ligand-substituted regenerated cellulose membranes would be very attractive for industrial-scale affinity separations provided the expendables could be made cost-competitive with agarose beads.

The issue is the tendency of such membranes to clog, not only on the membrane surface but within the body of the filter, which is not amenable to simple cleaning. Millipore has sold the Waters division (HPLC) but sold the MemSep line of *analytical* stacked disc for HPLC assay separately, to PerSeptive in Cambridge, MA. The MemSep line is available in the following chemistries: DEAE (diethylaminoethyl), QMA (quaternary), SP (sulfopropyl) and CM (carboxymethyl). There is a Protein A MemSep which likely is carbodiimide-coupled to the CM matrix. There are no plans at this time to scale up the MemSep line for preparative use.

Noncellulosic Woven and Nonwoven Fibers

Woven and nonwoven synthetic fibers include polyester, polyethylene, polystyrene and nylon. Exposure of the polyester to acid (dilute HCl, pH 1–2 for 1–5 hr) generates carboxylic acids on the fiber surface. The carboxylic ligands can be activated with a water-soluble carbodiimide and coupled to the $-NH_2$ of lysine on a protein or peptide at slightly acidic pH (e.g., 5.5) (Bauminger and Wilchek 1980). Exposure of polyethylene to conc HNO_3 for 1 hr generates carboxyl groups on the fiber surface. It is activated with a water-soluble carbodiimide (as above). Exposure of polystyrene to concentrated HNO_3 for 1 hr generates nitrate aromatic rings. The rings are reduced with Zn/HCl to $-NH_2$, then diazotized. A reaction occurs with histidine or tyrosine residues on the protein. Nylon can be activated in several ways. My preference is to expose the fibers to dilute HCl (pH 1) for 2 hr, then to wash and add a water-soluble carbodiimide at pH 5.5, wash with distilled water and add the protein to be coupled in water or saline (Olson 1983), as appropriate. Alternatively, nylon can be reacted with hydrazine, which converts the amides on the nylon fiber surface to hydrazides. Subsequent reaction with *nitrous* acid yields an azide that will react with amines (e.g., lysine).

I am not particularly fond of any of these matrices as they all tend to bind proteins nonspecifically. For example, nylon has some free amino groups and free carboxyllic acid groups that impart a pattern of positive and negative charges to the fiber surface. However, synthetic matrices can be particularly appealing as supports for enzymes. Should the reader wish to use one of these because of the strength of the materials, I recommend polyethylene.

Novel Systems

The unique systems most like existing matrices are the so-called *Perfusion Chromatography*™ matrices (POROS™) of PerSeptive Biosystems (Cambridge, MA) (Afeyan et al. 1990). These are beaded media, either of 20 μm diameter (POROS 20) or 50 μm diameter (POROS 50) of styrene-divinylbenzene copolymers, substituted with strong exchanger groups, or with Protein A or Protein G. The advantage of the medium is that bulk flow occurs through as well as around the medium, which makes for short diffusion distances.

The matrix is very strong and the flow is high because the matrix beads consist of very small beadlets. This medium should be excellent for production-scale purification of immunoglobulins and for ion exchange, but should be unsuitable for size-exclusion chromatography (SEC; gel filtration), which requires longer diffusion distances to be effective. The polystyrene-divynylbenzene matrix is coated with a polyol and subsitution of ion exchange or other groups to the matrix is via the polyol. Adsorption should be minimal, but I have not tested these materials. The POROS media merit consideration, especially for ion exchange.

What requires special attention is the process analysis of Fulton et al. (1992), who claim that, because the diffusion distances are so short in the POROS medium, small columns can be developed very rapidly. They claim that as many as 30 cycles may be processed for the generation of one batch of product on one small POROS column in the time that it takes to process one cycle per batch on a conventional large column. Consequently, since a much smaller system is used, equipment and media costs are much reduced. As yet, I have seen no confirming literature.

Sepracor, Inc. (Marlborough, MA) also claims high flow rates by virtue of matrix design (Holton et al. 1993). The Sepracor matrix, called Hyper D Media, has a "gel-in-a-shell" structure wherein a hydrogel of high porosity and binding capacity is contained within the pores of a rigid bead.

In an entirely different system, liquid droplets of fluorocarbons make possible very rapid, high surface-area interactions between ligands that have been perfluoroalkylated. The ligands stabilize the 10 μm or so fluorocarbon droplet surface, and are immediately available to solutes in the aqueous phase (Breillatt and Eveleigh 1987). Liquid-liquid systems like this pose special advantages for they are relatively easy to scale up, and they overcome the hydrodynamic weakness of gels in columns. The fluorocarbon liquid supports are relatively inert, chemically and biologically; they tend to be nonadsorptive, and because they have a specific gravity of 2–3, can be recovered easily by sedimentation in a tank or by centrifugation. The performance of this system, utilizing Procion Blue MX-R as an affinity ligand, was shown to be comparable to the same triazine dye on an agarose support (Stewart et al. 1989).

Aqueous two-phase systems have been used on occasion for the partial purification of proteins but more commonly are used in the purification of cells. However, if one selects a polymer-polymer (polyethylene glycol-dextran) or salt-polymer (phosphate or sulfate-polyethyleneglycol) system in which the protein of interest partitions *away* from a particular polymer, then coupling an affinity ligand to that polymer offers the possibility of a rapid purification method that is scaleable (Johansson and Tjerneld 1989).

Large-scale purification of formate dehydrogenase has been done directly from cell homogenates of *Candida boidinii* by means of poly(ethylene glycol) (PEG) and dextran, and with PEG-bound Procion red HE-3B-dextran. A PEG–hydoxypropylated starch (HPS) and Procion yellow HE-3G-PEG-HPS system was used by Tjerneld et al. (1987) to purify lactate dehydrogenase on a larger scale (Figure 7.5). Some very useful PEG derivatives for ligand coupling are available from Shearwater Polymers, Inc. (Huntsville, AL) including PEG succinimidyl succinate and the succinimidyl derivative of PEG–propionic acid. Less reactive for coupling is PEG–tresylate, and less reactive yet is PEG–glycidyl ether (PEG–epoxide).

Another ingenious approach is the attachment of an affine ligand (e.g., one of the Procion dyes, or an antibody) to a polymer that can be made soluble or insoluble as required (Fujii and Taniguchi 1991). The polymer-ligand is added to a solution of the material to be purified, conditions are altered to render the added polymer insoluble, and the suspension is centrifuged to recover polymer-ligand-protein. Such a system is readily scalable. A wide variety of polymers, many of them enteric coatings, can be used. Alginic acid is typical of many polymers that are soluble at pHs of 3–4 or less, which, in my opinion, can pose problems for bound macromolecules and for the integrity of the complex. However, 75 percent deacetylated chitosan is insoluble at pHs

Figure 7.5. Two-phase systems employing Procion dyes for the large-scale purification of lactate dehydrogenase, after Tjerneld et al. (1987).

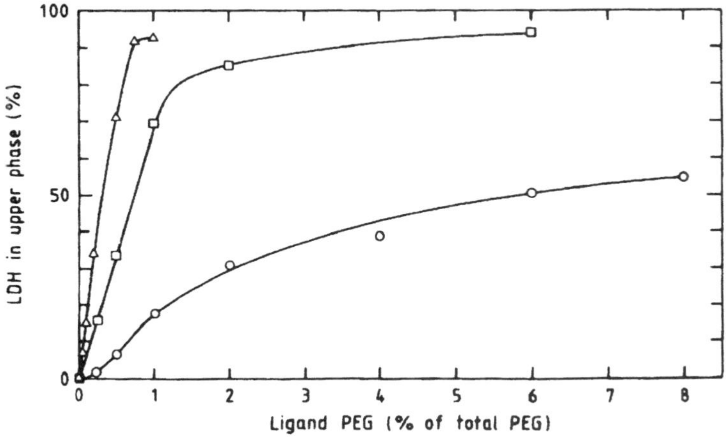

< 5.5 and soluble at pHs > 8.5 (the solubility of chitosan varies with the degree of deacetylation). Potentially more useful are polyelectrolyte complexes, better known to colloid chemists as complex coacervates. One such complex that seems useful is the complex of poly(4-vinyl-*N*-ethyl pyridinium bromide) [PVE] with polymethacrylic acid (PMA), which is insoluble at pH 2.9–5.8 and soluble above and below that range. The PVE–PMA complex also is insoluble in 0.2–0.4 *M* NaCl and soluble outside that range. This system may offer advantages in the purification of therapeutic proteins under mild conditions. However, it is clear from the literature that such systems have not been optimized, and that yield improvement would be essential if such systems are to be used in process.

Affinity separations can be done with cross-flow membrane systems (Herak and Merrill 1989). The affinity ligand (dye, Ab, etc.) is coupled to a macromolecule of very high molecular weight or to a very small particle, either of which is termed the *escort*. This approach clearly is scaleable to any dimension. In my view the issues that must be addressed are as follows:

- If the escort is particulate, the membrane system must not contain a woven or nonwoven upstream support or any other impediment to the flow of the escort.
- All proteins or polysaccharides in the source solution should be able to pass through the membrane filter; consequently, a first pass should be done in the absence of the escort and the filtrate should be used as source material.

In an application directly related to therapeutics, Luong et al. (1990) are utilizing the escort technique to purify urokinase from human urine. The original work of Luong et al. (1987), employing immobilized Cibacron Blue as the escort for the purification of HSA, could be put to profitable use in industry today, were the regulatory barriers more modest. In any event, Graves and Wu (1974) are correct that only affinity pairs that have a K_d significantly less than 10^{-6} can be used in such a system. If the K_d is significantly higher than 10^{-6}, considerable product may be lost during the washing of the escort.

The *reversed* micelle system of Wolf et al. (1989), employing aqueous systems emulsified in an organic solvent, would not be particularly attractive for the purification of peptides and proteins intended for pharmaceutical applications. The potential for denaturation of the proteins is too high.

Liapis (1989) contends that, with conventional affinity columns, operation with two short columns is more efficient than with one long column, and that the columns in series should be alternated, with one first and, in a subsequent run, the other. My experience has been that stacked columns reduce problems related to channeling of the medium (i.e., continuous large gaps through a packed bed or fibrous matrix, through which liquid flow is rapid and unimpeded, resulting in zone broadening).

LIGAND IMMOBILIZATION CHEMISTRIES

One of the better reviews on immobilization methods is that of Scouten (1987), although good reviews abound (e.g., Clonis 1987; Hermanson et al. 1992).

Epoxy (Oxirane) Activation

Expoxide (oxirane) rings are highly reactive with alkali and with amino groups, thiols, or hydroxyls. As is true of much of affinity technology, Jerker Porath developed oxirane methods (Sundberg and Porath 1974) which are summarized in Figure 7.6. The two-stage activation and coupling is simple and suitable for immobilization of any RNH_2-bearing species, including proteins, usually at a pH of 8.5–12 over a period of 5–18 hr. Stability of the secondary amine that is formed is excellent. Unreacted epoxide groups can be reacted with 1 *M* ethanolamine. Porath's coupler is 1,4-butanediol diglycidyl ether which, in addition to matrix activation, provides a spacer arm of more than six carbons. Other than epichlorohydrin, it is the least expensive oxirane activator (Clonis 1987). Epichlorohydrin is not a *bis*-epoxide, but behaves like one and is used as a Sephadex cross-linker. I have

Figure 7.6. Oxirane activation and coupling.

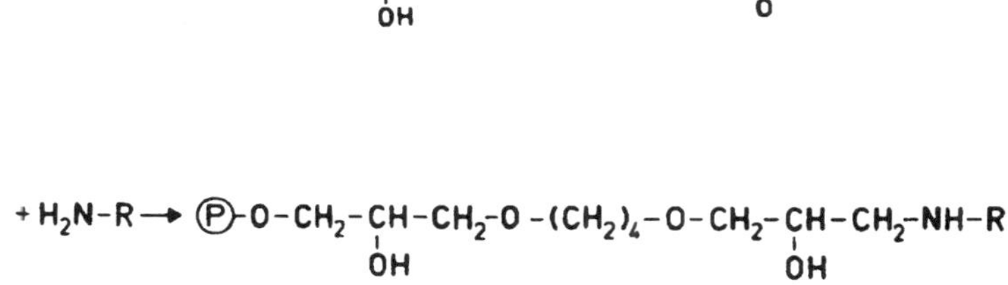

used the *bis*-oxirane method for coupling alkyl groups to nylon fibers (Olson 1983) and find that it works well.

Succinic or Glutaric Anhydride

The intent of the use of succinic or glutaric anhydrides is the generation of a carboxyl group on the matrix, which is C_4 or C_5 long. Succinylation prepares a matrix for carbodiimide activation. Nylon, which poses adsorptive problems due to the prevalence of amino and carboxyl groups on the partially hydrolyzed surface, can be succinylated (masking amino groups) and carbodiimide activated, then coupled.

Sulfonyl Chlorides

Organic sulfonyl chlorides for the activation of hydroxyl-bearing matrices were introduced by Nilsson and Mosbach (1980). Tosyl chloride (*p*-toluenesulfonyl chloride) and tresyl chloride (2,2,2-trifluoroethanesulfonyl chloride) generate active sulfonates on agarose or cellulose; activation times range from about 5 min to 1 hr. During reaction with nucleophiles such as amines or thiols, the sulfonates are the leaving groups. Tresyl forms sulfonates more readily and is easily displaced at pH 7 and 4–5°C (Nilsson and Mosbach 1981). With amino groups these sulfonates form secondary amines that have excellent stability.

A similar reagent that is a good activator and coupler is 2-fluoro-1-methylpyridinium toluene sulfonate (available from BioProbe International, Tustin, CA). The coupling is via nucleophilic substitution of the pyridinium ion (the pyridinium ion acts as an electron sink); the 1-methylpyridinol sulfonate is an excellent leaving group and the secondary amine formed between the matrix hydroxyl and the protein has good stability.

Divinylsulfone

The divinylsulfone (DVS) coupling method is another of Jerker Porath's inventions. Activation is rapid (0.5–2 hr) and the coupling of ligand is very rapid (0.5–2 hr) at pH 8–10. Adducts are very unstable at pH > 10, but fairly stable at pH < 7. Reaction is primarily with amines to form a secondary amine. Some ligand leakage should be expected in use. Although the method has been touted, the product is no more stable than that of the CNBr-activated matrices, and is no better than CNBr for the preparation of injectables.

Carbodiimide

An excellent review on carbodiimide reactions, which can be applied to the immobilization of a wide variety of ligands, is that of Bauminger and Wilchek (1980), although many other references can be found in the literature. Peptide bonds are readily generated in the presence of carbodiimides (CDs). If a carboxyl-bearing matrix is CD-activated (e.g., CD addition at a pH of 4.5–5.5), rinsed to remove the uncomplexed CD, and a lysine-bearing protein is added to the matrix, covalent coupling results. The resulting *N*-substituted carbamate is stable at pH < 10. Whether the coupling is to agarose, cellulose, or another matrix, –COOH groups are required on the matrix, and it is preferable that the carboxyl groups are some distance from the surface of the matrix (e.g., C_6 distant, as with the carboxyhexyl agarose available from Pharmacia). Carboxyl groups that have not reacted with the protein to be immobilized are masked with ethanolamine. Do not add a carbodiimide directly to a protein in solution; cross-linking will result.

The carbodiimide method is far preferable as a general coupling technique to the CNBr method, which is less stable and generates an anion-exchange background.

Chloroformates

A series of chloroformates with good aqueous solubility, most notably *p*-nitrophenylchloroformate, are quite good activators of hydroxyl groups. When the activated sites react with nucleophilic groups on the proteins being coupled, leaving groups are generated with high absorbance at specific wavelengths. This means that the extent of protein coupling can be measured spectrophotometrically. For example, *p*-nitrophenol is an obvious yellow and readily quantitated in the visible range.

Triazines, Including Dyes

Trichloro-*S*-triazine will react, more-or-less completely, with an hydroxyl-bearing matrix in 0.5–2 hr and will form a triazinyl link with amino, thiol, or hydroxyl groups over a period of 4–16 hr at pH 7.5–9. The stability of the linkage is quite good, but pi-bonding may occur with aromatics, causing nonspecific interactions.

The triazine dyes (e.g., Cibacron Blue F3G) were developed for the coloring of textiles. Triazine dyes (trade names: Procion M of ICI), like trichloro-*S*-triazine, react well with hydroxyl groups and amines. Although the pharmaceutical manufacturer may prefer to purchase one of the Procion dyes from ICI (United Kingdom), one should beware as the commercial dyes are mixtures (Hanggi and Carr 1985). The Procion-matrix systems available from Amicon (Lexington, MA) are prepared with purified dyes and, ultimately, are more practical than do-it-yourself.

The dyes react with dehydrogenases and other proteins, but dichlorotriazines also are useful as coupling agents for proteins (Figure 7.7). If you activate a matrix with a triazine and only one Cl remains on the ring, the half-life is rather long (24 hr in water at 40°C, and with quite good stability in the dry state). Therefore, dichlorotriazine-activated cellulosic or agarose-based matrices can readily be generated in-house. Coupling of lysine-bearing proteins or most H_2NR is very stable (Pierce and Rattee 1967).

Iminodiacetic Acid for Chelated Metals

The most interesting use of the interaction of peptides with immobilized chelated metals is as a coupling technique. A partially successful purification of F-VIII, based on binding to immobilized Ca^{2+}, has been reported (Mantovaara et al. 1991). However, the method proved to be noncompetitive with immunosorbent purification.

Chelated metals are not simple anion exchangers. Some of the interactions appear to be somewhat specific (e.g., Ni^{2+} with histidine). A

Figure 7.7. Dichlorotriazine activation and coupling.

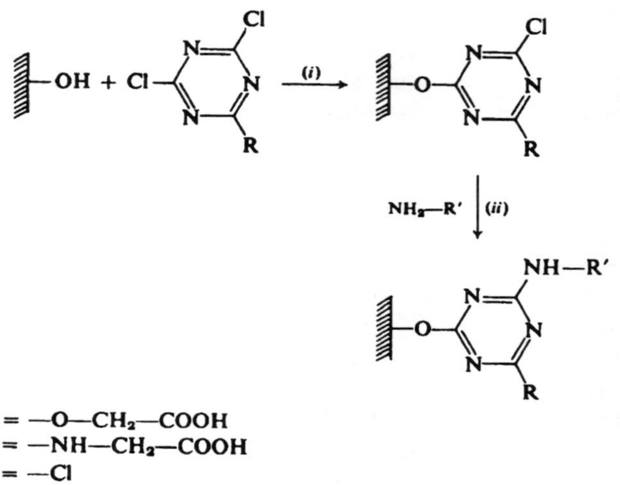

R = —O—CH_2—COOH
R = —NH—CH_2—COOH
R = —Cl

special advantage of the metal ion resins is that one should be able to clean and regenerate them under rigorous conditions, as is the case with many ion exchangers. Cleaning probably should commence with dilute acid to displace metallic cations, so that subsequent cleaning conditions do not generate insoluble hydroxides.

Glutaraldehyde and $NaIO_4$ Activation

Bis-glutaraldehyde is a classic bifunctional coupling agent. Activation time for the matrix varies from 1–8 hr and coupling at pH 6.5–8.5 can require 6–16 hr. Reaction is only with amines and, as the range of pH for reaction indicates, takes place under mild conditions. Highly reactive aldehydes such as glutaraldehyde form strong Schiff bases that cannot be completely reduced to a secondary amine with $NaCNBH_4$; they slowly release immobilized ligands or proteins. Nonetheless, stability of the immobilized group must be considered good. Grandics et al. (1990) claim development of an unidentified aldehyde coupling group that will tolerate steam autoclaving, and that does not release detectable amounts of bound ligand for periods up to one year. The review by Reichlin (1980) remains relevant.

Hou et al. (1990) had a very useful approach to glutaraldehyde activation. They reacted hydroxyl groups on a support with a diglycidyl ether (*bis*-oxirane) and, after the grafting, incubated the substituted matrix in dilute HCl (pH 0.5) overnight, followed by 1 hr at room temperature with 1.5 percent $NaIO_4$. The peptide to be coupled was made up at 5 mg/ml in 0.1 *M* sodium phosphate (pH 7.6) and the ligand was recirculated through the matrix for 3 hr at room temperature. Uncoupled ligand was removed by rinsing and the remaining active sites on the matrix were blocked with 2 percent glycine ethyl ester in 0.1 *M* borate buffer (pH 8.2). Stabilization of the ligand on the matrix was with 25 mg/ml sodium borohydride as a reducing agent, overnight.

Periodate oxidation of polysaccharide matrices generates aldehydes, usually at C_2 and C_3. Under mild conditions (pH 6–8), these form stable, uncharged carbamate linkages with amino groups (e.g., of lysine). A variation of this activation and coupling is used to couple antibodies to matrices via the carbohydrate groups on the Fc portion of the Ab.

Grandics et al. claim a solution for elution (the example is for the elution of Abs from immobilized anti-Abs) that does not cause appreciable denaturation of the immobilized protein (Ab, in this case). Particulars are withheld, and the secrecy may mean that the aldehydic function chosen for coupling is known and not patentable. The same reasoning applies to the proprietary eluting solution.

For those who prefer to use glutaraldehyde for immobilization of a binding protein, but who cannot use unspecified technology, the simplest approach is that of the Method M and the Scripps Clinic immunosorbent methods—use a strong exchanger on the downstream side of the column to trap either the eluted protein or the unimmobilized protein/ligand.

N-Hydroxysuccinimide Esters

N-hydroxysuccinimide (NHS) esters activate carbonyl groups bound to supports such as agarose (Cuatrecasas and Parikh 1972). The activated supports react with the amino groups of proteins spontaneously at pH 6.5–8.5 and 4°C, resulting in stable bonds. However, the active ester readily hydrolyzes, so if a commercially prepared NHS-activated matrix is used, coupling should commence shortly after hydration of the matrix. Pharmacia (Uppsala, Sweden and Piscataway, NJ) produces activated CH-Sepharose 4B and NHS–activated Superose. Bio-Rad (Hercules, CA) produces Affi-gel (activated agarose) and Affi-Prep.

Activated Matrices Available Commercially

The primary suppliers of activated matrices are Pharmacia (Uppsala, Sweden and Piscataway, NJ), Bio-Rad (Hercules, CA) and Pierce Chemical (Rockford, IL), although many newer companies now provide specialty affinity-type matrices (e.g., Sepracor [Marlborough, MA]). A sampling of the activated and affinity matrices available commercially are presented in Table 7.4.

A matrix that may hold promise for scaleup of process is "Glutaraldehyde-P" (J. T. Baker Inc., Phillipsburg, NJ). Spherical silica particles (40 μm diameter) with a wide range of pore sizes have been covalently coated with polyethyleneimine, which then has been reacted with 25 percent glutaraldehyde for 4 hr at room temperature, reduced with sodium cyanoborohydride, and dried. One of the advantages of glutaraldehyde is that it cross-links and inactivates antigens in pig heart valves used in humans. It is a cross-linker that has been used successfully for a long time (Richards and Knowles 1968) with implantable medical devices.

Ligand-Coupled Matrices Available Commercially

Table 7.5 details a proportion of the matrices presently available for the bulk purification of proteins.

Table 7.4. A Few of the Activated Matrices Available Commercially

Commercial Name	Matrix and Derivative	Source
Active thio-Sepharose	Glutathione-2-mercapto pyridine derivative of agarose	Pharmacia
Activated Sepharose-CH	Aminohexanoic acid *N*-hydroxy-succinimide ester of agarose	Pharmacia
Affi-Gel 10	Agarose, carboxylic acid-*N*-hydroxysuccinimide ester	Bio-Rad
Agarose adipic acid hydrazide	Self-explanatory	Pharmacia
CNBr Sepharose	CNBr derivative of agarose	Pharmacia
CPG-hydroxy-succinimide	Aminopropylsuccinimidyl *N*-hydroxysuccinimide-controlled pore glass	Pierce
Enzacryl-PTL	Cysteinyl thioacetyl derivative of poly(acrylamide)	Koch-Light
Epoxy-activated Sepharose	Bisoxirane derivative of agarose	Pharmacia
FMP-Trisacryl	2-fluoro-1-methylpyridinium toluene sulfonate derivative of acrylamide	BioProbe
Fractogel CDI	Carbiimidazole derivative of polyvinyl	Pierce
Glycidoxy CPG	Bisoxirane derivative of controlled pore glass	Pierce
NHS Superose	*N*-hydroxysuccinimide derivative	Pharmacia

Continued on next page

Continued from previous page

Commercial Name	Matrix and Derivative	Source
N-methyl-pyridinium trisacryl	*N*-methylpyridinol derivative of acrylamide	BioProbe
POROS™ AL-Aldehyde	Aldehyde polystyrene-divinylbenzene	PerSeptive Biosystems
POROS™ EP-Epoxy	Oxirane of polystyrene-divinylbenzene	PerSeptive Biosystems
POROS™ HY-Hydrazide	Hydrazide of polystyrene-divinylbenzene	PerSeptive Biosystems

Table 7.5. A Sampling of the Affinity Matrices Available Commercially, by Name, Affine Ligand, and Producer

Commercial Name	Matrix and Derivative	Source
Affi-Gel Blue Gel	Cibacron Blue F3G-A on beaded agarose	Bio-Rad
Affi-Gel Heparin Gel	Heparin on beaded agarose	Bio-Rad
Affi-Gel Protein A Agarose	Protein A coupled by a peptide bond to agarose	Bio-Rad
Arginine Sepharose 4B	Self-explanatory	Pharmacia
Benzamidine Sepharose 6B	Self-explanatory	Pharmacia
Blue Sepharose CL6B	Cibacron Blue F3G-A on cross-linked 6% beaded agarose	Pharmacia
Calmodulin Sepharose 4B	Self-explanatory	Pharmacia

Continued on next page

Continued from previous page

Commercial Name	Matrix and Derivative	Source
CM Affi-Gel Blue Gel	Cibacron Blue on carboxymethyl agarose	Bio-Rad
Con A Sepharose	Concanavalin A-Sepharose 4B	Pharmacia
DEAE Affi-Gel Blue Gel	Cibacron Blue on diethylamino-ethyl agarose	Bio-Rad
Gamma-binding Sepharose	Protein G-agarose	Pharmacia
Gelatin Sepharose 4B	Self-explanatory	Pharmacia
Glutathione Sepharose 4B	Self-explanatory	Pharmacia
Helix pomatia lectin Sepharose 6MB	Self-explanatory, save that the M designates a special grade of agarose	Pharmacia
Heparin Sepharose CL6B	Self-explanatory, save that CL6B designates cross-linked 6% beaded agarose	Pharmacia
Lentil lectin Sepharose 4B	Self-explanatory	Pharmacia
Lysine Sepharose 4B	Self-explanatory	Pharmacia
POROS 50 A	Protein A on 50 μm diameter polystyrene-divinylbenzene copolymer	PerSeptive Biosystems
POROS 50 G	Protein G on same matrix as POROS 50 A	PerSeptive Biosystems
Protein A HyperD	Protein A on "HyperD" cross-linked 6% beaded agarose	Sepracor

ELUTION STRATEGIES

Operation of columns is most efficient with two short columns in series rather than one long column (Liapis 1989). My experience has been that the zone broadening that results from unevenness in column packing or the inadvertent formation of channels in a column is reduced considerably by columns in series. Two or three columns can be alternated in position.

Addition of the feed and the wash step(s) should be in one direction, but elution of product should be in the opposite direction because product will be in highest concentration at the top if the feedstream is added at the top; elution of the product in the same direction will add diluent as a result of zone broadening. Concentrated product is preferred as it usually is more stable than dilute product and requires less additional processing.

ANALYSIS OF THE PRODUCT

All analyses relate to identity, safety, and efficacy. With regard to efficacy, testing also may relate to the delivery of a therapeutically effective dose. Testing might be summarized briefly as follows:

Safety

Mouse and Guinea Pig Safety

The test animals, injected into the peritoneum with product, must live and, in the case of the mice, gain weight. Some drugs may induce a temporary anorexia.

USP Sterility Test

Articles must be sterile, as labeled. The USP test is notoriously insensitive (Olson 1987a). Validation of all of the critical steps in process is more effective in ensuring sterility than the sterility test, but the test remains for the detection of gross contamination.

Absence of Viruses or DNA from the Host Cell or Animal

The concern with DNA is the possibility of introducing an oncogene or the activator of an oncogene into the recipient. Animal viruses may in some instances be transmissible to man, hence the concern for the most common of the viruses in the cell line used for manufacture of the product.

Absence of Immobilized Ligand and Matrix

If the immobilized ligand is a nonhuman protein (e.g., Protein A or Protein G from a staphylococcus), it is important that the eluted product is free from such ligands or fragments thereof, as they may be immunogenic. This is one of the reasons that protein-free matrices are preferred. A product that contains an undetectable trace of a dye, or a short piece of agarose, is less likely to be immunogenic to the recipient than a product containing a nonhuman protein.

Efficacy/Identity

Product efficacy is demonstrated in Phase II-Phase IV testing. Efficacy, *sensu strictu*, is not tested on each individual lot of product. Usually, some test of a *biological* attribute of the drug (e.g., the ability to bind to a particular receptor) is correlated with a physical or chemical attribute (e.g., reaction with a highly specific monoclonal antibody). Thereafter, a quantitative test employing either the receptor or the MAb is utilized as an indirect test of drug identity, hence efficacy.

Quantitation of the Purified Therapeutic Entity

A therapeutic protein usually is best identified by means of a quantitative immunochemical analysis [e.g., the Laurell (1966) rocket technique]. An outline of various quantitative immunochemical assays is provided in Stites (1978). Validation of the identity test includes a correlation with a bioassay that may, or may not, produce a quantitative response.

REFERENCES

Adair, W. L., and S. Kornfeld. 1974. Location of the receptors of wheat germ agglutinin and the *Ricinus communis* lectins from human erythrocytes using affinity chromatography. *J. Biol. Chem.* 249:4696–4704.

Afeyan, N. B., S. P. Fulton, N. F. Gordon, I. Mazsaroff, L. Varady, and F. E. Regnier. 1990. Perfusion chromatography an approach to purifying biomolecules. *Bio/Technology* 8:203–206.

Allison, M. A. K., S. E. Jones, and P. McGuffey. 1989. Phase II trial of outpatient interleukin 2 in malignant lymphoma, chronic lymphocytic leukemia and selected solid tumors. *J. Clin. Oncol.* 7:78–80.

Andersson, L.-O., H. Borg, and M. Miller-Andersson. 1975. Purification and characterization of human factor IX. *Thromb. Res.* 7:451–459.

Andrew, S. M., and J. A. Titus. 1992. Affinity chromatography using Protein A-Sepharose. In *Current protocols in immunology,* edited by J. E. Coligan, A. M. Kruisbeek, D. H. Margulies, E. M. Shevach, and W. Strober. NY: Green Publishing Assoc. and Wiley-Interscience.

Arnold, F. H. 1991. Metal-affinity separations: a new dimension in protein processing. *Bio/Technology* 9:151–156.

Axen, R., J. Porath, and S. Ernback. 1967. Chemical coupling of peptides and proteins to polysaccharides by means of cyanogen halides. *Nature* 214:1302–1304.

Baars, J. W., J. C. M. Fonk, R. J. Scheper, B. M. E. van der Filer, H. Bril, P. van der Valk, H. M. Pinedo, and J. Wagstaff. 1992. Treatment with tumor infiltrating lymphocytes and interleukin-2 in patients with metastatic melanoma: A pilot study. *Biotherapy* 4:289–297.

Barrett, P. N., A. Mitterer, W. Mundt, J. Eibl, M. Elbl, F. Dorner, R. C. Gallo, and B. Moss. 1989. Large-scale production, purification, and immunological analysis of vaccinia recombinant-derived HIV-1 gp160. In *Vaccines 89: Modern approaches to new vaccines including prevention of AIDS,* edited by R. A. Lerner, H. Ginsberg, R. M. Chanock, and F. Brown. Cold Spring Harbor, NY: Cold Spring Harbor Laboratory.

Bauminger, S., and M. Wilchek. 1980. The use of carbodiimides in the preparation of immunizing conjugates. *Methods Enzymol.* 70:151–165.

Beer, H. D., H. E. Maschke, and K. Schuegerl. 1992. Continuous production of restricion endonucleases: Continuous two-stage cultivation with *E. coli* JM103; continuous cell disintegration and purification by affinity chromatography. *Appl. Microbiol. Biotechnol.* 38:220–225.

Berg, K., I. Heron, and R. Hamilton. 1978. Purification of human interferons by antibody affinity chromatography using highly adsorbed anti-interferon. *Scand. J. Immunol.* 8:429–436.

Bog-Hansen, T. C., P. Prahl, and H. Lowenstein. 1978. A set of analytical electrophoresis experiments to predict the results of affinity chromatographic separations: fractionation of allergens from cow's hair and dander. *J. Immunol. Methods* 22:293–307.

Boschetti, E., P. Girot, J. P. Secheresse, and J. Saint-Blancard. 1981. Study of the molecular interaction between lysozyme and heparin and application in affinity chromatography. *J. Chromatog.* 210:469–475.

Breillatt, J. P., and J. W. D. Eveleigh. 1987. Bioaffinity and ion-exchange separations with liquid exchange supports. European Patent EP 246 103, 19 November.

Brocklehurst, K., J. Carlsson, M. P. J. Kierstan, and E. M. Crook. 1973. Covalent chromatography. Preparation of fully active papain from dried papaya latex. *Biochem. J.* 133:573–584.

Campbell, D. H., E. Leuscher, and L. S. Lerman. 1951. The use of an immunologically specific adsorbent for purification of antibody. *Proc. Natl Acad. Sci. USA* 37:575–578.

Chaga, G., L. Andersson, B. Ersson, and J. Porath. 1989. Purification of two muscle enzymes by chromatography on immobilized ferric ions. *Biotechnol. Appl. Biochem.* 11:424–431.

Clemetson, K. J., S. L. Pfueller, E. F. Suscher, and C. S. P. Jenkins. 1977. Isolation of the membrane glycoproteins of human blood platelets by lectin affinity chromatography. *Biochim. Biophys. Acta* 464:493–508.

Clonis, Y. D. 1987. Large-scale affinity chromatography. *Bio/Technology* 5:1290–1293.

Corbi, A. L., and J. Carreira. 1985. Distribution of allergen activity and identification of the main IgE-binding glycopeptide of *Paretaria judaica* pollen. *Int. Arch. Allergy Appl. Immunol.* 76:151–161.

Cosman, D., D. P. Cerretti, A. Larsen, L. S. Park, C. March, S. Dower, S. Gillis, and D. Urdal. 1984. Cloning, sequence, and expression of human interleukin-2 receptor. *Nature* 312:768–771.

Cuatrecasas, P., and C. B. Anfinsen. 1971. Affinity chromatography. *Annu. Rev. Biochem.* 40:259–275.

Cuatrecasas, P., and I. Parikh. 1972. Adsorbents for affinity chromatography. Use of N-hydroxysuccinimide esters of agarose. *Biochemistry* 11:2291–2299.

Davey, M. W., E. Sulkowski, and W. A. Carter. 1976. Binding of human fibroblast interferon to concanavalin A-agarose. Involvement of carbohydrate recognition and hydrophobic interaction. *Biochemistry* 15:704–713.

DeMaeyer-Guignard, J., M. G. Tovery, J. Gresser, and E. DeMaeyer. 1978. Purification of mouse interferon by sequential affinity chromatography on poly(U)- and antibody-agarose columns. *Nature* 271:622–625.

Easterday, R. L., and I. M. Easterday. 1974. Affinity chromatography of kinases and dehydrogenases on Sephadex and Sepharose dye derivatives. In *Immobilized biochemicals and affinity chromatography,* edited by R. B. Dunlap. New York: Plenum.

Edy, V. G., A. Billiau, and P. DeSomer. 1977. Purification of human fibroblast interferon by zinc chelate affinity chromatography. *J. Biol. Chem.* 252:5934–5935.

El-Rassi, Z., and C. Horvath. 1986. Metal-chelate interaction chromatography of proteins with iminodiacetic acid-bonded stationary phases on silica support. *J. Chromatog.* 359:241–253.

Flanagan, S. D., and S. H. Barondes. 1975. Affinity partitioning, a method for purification of proteins using specific polymer ligands in aqueous polymer two-phase systems. *J. Biol. Chem.* 250:1484–1489.

Ford, C. F., I. Suominen, and C. E. Glatz. 1991. Fusion tails for the recovery and purification of recombinant proteins. *Protein Expression Purif.* 2:95–107.

Francis, R., J. Bonnerjea, and C. R. Hill. 1990. Validation of the re-use of protein A Sepharose for the purification of monoclonal antibodies. In *Separations for biotechnology,* Vol. 2, edited by D. L. Pyle. London: Elsevier Applied Science.

Fujii, M., and M. Taniguchi. 1991. Review: Application of reversibly soluble polymers in bioprocessing. *Trends Biotechnol.* 9:191–196.

Fujikawa, K., K. Kurachi, and E. W. Davie. 1977. Characterization of bovine factor XIIa (activated Hageman factor). *Biochemistry* 16:4182–4188.

Fulton, S. P., A. J. Shahidi, N. F. Gordon, and N. B. Afeyan. 1992. Large-scale processing & high-throughput perfusion chromatography. *Bio/Technology* 10:635–639.

Ginjaar, H. B., H. B. M. van Paassen, J. T. den Dunnen, N.-t. Man, G. E. Morris, A. F. M. Moorman, and G. J. B. van Ommen. 1992. Construction of dystrophin fusion proteins to raise targeted antibodies to different epitopes. *FEBS Lett.* 308:293–297.

Ginsberg, M. H., and E. F. Plow. 1989. Fibronectin: a contender in platelet adhesive functions. In *Fibronectin,* edited by D. F. Mosher. San Diego: Academic Press.

Ginsburg, V. 1987. *Complex carbohydrates, Part E.* San Diego: Academic Press.

Goodheart, C. R., and R. H. Silverman. 1991. Cellular and plasma fibronectin. *SIM Industrial Microbiology News* 41:266–272.

Grandics, P., Z. Szathmary, and S. Szathmary. 1990. A novel immunoaffinity chromatography system for the purification of therapeutic proteins. *Ann. N.Y. Acad. Sci.* 589: 148–156.

Graves, D. J., and Y.-T. Wu. 1974. On predicting the results of affinity chromatography procedures. *Methods Enzymol.* 34:140–152.

Greten, H., and B. Walter. 1973. Purification of rat adipose tissue lipoprotein lipase. *FEBS Lett.* 35:36–40.

Haavik, S., B. Smestad Paulsen, and J. K. Wold. 1985. Glycoprotein allergens in pollen of timothy. I. Investigation of carbohydrates extracted from pollen of timothy (*Phleum pratense*) and purification of a carbohydrate-containing allergen by affinity chromatography on concanavalin A-Sepharose. *Int. Arch. Allergy Appl. Immunol.* 78:197–205.

Hakes, D. J., and J. E. Dixon. 1992. New vectors for high level expression of recombinant proteins in bacteria. *Anal. Biochem.* 202:293–296.

Hanggi, D., and P. Carr. 1985. Analytical evaluation of the purity of commercial preparations of Cibacron Blue F3GA and related dyes. *Anal. Biochem.* 149:91–104.

Hatchikian, E. C., and P. Monsan. 1980. Highly active immobilized hydrogenase from *Desulfovibrio gigas. Biochem. Biophys. Res. Commun.* 92:1091–1096.

Hedman, K., and A. Vaheri. 1989. Fibronectin and the pericellular matrix. In *Fibronectin,* edited by D. F. Mosher. San Diego: Academic Press.

Hellstrom, U., M.-L. Dillner, S. Hammarstrom, and P. Perlmann. 1976a. The interaction of non-mitogenic and mitogenic lectins with T lymphocytes: association of cellular receptor sites. *Scand. J. Immunol.* 5:45–54.

Hellstrom, U., S. Hammarstrom, M.-L. Dillner, H. Perlmann, and P. Perlmann. 1976b. Fractionation of human blood lymphocytes on Helix pomatia A hemagglutinin coupled to Sepharose beads. *Scand. J. Immunol.* 5, Suppl. 5:45–55.

Herak, D. C., and E. W. Merrill. 1989. Affinity cross-flow filtration: experimental and modeling work using the system of HSA and Cibacron-Blue agarose. *Biotechnol. Bioeng.* 5:63–68.

Hermanson, G. T., A. K. Mallia, and P. K. Smith. 1992. *Immobilized affinity ligand techniques*. San Diego: Academic Press.

Hochuli, E., and S. Piesecki. 1992. Interaction of hexahistidine fusion proteins with nitrilotriacetic acid-chelated Ni^{2+} ions. *Methods* 4:68–73.

Hodgson, J. 1992. Elective affinities: the art of chromatography. *Bio/Technology* 10:997–1000.

Hoffman, D. L. 1989. Purification and large-scale preparation of antithrombin III. *Am. J. Med.* 87 (suppl 3B):23S–26S.

Holton III, O. D., C. D. Blizzard, J. J. Vicalvi Jr., and A. J. Pearson. 1993. Hyperdiffusion chromatography for hi-speed, hi-capacity antibody purification. *Genetic Eng. News* January: 10.

Hou, K. C., S. Roy, R. Zaniewski, and E. Shumway. 1990. A method for extracorporeal heparin removal from blood by affinity chromatography. *Artif. Organs* 14:436–442.

Huang, J. W., M. W. Davey, C. J. Hejna, W. von Muenchhausen, E. Sulkowski, and W. A. Carter. 1974. Selective binding of humanm interferon to albumin immobilized on agarose. *J. Biol. Chem.* 249:4665–4667.

Huang, S. H., S. Roy, K. C. Hou, and G. T. Tsao. 1988. Scaling-up of affinity chromatography by radial-flow cartridges. *Biotechnol. Progr.* 4:159–165.

Hutchins, T. W., J. S. Magnuson, and T.-T. Yip. 1990. Secretory IgA, IgG, and IgM immunoglobulins isolated simultaneously from colostral whey by selective thiophilic adsorption. *J. Immunol. Methods* 128:89–99.

Hutchins, T. W., and J. Porath. 1987. Protein recognition in immobilized ligands: promotion of selective adsorption. *Clin. Chem.* 33:1502–1508.

Inogamov, U. K., S. K. Kasymov, A. V. Karas, V. T. Skvortsov, S. I. Salikhov, and A. E. Gurvich. 1991. Preparation of cellulose sorbent for affinity chromatography using epichlorohydrin and sodium borohydride. *Otkrytiya, Izobret.* No. 30:78. CA 117:51128q, issue 16, 1992.

Iverius, P.-H. 1972. The interaction between human plasma lipoproteins and connective tissue glycosaminoglycans. *J. Biol. Chem.* 247:2607–2613.

Jankowski, W. J., W. von Muenchhausen, E. Sulkowski, and W. A. Carter. 1976. Binding of human interferon to immobilized Cibacron Blue F3G-A. The nature of the molecular interaction. *Biochemistry* 15:5182–5187.

Janson, J.-C. 1982. Scaling-up of affinity chromatography, technological and economical aspects. In *Affinity chromatography and related techniques*, edited by T. C. J. Gribnau, J. Visser, and R. J. F. Nivard. Amsterdam: Elsevier.

Johansson, G., and F. Tjerneld. 1989. Affinity partition between aqueous phases—a tool for large-scale purification of enzymes. *J. Biotechnol.* 11:135–142.

Jordan, R. E., T. Zuffi and D. D. Schroeder. 1982. Isolation of antithrombin, high affinity heparin, or antithrombin:high affinity heparin complex by sequential affinity chromatography. In *Affinity chromatography and related techniques*, edited by T. C. J. Gribnau, J. Visser, and R. J. F. Nivard. Amsterdam: Elsevier.

Karlstam, B., and B. Nilsson. 1982. Characterization of grass pollen allergens by affinity chromatography on concanavalin A-Sepharose. *J. Immunol. Methods* 54:119–128.

Kerby, G. P., and G. S. Eadie. 1953. Inhibition of lysozyme by heparin. *Proc. Soc. Exp. Biol. Med.* 83:111–113.

Kiefer, M. C., C. Schmid, M. Waldvogel, I. Schlaopfer, E. Futo, F. R. Masiarz, K. Green, P. J. Barr, and J. Zapf. 1992. Characterization of human recombinant insulin-like growth factor binding proteins 4, 5, and 6 produced in yeast. *J. Biol. Chem.* 267:12692–12699.

Kisiel, W., and E. W. Davie. 1975. Isolation and characterization of bovine factor VII. *Biochemistry* 14:4928–4934.

Klapper, M. H. 1977. The independent distribution of amino acid near neighbor pairs into polypeptides. *Biochem. Biophys. Res. Commun.* 78:1018–1024.

Kurachi, K., and E. W. Davie. 1977. Isolation and characterization of bovine factor XI (plasma thromboplastin antecedent). *Biochemistry* 16:5831–5839.

Laurell, C. B. 1966. Quantitative estimation of proteins by electrophoresis in agarose gel containing antibodies. *Anal. Biochem.* 15:45–52.

Lebing, W. R., D. J. Hammond, J. E. Wydick III, and G. A. Baumbach. 1994. A highly purified antithrombin III concentrate prepared from human plasma fraction IV-1 by affinity chromatography. *Vox Sang.* 67:117–124.

Leonard, W. J., J. M. Depper, G. R. Crabtree, S. Rudikoff, J. Pumphrey, R. J. Robb, M. Kronke, P. B. Svetlik, N. J. Peffer, T. A. Waldmann, and W. C. Greene. 1984. Molecular cloning and exression of cDNAs for the human interleukin-2 receptor. *Nature* 311:626–631.

Letarte-Muirhead, M., A. N. Barclay, and A. F. Williams. 1975. Purification of the thy-1-molecule, a major cell-surface glycoprotein of rat thymocytes. *Biochem. J.* 151:685–697.

Liapis, A. I. 1989. Theoretical aspects of affinity chromatography. *J. Biotechnol.* 11:143–160.

Ljungquist, C., A. Breitholtz, H. Brink-Nilsson, T. Moks, M. Uhlen, and B. Nilsson. 1989. Immobilization and affinity purification of recombinant proteins using histidine peptide fusions. *Eur. J. Biochem.* 186:563–569.

Ljungquist, C., J. Lundeberg, A. M. Rasmussen, E. Hornes, and M. Uhlen. 1993. Immobilization and recovery of fusion proteins and B-lymphocyte cells using magnetic separation. *DNA Cell Biol.* 12:191–197.

Lotan, R., N. Sharon, and D. Mirelman. 1975. Interaction of wheat-germ agglutinin with bacterial cells and cell-wall polymers. *Eur. J. Biochem.* 55:257–262.

Lowe, C. R., S. J. Burton, N. P. Burton, W. K. Alderton, J. M. Pitts, and J. A. Thomas. 1992. Designer dyes: "biomimetic" ligands for the purification of pharmaceutical proteins by affinity chromatography. *Trends in Biotechnol.* 10:442–448.

Luong, J. H. T., K. B. Male, and A. L. Nguyen. 1990. Isolation of urokinase by affinity ultrafiltration. *Biotechnol. Bioeng.* 35:87–93.

Luong, J. H. T., A. L. Nguyen, and K. B. Male. 1984. Affinity cross-flow for purifying biomolecules. *Bio/Technology* 5:564–567.

Mantovaara, T., H. Pertoft, and J. Porath. 1991. Purification of factor VIII:c coagulant activity from rat liver nonparenchymal cell culture medium by immobilized metal ion affinity chromatography. *Biotechnol. Appl. Biochem.* 13:120–126.

Mantovaara-Jonsson, T., H. Pertoft, and J. Porath. 1989. Purification of human serum amyloid component (SAP) by calcium affinity chromatography. *Biotechnol. Appl. Biochem.* 11:564–571.

Mattiasson, B., B. Danielsson, C. Hermansson, and K. Mosbach. 1978. Enzyme thermistor analysis of heavy metal ions with use of immobilized urease. *FEBS Lett.* 85:203–206.

Menache, D., B. J. Grossman, and C. M. Jackson. 1992. Antithrombin III: Physiology, deficiency, and replacement therapy. *Transfusion* 32:580–588.

Merkle, R. K., and R. D. Cummings. 1987. Lectin affinity chromatography of glycopeptides. *Methods Enzymol.* 138:232–259.

Miller-Andersson, M., H. Borg, and L.-O. Andersson. 1974. Purification of antithrombin III by affinity chromatography. *Thromb. Res.* 5:439–452.

Mohapatra, S. S. 1992. Expression and thrombin cleavage of Poa p IX recombinant allergens fused to glutathione S-transferase. *Int. Arch. Allergy Immunol.* 98:343–348.

Nagai, K., and H. C. Thogersen. 1987. Synthesis and sequence-specific proteolysis of hybrid proteins produced in *Escherichia coli*. *Methods Enzymol.* 153:461–481.

Nethery, A., R. L. Raison, and S. B. Easterbrook-Smith. 1990. Single step purification of immunoglobulin M on C1q Sepharose. *J. Immunol. Methods* 126:57–60.

Nicola, A. N., A. W. Burgess, D. Metcalf, and F. L. Battye. 1978. Separation of mouse bone marrow cells using wheat germ agglutinin affinity chromatography. *Aust. J. Exp. Biol. Med. Sci.* 56:663–679.

Nikaido, T., A. Shimizu, N. Ishida, H. Sabe, K. Teshigawara, M. Maeda, T. Uchiyama, J. Yodoi, and T. Honjo. 1984. Molecular cloning of cDNA encoding human interleukin-2 receptor. *Nature* 631–635.

Nilsen, B. M., B. S. Paulsen, Y. Clonis, and K. S. Mellbye. 1990. Purification of the glycopeprotein allergen Ag7 from mugwort pollen by concanavalin A affinity chromatography. *J. Biotechnol.* 16:305–316.

Nilsson, K., and K. Mosbach. 1980. *p*-Toluenesulfonyl chloride as an activating agent of agarose for the preparation of immobilized affinity ligands and proteins. *Eur. J. Biochem.* 112:397–402.

Nilsson, K., and K. Mosbach. 1981. Immobilization of enzymes and affinity ligands to various hydroxy group carrying supports using highly reactive sulfonyl chlorides. *Biochem. Biophys. Res. Commun.* 102:449–457.

Nopper, B., F. Kohen, and M. Wilchek. 1989. A thiophilic adorbent for the one-step high-performance liquid chromatography purification of monoclonal antibodies. *Anal. Biochem.* 180:66–71.

O'Brien, R. D. 1960. *Toxic phosphorus esters: Chemistry, metabolism and biological effects.* NY: Academic Press.

Ohta, M., M. Okada, I. Yamashina, and T. Kawasaki. 1990. The mechanism of carbohydrate mediated complement activation by the serum mannan-binding protein. *J. Biol. Chem.* 265:1980-1984.

Olson, W. P. 1992. Targeted drug delivery based on the mode of action of bacillus Calmette-Guérin in bladder cancer. Ph.D. dissertation, University of Illinois at Chicago, Health Science Center; Ann Arbor, MI: UMI Dissertation Services.

Olson, W. P. 1987a. Sterility testing. In *Aseptic pharmaceutical manufacturing: Technology for the 1990s,* edited by W. P. Olson and M. J. Groves. Buffalo Grove, IL: Interpharm Press.

Olson, W. P. 1987b. Viruses in therapeutics from human plasma: upstream and downstream process strategies. In *Aseptic pharmaceutical manufacturing: Technology for the 1990s,* edited by W. P. Olson and M. J. Groves. Buffalo Grove, IL: Interpharm Press.

Olson, W. P. 1983. Particle adsorption. U.S. Patent 4,411,795, assigned to Baxter-Travenol Corp.

Parkinson, D. R., J. S. Abrams, P. H. Wiernik, A. A. Rayner, K. A. Margolin, D. A. Ven Echo, M. Sznol, J. P. Dutcher, F. R. Aronson, and J. H. Doroshow. 1990. Interleukin 2 therapy in patients with metastatic malignant melanoma: A phase II study. *J. Clin. Oncol.* 8:1650-1656.

Pierce, J. H., and I. D. Rattee. 1967. Acid-catalyzed hydrolysis of type 3 dye-fibre bonds between cellulose and triazinyl reactive dyes. *J. Soc. Dyers Colourists* 83:361–368.

Porath, J., J. Carlsson, P. Olsson, and G. Belfrase. 1975. Metal-chelate affinity chromatography, a new approach to protein fractionation. *Nature* 258:598–599.

Porath, J., F. Maisano, and M. Belew. 1985. Thiophilic adsorption—a new method for protein fractionation. *FEBS Lett.* 185:306–310.

Reichlin, M. 1980. Use of glutaraldehyde as a coupling agent for proteins and peptides. *Methods Enzymol.* 70:159–165.

Richards, F. M., and J. R. Knowles. 1968. Glutaraldehyde as a protein crosslinking agent. *J. Mol. Biol.* 231–233.

Robben, J., G. Massie, E. Bosmans, B. Wellens, and G. Volckaert. 1993. An *Escherichia coli* plasmid vector system for high-level production and purification of heterologous peptides fused to active chloramphenicol acetyltransferase. *Gene* 126:109–113.

Rodriguez, P. L., and L. Carrasco. 1995. Improved factor Xa cleavage of fusion proteins containing maltose binding protein. *BioTechniques* 18:238–243.

Rosenberg, S. A., M. T. Lotze, L. M. Muul, A. E. Change, F. P. Avis, S. Leitman, W. M. Lineban, C. N. Robertson, R. E. Lee, J. T. Rubin, C. A. Seipp, C. G. Simpson, and D. E. White. 1987. A progress report on the treatment of 157 patients with advanced cancer using lymphokine-activated killer cells and interleukin 2 or high dose interleukin 2 alone. *N. Engl. J. Med.* 316:889–897.

Saint-Blancard, J., J. M. Kirzin, P. Riberon, and F. Petit. 1982. A simple and rapid procedure for large scale preparation of IgGs & albumin from human plasma by ion exchange and affinity chromatography. In *Affinity chromatography and related techniques*, edited by T. C. J. Gribnau, J. Visser, and R. J. F. Nivard. Amsterdam: Elsevier.

Sawhney, A. S., C. P. Pathak, and J. A. Hubbell. 1993. Bioerodible hydrogels based on photopolymerized poly(ethylene glycol)-*co*-poly(alpha-hydroxy acid) diacrylate monomers. *Macromolecules* 26:581–587.

Scatchard, G. 1949. The attraction of proteins for small molecules and ions. *Ann. N.Y. Acad. Sci.* 51:660–672.

Schneider, C., R. A. Newman, D. R. Sutherland, U. Asser, and M. F. Greaves. 1982. A one step purification of membrane proteins using a high efficiency immunomatrix. *J. Biol. Chem.* 257:10766–10769.

Scouten, W. H. 1987. A survey of enzyme coupling techniques. *Methods Enzymol.* 135:30–65.

Secher, D. S., and D. C. Burke. 1980. A monoclonal antibody for large-scale purification of human leukocyte interferon. *Nature* 285:446–450.

Singer, I. I. 1989. Fibronectin-cytoskeleton relationships. In *Fibronectin*, edited by D. F. Mosher. San Diego: Academic Press.

Smith, M. C., T. C. Furman, and C. Pidgeon. 1987. Immobilized iminodiacetic acid metal peptide complexes. Identification of chelating peptide purification handles for recombinant proteins. *Inorg. Chem.* 26:1965–1969.

Steinemann, A., and L. Stryer. 1973. Accessibility of carbohydrate moiety of rhodopsin. *Biochemistry* 12:1499–1502.

Stewart, D. J., P. Hughes, and C. R. Lowe. 1989. Affinity chromatography on triazine dyes immobilized on novel perfluorocarbon supports. *J. Biotechnol.* 11:253–266.

Stites, D. P. 1978. Clinical laboratory methods for detection of antigens and antibodies. In *Basic and clinical immunology,* 2nd ed., edited by H. H. Fudenberg, D. P. Stites, J. L. Caldwell, and J. V. Wells. Los Altos, CA: Lange Medical Publications.

Sulkowski, E. 1989. The saga of IMAC and MIT. *BioEssays* 10:169–175.

Sundberg, L., and J. Porath. 1974. Preparation of adsorbents for biospecific affinity chromatography. I. Attachment of group-containing ligands to insoluble polymers by means of bifunctional oxiranes. *J. Chromatog.* 90:87–98.

Tjerneld, F., G. Johansson, and M. Joelsson. 1987. Affinity liquid-liquid extraction of lactate dehydrogenase in large scale. *Biotechnol. Bioeng.* 30:809–816.

Tsay, G. C. 1991. Lectin affinity chromatography of recombinant blood-coagulation factor VIII for glycoprotein removal. U.S. Patent 4,981,951, 1 January, 1991, assigned to Miles, Inc.

Turkova, J., K. Blaha, J. Horacek, J. Vajcner, A. Frydrychova, and J. Coupek. 1981. Hydroxyalkyl methacrylate gels derivatized with epichlorohydrin as supports for large scale and high performance affinity chromatography. *J. Chromatog.* 215:165–179.

Veldhoven, C. H. A., J. F. Meilof, J. G. Huisman, and R. J. T. Smeenk. 1992. The development of a quantitative assay for the detection of anti-Ro/ss-A and anti-La/SS-B auatoantibodies using purified recombinant proteins. *J. Immunol. Methods* 151:177–189.

Vinazzer, H. 1987. Clinical use of antithrombin III concentrates. *Vox Sang.* 53:193–198.

West, W. H. 1989. Continuous infusion recombinant interleukin 2 (rIL2) in adoptive cellular therapy of renal cell carcinoma and other malignancies. *Cancer Treat. Rev.* 16:83–89.

White, A., P. Handler, and E. L. Smith. 1973. *Principles of biochemistry,* 5th ed. NY: McGraw-Hill Book Co.

Whitehead, R. P., K. J. Kopecky, M. K. Samson, J. J. Costanzi, R. B. Natale, L. G. Feun, E. M. Hersh, and J. J. Rinehart. 1991. Phase II study of intravenous bolus recombinant interleukin 2 in advanced malignant melanoma. Southwest Oncology Group Study. *J. Natl Cancer Inst.* 83:1250–1252.

Whiteside, T. L., E. Letessier, H. Hirabayashi, D. Vitolo, J. Bryant, L. Barnes, C. Snyderman, J. T. Johnson, E. Myers, R. B. Herberman, J. Rubin, J. M. Kirkwood, and D. R. Vlock. 1993. Evidence for local and systemic activation of immune cells by peritumoral injections of interleukin 2 in patients with advanced squamous cell carcinoma of the head and neck. *Cancer Res.* 53:5654–5662.

Wolf, J. M., T. A. Hatton, and M. L. Yarmush. 1989. Bioaffinity separations using reversed micellar extraction. *Biotechnol. Progr.* 5:57–62.

Yip, T.-T., and T. W. Hutchins. 1992. Immobilized metal ion affinity chromatography. In *Practical protein chromatography*, edited by A. Kenney and S. Fowell. Totowa, NJ: Humana Press.

Yoshimoto, T., and D. Tsuru. 1974. Affinity chromatography of lysozyme on a Sepharose derivative coupled with lysozyme lysate of bacterial cell wall. *J. Biochem.* 76:887–889.

Zhang, S., W. Zhu, K. Zhou, D. Xia, and H. Liu. 1992. Removal of endotoxin in various solution by polymyxin B-sepharose 4B chromatography. *Shanghai Yike Daxue Xuebao* 19:9–12.

8

Monoclonal Antibodies

Gary Christiansen

The manufacture of predefined specific antibodies by means of permanent tissue culture cell lines is of general interest . . . Such cultures could be valuable for medical and industrial use.

These are the first and last statements in a paper published by Köhler and Milstein (1975). These were prophetic understatements to say the least.

Monoclonal antibodies produced from hybridoma cells are a significant part of the biotechnology boom that began in the mid 1970s. It was quickly recognized by immunologists and other scientists that there were far-reaching implications and tremendous opportunities of this technology (Hurrell 1982, 4–54; Lyle 1982; Strelkauskas 1987). Monoclonal antibodies have two very important advantages over conventional polyclonal antibodies:

1. High specificity
2. The ability to be produced in an unlimited supply and with a constant specificity and affinity

These characteristics have led to a number of applications for monoclonal antibodies, including cell-surface studies, diagnostic testing (ELISA, RIA, IRMA, etc.), tumor-associated antigen studies, histocompatibility studies, studies of infectious diseases, immunotherapy (radioimmunotherapy, chemotherapy, etc.) and purification of antigenic materials.

DEFINITIONS

Some useful definitions are given here to help in understanding this chapter.

Murine means derived from an animal belonging to the *Muridae* family which includes mice and rats.

Master cell bank (MCB) means the original cell line on which production is based. It may or may not necessarily have been produced by the manufacturer.

Manufacturers Working Cell Bank (MWCB) means a homogeneous suspension of seed cells derived from the MCB by a finite passage level, aliquoted into individual containers for storage (e.g., in a liquid nitrogen refrigerator). All containers are treated identically during storage and once removed from storage, the containers are not returned to the seed stock.

Hybridomas are cell lines made by fusion of lymphocytes from an immunized donor with myeloma cells.

Process validation is establishing documented evidence which provides a high degree of assurance that a specific process will consistently produce a product meeting its predetermined specifications and quality characteristics.

HISTORICAL NOTES

Köhler and Milstein (1981; 1976; 1975) and Milstein (1980) were studying mechanisms of gene expression and chose the synthesis of antibodies by lymphoid cells as their experimental model. They found that the cell fusion techniques they described were powerful tools to produce a specific antibody directed against a selected antigen. In their experiments they demonstrated the methods to isolate hybrid lines producing different antibodies directed against the same antigen and carrying different effector functions. Their work demonstrated that the selection of hybridoma lines in which only the specific antibody chains were produced was reasonably simple. They concluded that these cells could be grown in vitro in massive cultures to provide the specific antibody in large quantities.

THE PRODUCTION OF MONOCLONAL ANTIBODIES

The Fusion Process

In order to understand the purification process of monoclonal antibodies, it is necessary to understand the origins and production process of monoclonal antibodies. The specific production process will help to define the purification process by identifying certain impurities that must be removed by the purification process. The source of the hybridoma cell also defines the impurities that must be removed during the purification process.

The process for producing monoclonal antibodies (Goding 1986, 56–89; Hurrell 1982; Lyle 1982) starts with the injection of an antigen into a mouse. Methods have also been developed and reported for in vitro immunization to produce specific antigen-activated lymphocytes for fusion to produce hybridomas (Luben and Mohler 1980). For this purpose the antigen does not need to be highly purified as in traditional antiserum production methods. After a few days, the lymphocytes proliferate in response to the antigen. Antibodies are produced in cells by a group of cells called B cells. These cells recognize antigens through surface receptors that become copies of the serum antibodies that are found later, after the immune response is developed. The foreign antigen combines with the receptors and causes the cells to proliferate. Eventually the B cells become mature cells that secrete antibodies into the serum. The antibodies are specific for the initial stimulating antigenic determinant. These B cells, however, are highly diverse and each B cell will produce antibodies that have a unique way of binding to the antigenic molecule. Each cell produces millions of copies of the one antibody, but it does have a finite lifetime and does not grow well under culture conditions. At this time the spleen, which is rich in lymphocytes, is extracted and the entire population of the lymphocytes obtained from the spleen are mixed with the myeloma cells.

Myeloma cells, on the other hand, grow readily in culture and are essentially immortal. The most common myeloma cells used are of mouse origin. These malignant cells have the ability to translate the genetic information from the immunized animal's spleen cell into functional antibodies. Myeloma cells can, in fact, make antibodies on their own, but these antibodies usually have no identifiable specificity. Some myeloma cells make light chain portions of the antibody and others make no identifiable immunoglobulin portions. By fusion of the spleen cells with myeloma cells that produce or secrete no immunoglobulin on their own, the resultant cell line may produce antibodies having the characteristics of the cells from the immunized animal only.

Cells are fused most commonly by adding polyethylene glycol (PEG) to a concentrated mixture of the two cell types. The PEG causes agglutination of the cells. The mixture is then diluted and fusion of cell pairs takes place randomly (Knutton and Pasternak 1979). Since the lymphocyte cells cannot grow in tissue culture, the myeloma cells will quickly overgrow the fused cells. Myeloma cells that are enzyme deficient mutants are used in the fusions and this characteristic is used to eliminate the unfused myeloma cells (Paul 1975, 394–396). The cells are collected by centrifugation and resuspended in tissue culture media containing animal sera. Virtually all hybridoma procedures are based upon the growth of these cells in a selective medium. The medium contains hypoxanthine, aminopterin, and thymidine (HAT). The cells are distributed to tissue culture plates. Certain myeloma cell lines have a defective enzyme, hypoxanthineguantine phosphoribosyl transferase (HGPRT), and will die when placed in this medium. The cells are grown for the next few days in HAT medium and the myeloma cells and the lymphocyte cells die, leaving only the fused cells. The culture now contains a mixture of rapidly growing hybridoma cells.

The cells are isolated and screened for the production of the antibody of the desired characteristics. Once an individual clone is selected, the cell may be placed in tissue culture media and grown by conventional tissue culture means. The fusion product or hybridoma does not adhere to glass or plastic so it may be conveniently maintained as a suspension culture. The selected hybridoma cells are grown in larger quantities by growing them in culture fluid (e.g., in spinner flasks). The cells can be preserved by freezing in liquid nitrogen for long periods of time or merely sustained in tissue culture fluid. From these cells the seed lot can be established. The MCB and the MWCB are established and characterized. These cells are used for the production of future lots of monoclonal antibodies. Production of the monoclonal antibody product can be prepared either in vivo and isolated from ascites fluid or in vitro and isolated from the cell culture supernatant.

In Vivo Production

For large quantities of antibody, the cells are injected into the peritoneal cavities of mice. Often the mice have been primed by injection with a mineral oil such as pristane (tetramethyl pentadecane) to make them more prone to the development of tumors. The cells grow and produce large quantities of antibody, which can be obtained from the resultant ascitic fluid. This procedure will produce several milligrams per milliliter quantities of antibody. Depending on various conditions, large quantities of antibody can be obtained from a single mouse. Each production batch should originate from a fresh vial of the MWCB. The number of allowable serial passages of the cells should be restricted and defined. The ascites fluid is harvested from the animals and the purification steps begin. This procedure, though widely used, does have some drawbacks. A number of contaminants are present in the ascites fluid including irrelevant mouse immunoglobulins, proteases and nucleases, and viral contamination. Variability in the animals and the content of the ascites fluid all add to the difficulties of purification.

In Vitro Production

Alternatively, and in most cases, preferably, large quantities of antibody can be produced by tissue culture methods. There are many variations on the methods of tissue culture propagation. Again, each production batch should originate from a fresh vial of the MWCB. Details of the production method should include limits on the allowable number of cell population doublings. Criteria, such as titer of the monoclonal antibody, should be established. The cell culture supernatants must meet the criteria before it is acceptable for further processing or purification. Batchwise production methods are easier to control and standardize, but continuous culture production, which can pose considerable problems, is acceptable if it can be demonstrated that the method is advantageous and that all aspects are critically controlled. Depending on the nature of the culture medium employed, the purification process will be affected. The concentration of antibody in culture fluid supernatants is generally at a much lower concentration than the concentration in ascites fluid. Therefore, the purification schemes for antibodies from these different methods may call for quite different approaches, at least in the early steps of the purification.

A number of variations of the in vitro production process have been developed. The cells can be grown in T-flasks, roller bottles, and large fermentor tanks of various sizes from a single liter to hundreds of liters. The cells can be encapsulated or entrapped in a hollow fiber device. Variations in the cell-culture medium are many and diverse. The medium may contain animal sera, such as fetal calf serum, or it may be a specially developed low-serum or serum-free medium. Any culture medium is a complex mixture of components, including growth factors, hormones, transferrin, and other additives that are necessary or beneficial to cell growth. The more complicated the media, the more complicated will be the purification process. In developing the purification process, the various types of contaminants must be taken into account. The antibody concentration in the culture supernatant is generally quite low. Steps must be included in the purification scheme to concentrate the antibody to facilitate the final purification.

PURIFICATION OBJECTIVES AND GOALS

Challenges

Some of the challenges that must be faced when purifying a monoclonal antibody result directly from the nature of the antibody product. Monoclonal antibodies have variability in classes, subclasses, clones, variants, and batches. The amount and type of contaminants vary in different production systems. These differences are a challenging problem for both analytical and preparative purifications.

The problems are both technical and cost-effectiveness in nature. One must determine purification steps that are efficient, reproducible, and rapid. Ideally, the purification process will not be labor intensive. The challenges are to determine effective prepurification protocols that do not result in reduced yields. Preliminary steps may include concentration steps, delipidation steps, and so on. It is important that the monoclonal antibody is still active and stable after purification. Methods of purification must be selected that will minimize the loss of activity or reduction in stability. The ultimate scale of the purification will have an impact on the purification scheme selected. It is not desirable to develop the purification based on one set of purification parameters, only to discover that the scheme must be changed when the process is scaled up.

It is useful to understand the properties of the monoclonal antibody in question. The more knowledge available about the antibody, the easier will be the selection of purification steps. Even though two monoclonal antibodies are very similar in terms of class, subclass, and source species, they may differ significantly in their physical and chemical characteristics. These differences allow the separation of closely related antibodies. The method of choice for the separations can be difficult. Factors such as the isoelectric point (pI) of the monoclonal antibody, the stability at different pH or different salt concentration, and stability to different salts are all important factors in the selection of a purification method.

The actual requirements for purity of the final product will depend on the ultimate use of the monoclonal antibody. It is important to determine the degree of purity that is actually required as accurately as possible before developing and finalizing the purification scheme. There is no need to purify the monoclonal antibody beyond the actual needs. The time and resources can be very expensive. There may be cheaper or less specific methods that can be used (e.g., gel filtration instead of affinity chromatography) and still be sufficient.

Often the monoclonal antibody is a minor component in a complex mixture of similar proteins. The monoclonal antibody may coelute with some of the major contaminants. Antibody concentrations may vary from micrograms per milliliter to tens of milligrams per milliliter, depending on the source of the antibody and the procedure for manufacturing it. The subclass may not always be known before purification is attempted. Many factors must be considered. The antibody must be fully active when it is recovered. It cannot be affected or denatured by the purification process. The challenge for therapeutic or injectable antibodies is even greater because of the extremely high purity requirement for these products.

In addition, analytical methods capable of monitoring the degree of purity must be available. Scaleup from pilot- or research-scale purification to production scale is always a major challenge for any process.

The fact that the monoclonal antibodies are prepared in growth media poses a constant threat of contamination. Extreme caution in the process is required to prevent inadvertent contamination of the product during the processing. Processes must be reliable and reproducible and, therefore, must be capable of being validated.

Monoclonal antibodies pose additional problems compared to traditional polyclonal antibodies. The challenge is to identify the special properties of the monoclonal antibodies, such as pI, solubility; and tolerances to heat, pH, salt, or chaotropic conditions; and to take advantage of them in the development of the purification scheme. The purification process must be simple, fast, reliable, and cost-effective. The results of the process must be a product that is of high purity, high recovery, and good throughput. In addition, it is ideal if there is linear scaleup in the process.

Additional challenges in the purification process are imposed by regulatory requirements for the removal of virus contamination, DNA contamination, and pyrogen contamination. Further, the challenge to validate the effectiveness of the process is a major consideration.

Regulatory Requirements

The testing requirements for monoclonal antibody products are set forth as testing recommendations by the FDA in the Points to Consider Documents described below. The *Points to Consider in the Manufacturing and Testing of Monoclonal Antibody Products for Human Use* were reissued in 1987 (Food and Drug Administration 1987d). The *Points to*

Consider in the Characterization of Cell Lines Used to Produce Biologicals (Food and Drug Administration 1987c) covered the hybridoma cells used to produce biological products. The recommended testing needed to fully characterize a cell line is described in detail in this document. These requirements and the rational for the virus assays have been reviewed (Joner and Christiansen 1988).

Since there is the possibility that viable viruses might be present in the purified monoclonal antibody products, such viruses could infect humans, particularly patients who are immunocompromised by their disease states. Monoclonal antibody products derived from ascites produced in mouse peritoneal cavities are subject to contamination risks with viruses in the host mouse as well as from viruses propagated by the hybridoma cells. Murine and chimeric monoclonal antibody products produced in vitro can be contaminated by murine viruses. Hybridomas that are partially or entirely of human origin may have human viruses that contaminate the antibody. The generation of antibody brings the risk of contamination from personnel as well as tissue culture procedures. As the hybridomas age, there is the possibility that the hybridoma cell may express certain viruses. The *Points to Consider* recommend testing the murine antibody products for the following list of murine viruses:

- Sendai
- Ectromelia
- Mouse adenovirus
- Mouse hepatitis virus
- Minute virus of mouse
- Pneumonia virus of mouse
- Reovirus type 3
- Polyoma
- Hantaan
- Epizootic diarrhea of infant mice
- Murine cytomegalovirus
- Lymphocytic choriomeningitis virus (LCMV)
- Lactic dehydrogenase elevating virus (LDHV)

The presence of these viruses can be detected by the mouse antibody production (MAP) test. In general, mice are injected with the monoclonal antibody product through various routes. Serological tests are performed that determine whether the animals have any detectable antibodies to the viruses in question. Additional tests for viruses include tissue culture isolation methods followed by biochemical analysis to detect reverse transcriptase levels typical of these murine retroviruses, or to detect increases in lactic dehydrogenase levels in the animal serum 3 to 7 days after injection, which is indicative of the presence of LDHV. Another test that should be performed is the tissue culture safety test, a general screen for a variety of infectious viruses. This assay uses isolation assays on several sensitive cell lines. These cell lines exhibit reactions to several viruses, including herpesviruses, paramyxoviruses, parainfluenza viruses, poxviruses and rhabdoviruses. If the source of the monoclonal antibody product is of human origin, there are several human viruses that should be tested for. These would include the following:

- Epstein-Barr virus (EBV)
- Human cytomegalovirus
- Hepatitis B virus
- Retroviruses

Various assays are available to test for these.

Products from either ascites fluid or from tissue culture methods should be tested for viruses before and after purification. Validation studies should be performed that demonstrate that the adventitious agents are removed or inactivated by the purification process. These validation studies are important in characterizing the purification process and will be discussed later in the chapter.

There are additional recommendations in the *Points to Consider* documents that help to define the requirements of the production and purification process. The *Points to Consider* recommend strongly that purification steps beyond simple salt fractionation of culture supernatants or ascites fluids take place. Suggested steps are size-exclusion chromatography (SEC), affinity chromatography, ion-exchange chromatography (IXC), and/or ultracentrifugation. Other treatments are recommended to ensure that the final product will be free from contamination. These include high temperature incubations; pH treatments; and exposure to organic solvents, chelating agents, or nucleases.

The test requirements for the product also help to define the requirements for the purification process. For example, tests are recommended to detect aggregation, denaturation, or fragmentation of the immunoglobulin product; therefore, it is concluded that the purification process must be capable of removing these potential contaminants. The product must be sterile and free of mycoplasma contamination. The best method, of course, is to eliminate or minimize the possibility of any contamination during the process, but the purification process must be capable of removing any such contamination regardless. Any additives used in the propagation of the hybridoma should be removed during purification, since serum additives of animal origin are capable of producing an allergic response in human subjects. Tests for residual amounts of these additives should be performed to demonstrate the residual amount or complete absence of these additives. Since the final product is being tested for potency, the purification process should affect the activity of the antibody minimally. Tests must be performed for polynucleotide contamination that have a sensitivity on the order of 10 pg of DNA per dose. Validation procedures are recommended to establish that any potential viral and nucleic acid contaminants have been effectively excluded or inactivated from each bulk lot by the manufacturing process used.

In addition to the *Points to Consider* documents, there are guidelines issued by the Commission of the European Communities for the production of monoclonal antibody products (EEE Ad Hoc Working Party 1991). These guidelines offer additional insight to the expectations of the purification process. Many of the issues are the same or similar to those in the *Points to Consider* documents from the FDA. There are some additional points that are insightful. Specific controls are recommended for the production areas, including that the production area be decontaminated before being used for the manufacture of monoclonal antibody products and that access to the area be limited. Again, in these guidelines it is strongly recommended that the purification process be validated to determine the capacity of the process to remove contaminants, including unwanted immunoglobulins, DNA, viruses, and irritants (e.g., pristane). Any inactivation process should be shown to be effective and to not compromise the biological activity of the product. Analytical tests are recommended that include column chromatography (including HPLC), isoelectric focusing, SDS–PAGE under reducing conditions, and a DNA assay method.

Validation Requirements for Purification

Once the purification process is determined, it must be validated in order to comply with the requirements set forth in the FDA guideline for process validation (Food and Drug Administration 1987a) For a purification process there is certain information the regulatory authorities will demand. Some of the validation parameters that must be considered are as follows:

- Column qualification
- Acceptance criteria for chromatographic media

- Maintenance and sanitization
- Column lifetime
- Clearance of nucleic acids, endotoxins and viruses
- Leakage of contaminants from the chromatographic media

Column qualification will include consideration of a number of factors. SOPs for the use of the columns must be written that include the packing of the column, procedures for the equilibration of the column, performance determination, regeneration of the column, column maintenance, and proper storage conditions and times for the packed column.

The information should include the column size and the materials from which it is made. The location of the column should be defined (i.e., in a class 100 hood or a clean room). Documentation of the system should include a schematic diagram of the system. Additional parts and components of the system (such as tubing materials and size, pressure gauges, pumps, fittings, etc.) should be clearly stated.

The SOP for the packing of the column can usually be developed from information provided by the vendor of the column and packing material. A pressure/flow curve for each column should be developed and used to establish the optimal packing for each lot of chromatographic medium. The effectiveness of the packing procedure can be defined by the efficiency of the packing method by determining the height equivalent of the theoretical plates (HETP) and peak symmetry. Determination of peak homogeneity for HPLC was reported by Carter et al. (1982). Specifications for these factors should be established and measured as part of the production process. These specifications should be defined for the worst case (i.e., at both the upper and lower limits that might occur). The peak symmetry is calculated by the formula

$$S = b/a,$$

where S is the symmetry, b is the trailing peak half-width at 10 percent of the peak height, and a is the leading peak half-width at 10 percent of the peak height. Records of the column performance should be maintained to detect any changes or trends. Performance characteristics that must be determined during development are purity and yield data, flow rate ranges, sample load ranges, selectivity, and the capacity of the chromatographic medium. As part of the validation of the column, the effective lifetime of the column should be determined. This can be arrived at from the preventive maintenance data, evaluation of the buildup of material on the column, efficiency of the cleaning methods, and the stability of the medium to the cleaning agents. The column should be evaluated after each cleaning for column performance to assure that the medium can be regenerated to the original starting conditions. Data can be generated on a small scale, using an automated system that will generate great quantities of data in a short period of time, then the data can be collected on the large-scale columns in a database to assure that the performance of the column during its lifetime is adequate.

Once a system for purification is selected, the installation qualification (IQ) should be performed and documented. The IQ protocol should be written and approved in advance of performing the qualification. Items that would be included in the IQ are evaluation of the equipment design for the intended purpose, evaluation of equipment for function, and calibration. Procedures for cleaning, maintenance, and operation should be written and approved. The IQ also includes a description of the process (often a flowchart is useful), a description of the facility and location of the equipment (including air quality conditions), information of the electrical requirements and any other utility requirements, and testing and calibration protocols and data.

The operational qualification (OQ) would include such information as the proper operation of the pumps, detectors, and fraction collectors; documentation of the proper flow rates; and proper functioning of any valves and controllers. If the process is controlled by a computer program or a microprocessor, these should also be documented and qualified.

The performance qualification (PQ) would include such information as yield at each step, documentation of the removal of contaminants, removal and/or inactivation of potential viruses, the sterility and pyrogenicity at each step, and the final conformance to specifications of the product.

Virus Removal and Inactivation

The nature of the production of monoclonal antibodies, whether from ascites fluid or from tissue culture methodologies, makes the product susceptible to viral contamination. Viruses come from two sources: (1) endogenous viruses that are covalently linked to the host genome and (2) adventitious viruses that are external to the host genome. It is required that the MCB and culture medium are free from retrovirus or oncogenic material contamination. The characterization of the MCB and the MWCB will make that determination. It must be further demonstrated that viruses are not copurified with the product. Viral clearance and viral inactivation studies must be performed for monoclonal antibody products. Recommended procedures are in the EEC Guidelines (EEC Ad Hoc Working Party 1991). The removal and/or inactivation of model viruses is performed by spiking the product with model viruses that can be cultivated at high titers. Viral inactivation can occur during the chromatography steps as a result of the conditions, such as pH, salt concentration, or denaturing agents. Steps such as heat treatment and addition of detergents or PEG or other viral inactivation components can be included in the purification process. Virus removal can also be accomplished by chromatography methods, including gel filtration, IXC, and affinity chromatography. Carthew (1986) demonstrated the removal of rodent viruses during fractionation of monoclonal antibodies using gel filtration and affinity chromatography. Zolton et al. (1985) has demonstrated the removal of hepatitis B virus from immunoglobulins by anion exchange chromatography. Tsao et al. (1988) used affinity chromatography for the removal of viruses and virus infected cells.

The chromatography column and medium collect and concentrate the viral contamination. Therefore, the cleaning of the chromatography column must take into consideration the inactivation of viral contamination of the chromatography medium. This can be accomplished using pH treatment, PEG, detergents, and NaOH treatments. This must be considered as part of the validation of the chromatography column.

Each step of the purification procedure must be evaluated for the effectiveness of viral inactivation or removal. By determining the log reduction of the spiked viral contamination for each step, one can determine the total viral reduction capability of the purification process by adding together the log reductions for each step.

DNA and Nucleic Acid Removal

The requirements for DNA contamination limits are in the *Points to Consider* (Food and Drug Administration 1987c) and in World Health Organization (WHO) guidelines (1991). The WHO guideline states the upper limit at 100 pg/dose, whereas the *Points to Consider* set the limit at 10 pg/dose. The potential sources of nucleic acid contamination are from the host cell DNA and from retroviral RNA.

In a manner similar to the viral inactivation and removal, the removal of nucleic acids in the purification process can be demonstrated by clearance experiments with spiked DNA. Each purification step is evaluated for clearance of DNA. The spiking experiments can be performed at a small scale, representative of the large-scale purification using the same conditions. The size of the spiked DNA should be similar to that found in the host. The DNA levels may be determined by immunochemical methods that determine all DNA, regardless of source, or by probe hybridization methods that are more specific.

Techniques that have been found to be effective for the removal of DNA include anion exchange, DNAse treatment, and specific binding using immunoadsorbents. The total clearance factor can be determined as the sum of the clearance factors for each step. As the case for viruses, the cleaning procedure for the chromatographic medium must be validated for the removal of DNA that is concentrated on the column substrate.

Pyrogen Removal

The most common pyrogenic contaminants encountered are endotoxins. Other pyrogenic materials can include pyrogenic steroids, whole Gram-negative and Gram-positive organisms, peptidoglycans, fungi, and so on. Endotoxins are lipopolysaccharides from gram-negative bacteria. Some of the characteristics of the endotoxin that affect its removal from the product are that endotoxin is highly negatively charged, it is heat stable, relatively insensitive to pH changes, and tends to exist in high molecular weight aggregates.

Methods used to detect endotoxins are the *Limulus* amebocyte lysate (LAL) test, the rabbit pyrogen test and the human leukocytic pyrogen assay. The LAL test is sensitive and fast, but it only detects endotoxins. Many substances can interfere with the LAL test such as buffers and chromatography media. The test must be validated to account for any interference factors. Methods for the validation of the LAL assay are outlined in the FDA's *Guidelines on Validation of the Limulus Amebocyte Lysate Test as an End-Product Endotoxin Test for Human and Animal Parenteral Drugs, Biologicals, and Medical Devices* (Food and Drug Administration 1987b). Although the rabbit test is sensitive to nanogram levels, it is error prone due to the environment of the testing conditions, immune responses by the rabbits to the biologic materials being tested (especially if they are used for more than one test), and the development of a tolerance to endotoxins.

Endotoxin contamination can arise from human operators, buffers, salts, chromatographic media, and water. Raw materials should be tested before use to assure the absence of endotoxins. Bacterial growth in the equipment contributes to the endotoxin load. This can be minimized by avoiding threaded connections on the equipment, avoiding crevices in the piping or tubing, and changing filters frequently, all of which tend to harbor bacterial growth and are difficult to clean effectively. Practices that are useful in producing pyrogen free products are to use closed systems, to filter all solutions, to use pyrogen free water, to work in sterile areas whenever possible, to use aseptic techniques, and to store product at low temperatures to inhibit bacterial growth. Precautions must be taken to eliminate endotoxins in the process, including the depyrogenation by dry heat of components that can withstand dry heat depyrogenation conditions, washing with NaOH, and screening materials for endotoxins. Endotoxins can be removed from aqueous solutions by the use of depth filters, some detergent treatments, ultrafiltration, gel filtration, IXC, and affinity chromatography (Issekutz 1983). The elimination of endotoxins based on size differences would include UF (Sharma 1986), gel filtration, and density gradient centrifugation.

Quality Control of the Monoclonal Antibody

Ultimately, the purification of the monoclonal antibody must result in a product that meets the specifications for a safe and effective product. The quality control of the process, the intermediate stages and the final product must be capable of determining the conformance to the regulatory requirements and the specifications. Methods must be developed to detect potential contaminants at very low levels. Assays that are used to test for these trace contaminants include assays based on such technologies as immunoassays (ELISA, IRMA, RIA, etc.), HPLC, and hybridization techniques (DNA or gene probe assays). The development of these analytical methods can be a significant challenge.

The essential requirements of the quality control of pharmaceutical monoclonal antibodies are identity, purity, potency, efficacy, stability, and safety. Proper quality control of the development and characterization of the hybridoma cell (or other organism) producing the monoclonal antibody, quality control of the culture process, quality control of the purification processes, and quality control of the final product are all essential.

Tests for identity include PAGE gels, amino acid analysis, isoelectric focusing, HPLC, immunoassays, and immunoreactivity. Purity tests include PAGE gels, HPLC, and a variety of tests for specific impurities or contaminants. Determination of purity for monoclonal antibodies is difficult compared to standard pharmaceuticals. Chemically equivalent forms may not be acceptable (e.g., aggregates). Contaminants include bacteria, fungi, viruses, mycoplasma, endotoxins, residual solvents or cleaning agents. Impurities may include host cell components, cell culture media, nucleic acids, column materials, and so on. Bioassays may be used for potency. Safety will be evaluated by a number of procedures throughout the development cycle, but the General Safety Test, as described in 21 CFR 610.11, will be required for these biological products.

PURIFICATION METHODS

Miscellaneous Classical Methods

The first steps in a purification process are relatively crude procedures to either concentrate the antibody, to provide a gross separation from other materials, or to perform a buffer exchange. All of of which is dependent on the source of the antibody.

In the case of ascites-produced monoclonal antibodies, the initial step in the purification consists of the removal of lipids, clotted material, and any other solid material contained in the ascites fluid. Filtration through a 0.45 μm filter removes the solid debris and serves to reduce and maintain a low bioburden. Precipitation of the immunoglobulin fraction with either sodium sulfate or ammonium sulfate at approximately 50 percent saturation provides the initial isolation of the monoclonal antibody. Chromatographic steps provide the ultimate purification step to result in high purity product.

In the case of antibody produced in culture fluid, whether serum-containing media, low-serum media, or serum-free media, a concentration and/or a buffer exchange step is necessary, as is the initial separation of the cells and cell debris. The cells can be separated by centrifugation, by filtration, or by filtration through a 0.2 μm hollow fiber cartridge. Ultrafiltration can be used to concentrate the cell-free harvest and at the same time to exchange the buffer to the buffer of choice for the continued purification steps.

Dialysis is a relatively simple process that employs gentle conditions and is adaptable to a various batch sizes, from a few milliliters to hundreds of liters. Membranes are characterized by a nominal molecular weight cutoff (MWCO). The dialysis of a solution containing the product and a mixture of solutes will result in a retained solution that still contains all of the solutes, but at a different ratio, enriched in the higher molecular weight components. Ultrafiltration (UF) uses applied pressure to effect the separation. Ultrafiltration is relatively gentle as a process and produces similar results, although the retentate will have a greater concentration than the original solution because the solvent and inorganic salts permeate more rapidly than any of the solutes. The utility of this process is in the concentration of the product in the retained solution. Theoretical aspects of the methods are reviewed by Saltonstall (1992).

A variation of the membrane technology is affinity membrane filtration (Pfund and Charles 1988). In this technology, attachment of ligands (such as antibodies, enzymes, or other coatings) convert the membrane from a passive, size-only separation device to an active separation tool. Affinity membrane technology is available with Protein A specificity and Protein G specificity; and others are available or under development from companies such as Amicon (Beverly, MA), Millipore (Bedford, MA), Pall (East Hills, NY), Sepracor (Marlborough, MA), and so on. Other variations include hollow fiber devices, spiral membrane cartridges, and cross-flow or tangential flow ultrafiltration (TFU). Cross-flow or tangential flow refers to the principle direction of process flow relative to the membrane surface. This technology decreases fouling or clogging of the membrane. Both hydrophilic and hydrophobic membranes are available and were compared by Levy and Sheehan (1991).

Generally, the problems with membrane technology are fouling and clogging; variation in pore size; and susceptibility to degradation on repeated cleaning, sanitizing, and use. The advantages include large-scale application, gentle conditions, and relatively low cost for unsubstituted membranes.

The concentrated antibody solution is now ready for further purification. Again, chromatographic steps will provide the ultimate purification steps to yield the high purity product.

Chromatographic Methods

There are a number of chromatographic reagents available to purify the antibody. In selecting the proper chromatographic media it is necessary to define the purification goals. One must consider the regulatory requirements of the purified product that were discussed earlier. One must also consider the known contaminants that may be present and their characteristics. And, of course, a thorough knowledge of the properties of the antibody itself will help in selecting the proper chromatographic media. In order to obtain a therapeutic-grade antibody it is likely that more than one chromatographic medium will have to be used. The sequence in which the media are used is also of importance. The types of chromatographic media include cation/anion exchange, hydrophobic interaction, affinity, and gel filtration. The general types of chromatography and their modes of separation are summarized in Table 8.1. A wide variety of each type of chromatography media are available commercially.

Gel filtration media separate proteins on the basis of their size and molecular shape. These media are generally low cost with medium resolution but also have low capacity. Gel filtration or size exclusion chromatography (SEC) is a cost-effective method for desalting or for buffer exchange. High resolution media are available that have been optimized for increased resolution and stability due to a higher degree of cross-linking and small particle size. They are available in different fractionation ranges. Often the first chromatographic step in a purification scheme is SEC because it can be used in a variety of conditions,

Table 8.1. Basis of Separation of Various Chromatographic Media

Media	Basis of separation
Gel filtration	Molecular size and shape
Ion exchange	Net charge of molecule
Hydrophobic interaction	Hydrophobic nature of the molecule
Affinity chromatography	Biospecific interaction

eliminating the need for special pretreatment of the sample. Yields are excellent and predictable. It is best used before IXC, particularly with crude preparations, to reduce the number of components that must be dealt with on the next step. It is sometimes used for final cleanup. It can be used for the removal of transferrin after ion-exchange steps. With each step purity increases; but the yield also decreases, so it is important to select the purification steps rationally and so that there are as few steps as possible.

Ion-exchange media separate the molecules based on differences in the charge characteristics of the molecules. Various forms of ion-exchange media are available, including both weak and strong anion and cation exchangers. Hydroxyapatite is an example of this type of gel that separates on the basis of charge and charge density. Ion-exchange is capable of separating proteins with small differences in charge and is, therefore, capable of high resolution. Nucleic acids and nucleotides are strongly charged and are, therefore, readily separated by ion exchange chromatography. This makes ion chromatography a useful tool in the removal of DNA from the antibody. Ion-exchange chromatography is an excellent second step for two reasons; it concentrates the sample and it provides high resolution. Ion-exchange is an excellent method for monoclonal antibody purification, however, it does require careful sample pretreatment since some contaminating proteins will copurify. Transferrin is one of the most common contaminants remaining after ion-exchange purification since it has similar charge properties. Albumin is much easier to remove using IXC. The albumin is readily removed on the ion exchange step since the pI of bovine albumin (pI 4.8) is far removed from that of the MAb. Some of the advantages of IXC include gentle purification conditions, conditions that are easy to scale up, high capacity, and the ability to purify nearly any monoclonal antibody; and there are many documented and proven techniques and conditions available. Disadvantages may include frequent contamination with transferrin, extensive optimization required to result in high purity, and sometimes difficult to predict or reproduce elution profiles.

The most powerful and cost effective single purification step is ion exchange. Many of the beaded polymeric ion exchangers can be cleaned using rigorous conditions (e.g., 1 *M* NaOH) and sterilized by autoclaving. The ion exchangers are available in uniform lots in up to carload quantities.

The pretreatment of the sample prior to IXC would include equilibration with the starting condition buffer (i.e., the salts, ionic strength, and pH desired for loading the sample onto the ion-exchange resin). Buffer exchange can be done first on a highly cross-linked size-exclusion medium, e.g., Sephadex G-25 Coarse, equilibrated with the starting condition buffer for the first ion exchanger. Alternatively, the sample may be recirculated through a UF with the desired buffer added to the feed tank. Concentration of the samples prior to the addition to ion exchange or affinity columns is not necessarily required since these types of chromatogrqaphy are concentrating steps, but to reduce the buffer equilibration time and the quantity of buffer consumed, one may choose to concentrate the sample in the UF and then diafilter to equilibrate the sample in the new buffer.

A generic list of ion exchange media is shown in Table 8.2. The list includes the inorganic medium, hydroxyapatite [$Ca_{10}(PO_4)_6(OH)_2$], that despite its heavy loading of Ca^{2+} is anionic (i.e., a cation exchanger).

The preferred media are the beaded polymeric resins commercially available from several sources. The beaded polymeric resins are relatively stable, available in large quantities, demonstrate good batch to batch reproducibility and scalability.

The ion exchangers of choice are strong rather than weak ion exchangers. They have high capacity over a wide pH range and have sharp release profiles. Major release of the desired antibody does not occur until the eluting buffer comes close to the critical pH or salt molarity. Because strong exchangers tend to be tenacious, they can be loaded amost to capacity.

The following describes a sample purification of a monoclonal antibody (MAb) with a pI of 6.1. The pI of 6.1 has been determined from electrofocusing experiments or by equilibrating small samples to various pHs, and strong exchanger minicolumns in the same buffers and noting at which pH the MAb is or is not bound. The MAb is a murine IgG produced by a hybridoma in culture fluid. The culture media contains bovine serum. There is a large volume of culture media containing a low concentration of the desired MAb. The volume of growth media is first reduced and buffer exchanged in a UF system.

Table 8.2. Commonly Used Ion Exchangers

Exchanger	Polymer
Strong anion exchangers:	
QAE-Sephadex	cross-linked dextran
QAE-Sepharose	cross-linked agarose
QA-Cellulose	cellulose
Strong cation exchangers:	
SP-Sepharose	cross-linked dextran
SP-Sepharose	cross-linked agarose
Bio-Rad AG50	acrylamide copolymer
Hydroxyapatite	crystalline Ca-Phosphate

Since the MAb has a pI of 6.1 and the first ion exchange resin selected is a cation exchanger, with either a sulfoethyl (SE) or a sulfopropyl (SP) ligand, the MAb, if it is to bind to the resin, must be cationic and the pH of the starting buffer must be significantly lower than 6.1, perhaps 5.5 or even as low as 5.0. The buffer system of choice should be inexpensive and pharmaceutically acceptable. Citrate and phosphate buffers are the most common examples used in the industry. In this example, a phosphate buffer in the range of 0.05 to 0.10 *M* is selected for starting condition development. Preliminary experiments, development of the column may have been done with a linear NaCl gradient until the MAb was eluted, usually in a peak skewed slightly. But a linear gradient system is neither practical nor economic on a production scale. Therefore, one quantitates the pre- and postelution conditions that result in the recovery of about 98 percent or better. All intact copies of the MAb will have the same or at least very close to the same pI. Therefore, if the pH of the starting conditions on a strong ion exchange resin are within 1 pH unit of the pI of the desired species, there is an excellent prospect of eluting the MAb with a salt gradient. The gradient will not be linear, but stepwise. One would have starting conditions during which the proteins that are anionic at pH 5.1 elute from the column as waste. One runs a minimum of one and a maximum of two bed volumes of the starting buffer to flush out the nonbinding species. The elution is performed slowly with the phosphate buffer + salt that elutes the MAb. A gradient will form within the column and a relatively sharp MAb peak, observable in a flow-through UV monitor, will result in one bed volume. Occasionally, dependent upon the matrix and culture medium, up to 5 percent of the MAb is recovered at a slightly higher molarity of salt. The elution conditions may be altered to the higher molarity conditions or a second eluting step may be added to the process. The column is then flushed with a pH 10 buffer, followed by a pH 10 buffer containing, e.g., 3 *M* salt to convert any remaining proteins to anions (which are not bound to the column) and to elute adsorbed materials. One bed volume of 1 *M* NaOH is a good sanitizing agent. The column is then re-equilibrated to the starting condition buffer. The pH and ionic strength is verified at equilibrium.

The recovered MAb (pI 6.1) may be further purified on an anion exchanger. For the MAb to be anionic, the pH should be around pH 7.0. A suitable strong anion exchanger would be one of the quaternary resins. It may be desireable to do a relatively flat linear gradient or to follow the same approach used with the cation exchanger. A flat gradient will not spread the MAb peak much, but will allow a finer cut of the MAb if there are no co-chromatographing contaminants. Usually there are not many proteins in this pI range. Antibody clones may range from about 6 to a high 7. Transferrin (Tf) has a pI of 5.9 and may co-chromatograph with the MAb with pI 6.1. Since virtually all life forms require some Fe, Tf is often added to hybridoma culture as either a purified protein or as a component of bovine serum. However, since Tf is 85 kDa it may be readily eliminated from the partially purified MAb by size exclusion chromatography. If Tf is present as a molecular contaminant, it can be removed from a 1 or 2 percent (w/v) solution of the partially purified MAb by size exclusion chromatography on a column of a small-pored, highly cross-linked beaded agarose. At this stage, the MAb may be a highly purified product. An additional cation exchanger step may or may not be needed.

Loading conditions vary for different media. For size exclusion (gel filtration) resins, the basic relationship is

$$V_t = V_f + V_i + V_o$$

where V_t is the total volume of a column of the size exclusion resin, V_f is the volume of the resin and bound liquid, V_i is the volume of the liquid inside the resin beads, and V_o is the volume of the liquid outside the resin beads or between the resin beads. The sample volume is designated as V_s. When buffer exchange is done on a highly cross-linked resin, e.g., P-2 (Bio-Rad) or G-25 (Pharmacia), V_s must be no more than 25 percent of V_t. The sample concentration is not as important as the viscosity, for if the sample is very viscous it will ooze between the beads and not penetrate into the individual beads of the gel medium. Often, this can be seen on the column and it is obvious in terms of failure to achieve complete solvent exchange. Even if viscosity poses no problem, the sample, eluting in the void volume (V_0), will be diluted about 10 to 15 percent, seldom more unless the sample volume is significantly less than 25 percent, in which case the dilution factor can be considerable.

Generally, the more porous size exclusion resins intended for separations according to size and shape, such as beaded agarose, perform well only if maximum loading is 4 percent of the bed volume (i.e., $V_s = 0.04V_t$). However, where the proteins to be separated are considerably different in molecular weight, loadings can be greater.

The loading of ion exchange resins in a low-salt starting condition buffer is independent of volume. A reasonable rule-of-thumb is to load approximately 50 percent of the w/v capacity, which must be established experimentally. Macromolecules that are anionic at a given pH will tend to be excluded from a cation exchanger since they have the same charge as the exchanger. When these molecules pass through the exchanger, they do not indicate the loading is beyond the capacity of the exchanger. They indicate only that the pass-through molecules are not bound under the starting conditions.

Chromatofocusing is a separation technique that is pI dependent. The concept of chromatofocusing was described by Sluyterman and coworkers (Sluyterman and Elgersma 1978; Sluyterman and Wijdenes 1977, 463–466; Sluyterman and Wijdenes 1978). The concept of chromatofocusing is that a pH gradient is produced when a buffer at one pH is run through an ion-exchange column initially equilibrated to a second pH. When a pH gradient is used to elute proteins bound to the ion-exchange medium, the proteins will elute in order of their pIs. Further, a focusing effect occurs that results in band sharpening and sample concentration along with high resolution. Although chromatofocusing does give excellent resolution of complex mixtures, it is best used in combination with methods using other criteria for separation. The high resolving power is best utilized in the intermediate or later stages of purification.

Hydrophobic interaction chromatography (HIC) separates protein molecules on the basis of their surface hydrophobicity. These media are generally low cost and high capacity. The resolution is intermediate, and because of the structure of the gel, the flow rates are fast. These gels are stable both physically and chemically. Variations in the degree of substitution determines the range of proteins that are bound.

This can be a highly effective method, but the monoclonal antibody must be stable at high salt concentrations. There are weakly hydrophobic gels now available that are capable of purifying labile immunoglobulins. All immunoglobulins are hydrophobic to some extent, but the degree varies even within the same subclass. This procedure can be used effectively to remove host IgG from the monoclonal antibody. Advantages of HIC are that the technique can be used for most monoclonal IgG that are stable at high salt concentrations, and that high purity can be achieved in a single step. The separation of the monoclonal antibody from host IgG and the fact that very little sample pretreatment is required are additional advantages. The conditions are relatively easy to optimize and the results are easy to interpret. The disadvantages are that some monoclonal antibodies may be sensitive to the high salt concentrations encountered and that the method is not as readily scaled up. Hydrophobic interaction media are frequently used analytically, usually for reverse-phase analytical separations (RP–HPLC), but the use of organic solvents for industrial scale separations present problems such as worker hazards and high costs of disposal. Therefore, hydrophobic media usually are not used for large-scale purifications.

Affinity chromatography is the most selective purification method available. It is capable of achieving a high degree of purity in a single step. The cost tends to be high; however, the speed and resolution are also very high. This type of chromatography is based on the specific and reversible binding of the molecule being purified to a complementary binding substance immobilized on the medium. Protein A is a commonly used affinity chromatography media for antibodies. It is a group specific adsorbent and is used primarily to separate IgG antibodies. It is specific to the F_c region of the IgG molecules. The procedure has the advantages of being highly specific (i.e., high purity can be obtained in a single step) and rapid and easy to perform. The process is easily optimized. The chromatography medium has a high capacity and is easy to clean and regenerate. Very little sample pretreatment is required. Disadvantages would include the lack of ability to distinguish between the host IgG and the monoclonal IgG and that it does not bind all classes and subclasses from all species. There is also a tendency for leakage of Protein A into the elution buffer (Bloom et al. 1989). Some monoclonal antibodies are sensitive to low pH conditions that might be required to elute the product. Protein G is another ligand that can be used for the purification of antibodies, but the elution conditions necessary are generally stronger. Protein G does have utility in purifying monoclonal antibodies since it does bind a wide range species and subclasses. It binds mouse IgG_1, which happens to bind poorly to protein A. Specific antigens can be used as the immunoadsorbent for more specific purifications. By attaching specific ligands to an activated support, a wide variety of customized purification media can be prepared that can be used to purify the specific antibody.

It should also be pointed out that each of these chromatographic separation techniques can be used not only for purification but also for analysis of the resulting product. These chromatographic procedures are adapted to a variety of formats. Gel filtration, for example, may be used batchwise (i.e., simply mixed with the crude mixture and separated by filtration or decanting, open column procedures, or HPLC methodologies).

With all of these tools available, a variety of purification schemes can be selected. It is preferable to have more than one chromatographic step in the purification in order to attain the high level of purity and to assure with a high level of confidence the adequate removal of viruses, nucleic acids, pyrogens, and other adventitious agents.

HPLC Methods

All of the chromatographic methods described previously can be used in traditional open-column liquid chromatography. A number of new methods have been developed that are more reliable and faster. As the demands for more highly purified monoclonal antibodies grows, the need for fast, reliable, and automated methods for purification also grows. High performance liquid chromatography (HPLC) methods meet this need and are capable of yielding large quantities of high purity antibody preparations in a short time. In many cases these systems accomplish this with very little operator contact. Fully automated systems can play a significant role in developing the purification scheme.

Preparative methods tend to correlate very well with the analytical methods. The preparative elution profiles follow the prediction of the methods development separation profiles. Optimal conditions for the purification of monoclonal IgG from mouse ascites fluid were reported for 26 monoclonal antibodies of subclasses IgG_1 and IgG_2 using HPLC for the purification (Bruck et al. 1982). Another report by Izant et al. (1982) provided purification conditions for the isolation of purified monoclonal antibody from hybridoma suspension culture. There are literally hundreds of published references on the high performance purification of monoclonal antibodies in the literature today.

Systems for purification are available with capacities ranging from microliters to liters, and from micrograms to tens of grams. High performance media are commercially available as prepacked columns that provide high quality reproducibility and convenience. Systems must be selected for cleanability, ease of sanitization, and biocompatibility of the materials used for components.

One of the most important capabilities in the HPLC methods is the ability to rapidly develop purification methods. With systems available today, a variety of purification conditions can be tested in a short period of time using various software programs linked to the system. The test runs can be programmed and run with virtually no operator contact. A series of experiments can be programmed into the systems that will systematically analyze the effects of various parameters, such as elution pH, type of ion, ionic strength during elution, conductivity, pressure, loading capacity, and loading flow rate. The instruments will perform the experiments automatically, analyze the results and accomplish all of this rapidly. The process of purification development can be streamlined very effectively. The systems are able to check a number of parameters that otherwise would require too much time and resources. These method development systems are powerful tools for optimization of the final purification scheme. The systems, consisting of automated hardware and software, allow control over mobile phase blending, pressure and gradient slopes, and the ability to record relevant data such as pH, conductivity, UV absorbance, and pressure. Process developers are able to execute many experiments and manage the data generated quickly and efficiently. The BioCAD™ workstation (PerSeptive Biosystems, Cambridge, MA) is one example of this type of system and has been designed to work with the perfusion technology they developed (Spalding 1991).

The TRIO™ system (Sepracor, Inc., Marlborough, MA) also is fully automated and scalable from process design to production scale purification. This processing system integrates several purification technologies into one processing system. The system can incorporate the use of membranes for clarification and microfiltration and affinity membrane purifiers for purification. The affinity membranes can be used for the removal of endotoxins or other contaminants. The third technology incorporated is the column purification that can remove contaminants and/or perform buffer exchanges. Membranes can be hollow fiber membranes or tangential flow filters. The affinity membranes may be hollow fibers or Protein A- or Protein G-specific membranes. The column purification can be low pressure chromatography in a variety of column volumes and chromatography media, including ion exchange, hydroxyapatite, reverse phase, gel filtration, or immobilized dyes. This unit provides a single system for the purification steps that incorporate different commonly used technologies.

The BioPilot® System and the BioProcess™ System (Pharmacia-LKB Biotechnology, Piscataway, NJ) are fully automated chromatography systems for pilot and full production-scale purifications. The

BioPilot® system is designed for process development and production of multigram quantities whereas the BioProcess™ system is for larger-scale purifications. Process development is aided by the capability of the BioPilot® system to permit programmed selection of different buffers, samples, cleaning agents, and columns for optimization of the purification protocols.

Another HPLC system specifically designed for the purification of monoclonal antibodies is the MAPS™ (Monoclonal Antibody Purification System) Preparative System (Bio-Rad Laboratories, Richmond, CA). This system combines software-controlled gradient HPLC with automatic sampling and fraction collection for a fully integrated monoclonal antibody purification system.

Scaleup Considerations

Different strategies can be applied to the scaleup for industrial processes. Lebherz (1988) has reviewed many of the considerations for large-scale production of monoclonal antibodies. Typical stages of the development of a scaleup plan are shown in Figure 8.1. One strategy is to minimize the number of steps involved in the purification, which then requires a highly specific or high-selectivity step such as immunoaffinity chromatography. Another strategy is to use a properly optimized series of more conventional steps with lower yields and productivity, but with similar overall process economics. In order to accomplish the regulatory requirements, however, it appears that a series of steps are needed.

One of the advantages of chromatographic techniques is the ability to scale up from laboratory methods to pilot plant to full production. Separation parameters can be worked out at the laboratory bench and translated into the production scale. However, it should be considered that some methods, while excellent at the bench scale do not always scale up to preparative scale efficiently. For example, gel filtration can be a very effective initial step in small-scale procedures, but because of its low capacity, it may not be an efficient initial step in a large-scale method. Gel filtration is suitable for final-stage purifications when critical contaminants need to be removed from small volumes of product. Ion exchange is suitable for handling large quantities of proteins. Affinity chromatography and HIC are suitable at any stage of the purification. Chromatofocusing is more appropriate at final stages of purification of high value products. Optimization of large-scale chromatography has been reviewed by Curling and Cooney (1982). Generally, a chromatographic purification system can be scaled up without having to change the elution buffers. Flow rates may be affected when using large columns and when using compressible gels. Scaleup of a chromatography step is usually accomplished by

Figure 8.1. Stages in a purification process.

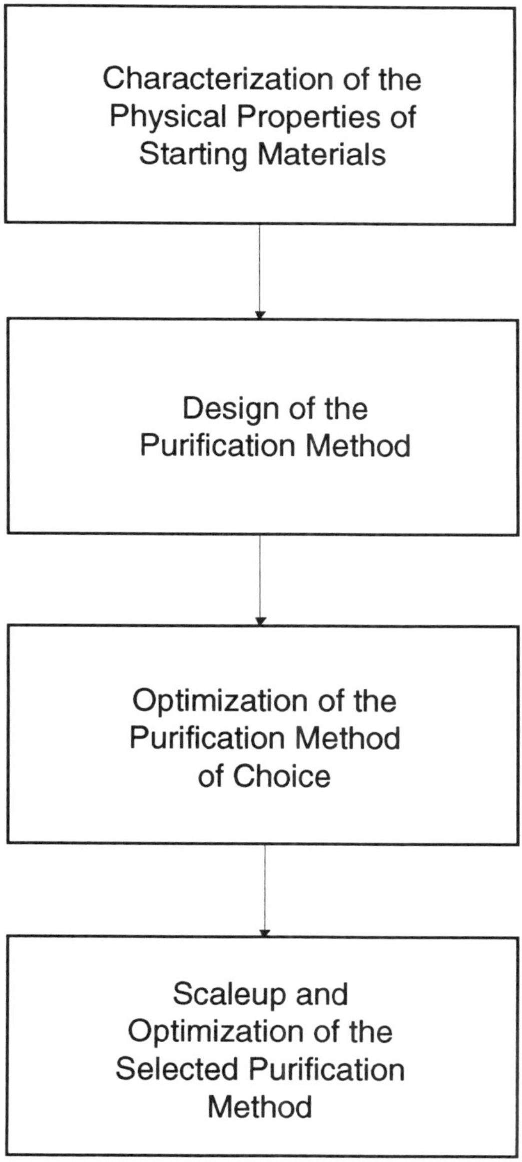

increasing the column diameter while holding constant the ratios of volumes of bed volume, feed, wash and elution buffers. With soft gels, scaleup is sometimes accomplished by increasing column diameter but not column length. The problems are with uniform sample application over the expanse of the resin and uniform packing of the bed. For these reasons, large columns in a series are used. The resolution is proportional to the square root of the bed height for gel filtration; therefore, increases in resolution must be accompanied by increases in column height. Ion-exchange chromatography on an industrial scale generally is performed as a step-wise gradient. The linear flow velocity and ratios of the gradient buffers are kept constant.

Case Studies

The approval of Orthoclone OKT® 3 (MUROMONAB-CD3) was a landmark for monoclonal antibodies (Rao et al. 1991; Food and Drug Administration 1986). In the summary basis for approval, the purification from ascites is described. Purification steps begin with the pooling of batches of frozen ascites and the clarification of the pooled ascites by filtration. The IgG is isolated by ammonium sulfate fractionation. Further purification is accomplished by anion exchange chromatography. The purified antibody is diluted and dialyzed against phosphate buffered saline (PBS). The solution is filtered and stored frozen until it is thawed, pooled, formulated, and sterilized by membrane filtration. A murine virus spiking study was performed to demonstrate six-log reduction of virus by the purification process. In addition, a six-log reduction of DNA from the purified monoclonal antibody was demonstrated.

An evaluation of methods for the purification of therapeutic grades of monoclonal antibodies was performed and reported by Duffy and coworkers (1989). In this case the monoclonal antibody was prepared in culture media. Criteria for the final product included retention of biological activity, freedom from high molecular weight aggregates that are more immunogenic than the monomer, lot-to-lot consistency and cost-effective production of large quantities (multikilograms) of highly purified antibody. In addition, the compliance with FDA guidelines for purity and the removal of viruses, DNA, and endotoxins was crucial. Comparisons of Protein A chromatography and ion-exchange chromatography were made. The purification scheme began with concentration of the cell culture in a plate-and-frame UF unit containing cellulose acetate membranes. Following the UF, two methods of purification were compared. One method included a protein A affinity chromatography step and the other included a series of IXC steps. The two pathways are shown in Figure 8.2. The series of ion-exchange steps include a cation exchange step, followed by an anion exchange step, and finally a second cation exchange step. Ammonium sulfate precipitation steps followed by desalting with gel filtration resulted in the final product.

Figure 8.2. Scheme for the purification of therapeutic-grade monoclonal antibodies.

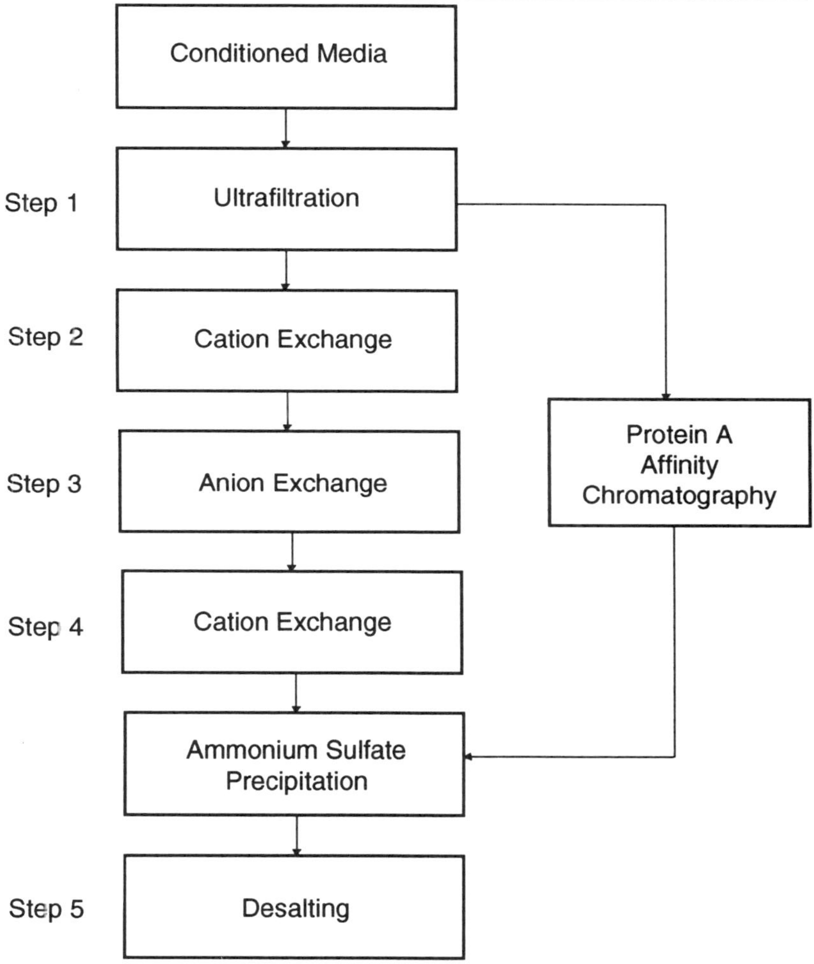

The advantages of the IXC and the protein A affinity chromatography steps were compared. Protein A chromatography provided a generic method for isolating antibodies and it resulted in a rapid higher-fold purification in a single step. However, the protein A was not the choice recommended for large scale purification by the authors. The limitations of protein A chromatography included copurification of serum-derived IgGs, leaching of protein A and its breakdown products, loss of capacity upon sanitization, harsh elution conditions, formation of aggregates in the product, high costs, and the inability to bind IgMs and certain IgG subclasses. The advantages of the IXC steps were that the high flow rates permitted fast high volume separations and that the immunoglobulins with pIs at >6 could be separated efficiently from most serum proteins with pIs at <5.5. Although the ion-exchange method must be customized for each monoclonal antibody, the method does exploit the ability to separate one particular protein and to minimize copurification of other serum-derived immunoglobulins. Ion-exchange resins are easily sanitized, effectively reducing costs. Protein A chromatography was viewed as a very effective analytical tool.

The development of serum-free media or low-serum media has simplified the purification process somewhat, but there are still problems that need to be addressed while developing the final purification process. A strategy was developed and reported by Hanna et al. (1991) without full characterization of the starting material. The media contained added protein components, such as bovine serum albumin (BSA), transferrin, and insulin. Potential contaminants included DNA; proteins; endotoxin; closely related proteins such as fragments, aggregates, and denatured product; BSA; transferrin; insulin; carbohydrates; lipids; and other contaminants derived from the culture media. In addition, contaminants are potentially introduced from the purification itself, including resin fragments and ligands, pyrogens, and buffer salts. A list of the physical properties of the known contaminants was prepared, which included the molecular weights and pIs. With these data the purification steps could be selected in a logical manner. The first step used in their purification of this T cell binding antibody was diafiltration to concentrate and prepare the protein solution for the first chromatography step.

Protein A affinity chromatography binds the antibody, while allowing the contaminants to pass through the column or to be readily washed off the column and accomplish a high degree of purity in this one step. Additional steps were added to remove the trace amounts of contaminants that remained. After comparing the pIs of the antibody and the known contaminants, the authors decided to use an anion exchange to further purify the antibody.

Gel filtration step removed antibody aggregates and fragments that may have similar charges but significantly different molecular weights. Gel filtration also exchanged the buffer to the final formulation buffer. To improve the effectiveness of the gel filtration step, a concentration step was inserted prior to the gel filtration to maintain a reasonable working volume without sacrificing resolution.

DNA removal was evaluated in the various chromatographic steps by spiking with radiolabeled DNA and determining the clearance factor. Protein A had a clearance factor of 10,000 (4 logs of DNA clearance) and anion exchange had a clearance factor of 1000. Clearance factors for other known contaminants were similarly determined. Assays for trace contaminants ultimately verified the removal of contaminants in the final purified antibody.

By logically examining the physical characteristics of each known contaminant, a purification scheme was developed that took advantage of the selectivity of the purification steps.

The purification steps also must be evaluated for virus removal and/or inactivation. The validation of the purification process is accomplished by challenging each purification step with spiked challenges of a representative panel of infectious model viruses. Results of a number of viral removal/inactivation steps were reported by Grun et al. (1992). The validation of virus removal is safer performed outside the production area since high titers of infectious viruses will be used in the studies. The process is scaled down, but represents the manufacturing process accurately in terms of temperature, buffers, flow rates, and elution profiles. Each virus will have its own characteristic log reduction for each step; therefore, the cumulative effect of the different purification steps may only be applied to each virus separately. Care must also be taken in using inactivation steps or removal steps so that the monoclonal antibody itself is not affected in terms of activity, stability, potency, or form.

RECENT ADVANCES AND FUTURE TRENDS

In order to avoid the potential immune response to murine antibodies, a variety of methods are being explored. One of these strategies is to prepare human monoclonal antibodies or "humanized" mouse antibodies. This process is accomplished by gene manipulation, splicing the antigen-binding region of the murine antibody gene to the nonantigen-binding region of human antibody. This creates a chimeric gene and this process has been reviewed extensively (Aguila et al. 1986; Reichmann et al. 1988). Because the antigenic portion of the antibody produced is now human, it has a lower tendency to produce an immunogenic response. These chimeric antibodies are produced in mammalian cell in vitro production. The cells used for this production require high nutrient serum for maintenance of cell viability and production of antibody. The presence of high levels of proteins and the fact that the chimeric antibodies are generally expressed in lower quantities than traditional murine monoclonal antibodies increases the problems in purification of these chimeric antibodies. The purification of a chimeric antibody is described by Moellering and Prior (1990). In their purification process the cell culture medium was concentrated using a plate and frame ultrafiltration device as the first step.

The concentrated protein mixture was pH adjusted and passed through an ion exchange column containing Q Sepharose Fast Flow™ (FFQ) (Pharmacia-LKB Biotechnology, Piscataway, NJ) followed by an ABx column (J. T. Baker, Phillipsburg, NJ). The antibody was then passed through an S-Sepharose Fast Flow™ column (Pharmacia). In this scheme the purification process was monitored with Protein A HPLC and a size-exclusion HPLC column. The Protein A HPLC was effective in demonstrating the removal of the bovine immunoglobulin from the chimeric antibody.

Another approach to minimize the immunogenic response to murine monoclonal antibodies is to utilize the smaller active fragments of the antibodies that will have a lower antigenic character. Gavit et al. (1992) describe the methods for producing these fragments (e.g., Fab molecules, have been produced by papain cleavage of whole immunoglobulins and consist of an Fd heavy chain and either a lamda or kappa light chain held together by an interchain disulfide bond and hydrophobic forces). The cleavage products tend to be a heterogeneous mixture because of the nonspecific cleavage. The combination of hybridoma and recombinant DNA technologies has allowed the production of Fabs from *E. coli.* The Fabs are produced by the *E. coli* in fermentation media and can be purified directly. The reduction of endotoxin from the fermentation media becomes a significant step in this purification scheme. The purification process begins with the separation of cells from the medium using a hollow fiber cartridge. The cell-free medium was then diafiltered and concentrated using an UF unit fitted with a spiral cartridge. A series of chromatography steps were used to achieve final purification. These steps included a DEAE cellulose column (DE-52, Whatman Inc., Clifton, NJ) followed by a CM-cellulose (CM-52, Whatman) and a CM-Sepharose Fast Flow™ (Pharmacia, Piscataway, NJ). The final chromatography step was another DEAE cellulose column. Concentration and buffer exchange was accomplished with UF and the final step was sterile filtration. These steps were considered to be readily scalable and effective in purifying the antibody fragment to regulatory standards. The methods used were shown to be effective at removing endotoxin, host cell protein, and DNA to acceptable levels.

The production of monoclonal antibodies has increasingly trended toward cell-culture production. Historically, the cell culture of mammalian cells has involved the use of culture media supplemented with up to 5 to 20 percent serum (Fike et al. 1991). The recent trend has been to eliminate the serum in the media to minimize the downstream processing problems and costs of removing the extraneous contaminants.

The elimination of serum proteins also reduces the possibility of introducing viruses, mycoplasma, and other adventitious agents. There are a great number of reports of the use of various defined media that are either low-serum or serum-free media. Direct comparisons of the antibody production in serum-free media have been found to be equal to or better than in serum-containing media. The adaptation periods of cells to serum-free media are minimal with proper formulation and there is significantly less product contamination.

The trend in purification of commercial quantities of monoclonal antibodies will be toward the use of automated process control. Process automation will provide the competitive edge by significant long-term cost reductions. Advantages of automated process control include improved reliability and reproducibility. There is the opportunity to increase productivity by the full-time use of the equipment. The product can be isolated from contact with the operators and the environment, providing product protection from contamination. The records can be more reliable and complete. Of course, the automated system must be validated in order to comply with cGMP regulations. Process automation is readily applicable to batch operations as well as to continuous operations.

A number of improvements have been made recently that have improved the performance of chromatography methodologies. The types of improvements include less compressible gels that accommodate scale-up more readily, better designs of pumps, and new column designs. Companies are working on more durable affinity chromatography gels in order to reduce the net cost by increasing the column lifetime. Improvements are being made on membrane filtration and other techniques. All of these technologies for separation and purification are evolving rapidly. Since a significant part of the cost of a product is in the purification these improved technologies will have a significant impact of the future costs of the monoclonal antibody products.

Perfusion Chromatography™ media have been introduced recently and represent a new development in purification technology (Alfeyan et al. 1990; Alfeyan et al. 1991, Alfeyan et al. 1990; Fulton et al. 1991; Krips et al. 1991; Salzman 1992). Perfusion Chromatography™ is a patented process and a trademark developed by PerSeptive Biosystems (Cambridge, MA). The media used in this process has pores that allow liquid to flow through the chromatographic particles, greatly increasing the binding capabilities of the media. This technology has allowed more rapid purifications with higher resolution without the reduction in media particle size that results in increased back pressure. Reduced residence times on the column maximize the retention of biological activity of the antibodies being purified.

CONCLUSIONS

Selecting a method for the purification of a monoclonal antibody is dependent on many factors. These factors include regulatory requirements, the ultimate use of the purified monoclonal antibody (i.e., the level of purity actually required), the production quantity of monoclonal antibody, and the antibody production method. Various criteria for the purification of monoclonal antibodies have been presented. The purification methods and the tools available are diverse. The combination of methods seems without limit. Constant improvements are being made by a number of companies on the equipment and supplies for purification. Requirements for the purity of these products tends to become tighter as more data becomes available. As analytical methods become more sophisticated for the determination of contaminants, so will the demands on the purification processes to assure the removal of these contaminants. The ability to purify monoclonal antibodies cost-effectively on a commercial scale will play a significant role in bringing the products to market and in the realization of the full potential of the technology.

REFERENCES

Afeyan, N. B., S. P. Fulton, N. F. Gordon, I. Mazsaroff, L. Varady, and F. E. Regnier. 1990. Perfusion chromatography: An approach to purifying biomolecules. *Biotechnology* 8:203–206.

Afeyan, N. B., S. P. Fulton, and F. E. Regnier. 1991. Perfusion chromatography packing materials for proteins and peptides. *J. Chromatogr.* 544:267–279.

Afeyan, N. B., N. F. Gordon, I. Mazsaroff, L. Varady, S. P. Fulton, Y. Yang, and F. E. Regnier. 1990. Flow-through particles for the high-performance liquid chromatographic separation of biomolecules: perfusion chromatography. *J. Chromatogr.* 519:1–29.

Aguila, H. L., R. R. Pollock, G. Spira, and M. D. Scharff. 1986. The production of more useful monoclonal antibodies. *Immuno. Today* 7 (12):380–383.

Bloom, J. W., J. F. Wong, and G. Mitra. 1989. Detection and reduction of Protein A contamination in immobilized Protein A purified monoclonal antibody preparations. *J. Immunol. Methods* 117:83–89.

Bruck, C., D. Portetelle, C. Glineur, and A. Bollen. 1982. One-step purification of mouse monoclonal antibodies from ascitic fluid by

DEAE Affi-gel Blue chromatography. *J. Immunol. Methods* 53:313–319.

Carter, G. T., R. E. Schiesswohl, and H. Burke. 1982. Peak homogeneity determination for the validation of high-performance liquid chromatography assay methods. *J. Pharm Sci.* 71 (3):317–321.

Carthew, P. 1986. Is rodent virus contamination of monoclonal antibody preparations for use in human therapy a hazard? *J. Gen. Virol.* 67:963–974.

Curling, J. M., and J. M. Cooney. 1982. Operation of large scale gel filtration and ion-exchange systems. *J. Paren. Sci. Technol.* 36:59–64.

Duffy, S. A., B. J. Moellering, G. M. Prior, K. R. Doyle, and C. P. Prior. 1989. Recovery of therapeutic-grade antibodies: Protein A and ion-exchange chromatography. *BioPharm* 2 (6):34–47.

EEC Ad Hoc Working Party on Biotechnology/Pharmacy. 1991. Validation of virus removal and inactivation procedures. *Biologicals* 19:247.

Fike, R. M., J. L. Pfohl, D. A. Epstein, D. W. Jayme, and S. Weiss. 1991. Hybridoma growth and monoclonal antibody production in protein-free hybridoma medium. *BioPharm* 4 (3):26–29.

Food and Drug Administration. 1987a. *Guideline on the general principles of process validation.* Rockville, MD: Center for Drugs and Biologicals and Center for Devices and Radiological Health.

Food and Drug Administration. 1987b. *Guidelines on validation of the limulus amebocyte lysate test as an end-product endotoxin test for human and animal parenteral drugs, biologicals, and medical devices.* Rockville, MD: Center for Drugs and Biologics and Center for Devices and Radiological Health.

Food and Drug Administration. 1987c. *Points to consider in the characterization of cell lines used to produce biological products.* Bethesda, MD: Office of Biologics Research and Review, Center for Drugs and Biologics.

Food and Drug Administration. 1987d. *Points to consider in the manufacture and testing of monoclonal antibody products for human use.* Bethesda, MD: Office of Biologics Research and Review, Center for Drugs and Biologics.

Food and Drug Administration. 1986. *Summary basis for approval for Orthoclone OKT®3 (MUROMONAB-CD3).* Bethesda, MD: Department of Health and Human Services, Center for Drugs and Biologics.

Fulton, S. P., N. B. Afeyan, N. F. Gordon, and F. E. Regnier. 1991. Very high speed separation of proteins with 20 μm reversed-phase sorbent. *J. Chromatog.* 547:452–456.

Gavit, P., M. Walker, T. Wheeler, P. Bui, S. Lei, and J. Weickmann. 1992. Purification of a mouse-human chimeric Fab secreted from E. coli. *BioPharm* 5 (1):28, 34–37.

Grun, J. B., E. M. White, and A. F. Sito. 1992. Viral removal/inactivation by purification of biopharmaceuticals. *BioPharm* 5 (9):22–30.

Goding, J. W. 1986. *Monoclonal antibodies: Principles and practice.* London: Academic Press.

Hanna, S., P. Pine, G. Reuzinsky, S. Nigam, and O. R. Omsted. 1991. Removing specific cell culture contaminants in a MAb purification process. BioPharm 4 (9):33.

Hurrell, J. G. R. 1982. *Monoclonal hybridoma monoclonal antibodies: Techniques and applications.* Boca Raton: CRC Press.

Issekutz, A. C. 1983. Removal of gram-negative endotoxin from solutions by affinity chromatography. *J. Immunol. Methods* 61:275–281.

Izant, J. G., J. A. Weatherbee, and J. R. McIntosh. 1982. A microtube-associated protein in the mitotic spindle and the interphase nucleus. *Nature* 295: 248–250.

Joner, E., and G. D. Christiansen. 1988. Hybridoma technology products: Required virus testing. *BioPharm* 1 (8):50–55.

Knutton, S., and C. A. Pasternak. 1979. The mechanism of cell-cell fusion. *Trends Biochem. Sci.* 4 (4):220–223.

Köhler, G. 1981. The technique of hybridoma production. *Immunol. Methods* 11:285–298.

Köhler, G., and C. Milstein. 1976. Derivation of specific antibody-producing tissue culture and tumor lines by cell fusion. *Eur. J. Immunol.* 6:511–519.

Köhler, G., and C. Milstein. 1975. Continuous cultures of fused cells secreting antibody of predefined specificity. *Nature* 256:495–497.

Krips, D. M., R. D. Sitrin, and C. N. Oliver. 1991. A very rapid two-minute Protein A HPLC assay for monoclonal antibodies. *FASEB J.* 5:A465.

Lebherz III, W. B. 1988. Batch production of monoclonal antibody by large-scale suspension culture. *BioPharm Manu.* 1 (2):22–32.

Levy, P. F., and J. J. Sheehan. 1991. Performance characteristics of polysulfone and cellulose membranes for the ultrafiltration of biological process streams. *BioPharm* 4 (4):24–33.

Luben, R. A., and M. S. Mohler. 1980. In vitro immunization as an adjunct to the production of hybridomas producing antibodies against the lymphokine osteoclast activating factor. *Mole. Immunol.* 17:635–639.

Lyle, L. 1982. Production and applications of monoclonal antibodies. *MD & DI* June:41–44, 83.

Milstein, C. 1980. Monoclonal antibodies. *Sci. Am.* 243:66–74.

Moellering, B. J., and C. P. Prior. 1990. Separating clinical-grade chimeric antibodies from serum-derived immunoglobulins. *BioPharm* 3 (1):34–38.

Paul, J. 1975. *Cell and tissue culture,* 5th ed. London: Churchill Livingstone.

Pfund, N., and K. G. Charles. 1988. Separations technology outlook, part II: Improved recovery and greater purity. *BioPharm Manuf.* 1 (1):29–33.

Rao, P., A. Williams, A. Baldwin-Ferro, E. Hanigan, D. Kroon, M. Makowski, E. Meyer, V. Numsuwan, E. Rubin, and A. Tran. 1991. C-terminal modification occurs in tissue culture produced OKT3. *BioPharm* 4 (10):38–43.

Reichmann, L., M. Clark, H. Waldman, and G. Winter. 1988. Reshaping human antibodies for therapy. *Nature* 332:323–327.

Saltonstall, C. W. 1992. Separations with dialysis and ultrafiltration: Theoretical and practical considerations. *Amer. Biotechnol. Lab.* 10 (8):18.

Salzman, A. T. 1992. Perfusion chromatography and beyond. *Biotech Forum Eur.* 7/8/92, 442–445.

Sharma, S. K. 1986. Endotoxin detection and elimination in biotechnology. *Biotechnol. Appl. Biochem.* 8:5–22.

Sluyterman, L. A., and O. Elgersma. 1978. Chromatofocusing: Isoelectric focusing on ion exchange columns. I. General principles. *J. Chromatog.* 150:17–30.

Sluyterman, L. A., and J. Wijdenes. 1977. Chromatofocusing on ion exchangers in the absence of an externally applied potential. In *Joint*

Proceedings of the International Symposium on Electrofocusing and Isotachophoresis, Edited by B. J. Radola and D. Grasslin. Berlin: de Gruyter.

Sluyterman, L. A., and J. Wijdenes. 1978. Chromatofocusing: Isoelectric focusing on ion exchange columns. II. Experimental verification. *J. Chromatog.* 150:31–44.

Spalding, B. J. 1991. Downstream processing: Key to slashing production costs 100 fold. *Biotechnology* 9:229–241.

Strelkauskas, A. J., ed. 1987. *Human hybridomas: Diagnostic and therapeutic applications.* New York: Marcel Dekker, Inc.

Tsao, I. F., C. Shipman, and H. Y. Wang. 1988. The removal of adventitious viruses and virus infected cels using a cellular adsorbent: A feasibility study. *Biotechnology* 6:1330–1333.

World Health Organization. 1991. Proposed requirements for monoclonal antibodies for clinical use in humans (requirements for biological substances No. 48). *Technical Report BS/91:1657.*

Zolton, R. P., J. V. Padvelskis, and P. M. Kaplan. 1985. Removal of hepatitis B virus infectivity from human gamma-globulin prepared by ion-exchange chromatography. *Vox. Sang.* 49:381–389.

9

Immunoaffinity Purification of Proteins for Injection, and the Issue of Animal Viruses

Wayne P. Olson

IMMOBILIZED ANTIGENS AND ANTIBODIES—A RESEARCH TOOL

Campbell et al. and Early Immunosorbent Work

Campbell et al. (1951) sorbed and diazo-coupled protein antigens (Ags) to various supports. Antibodies (Abs), in antiserum added to a column of the cellulose-Ag conjugate, bound to the Ag. Rinsing the cellulose eliminated non-Ab proteins. Polyclonal Abs (PAbs) then were eluted by means of a glycine-HCl buffer. The beauty of this approach was that the purified polyclonal Abs then could be coupled to a diazotized support as a means of purifying the Ag from complex solutions. However, the starting point was a purified Ag. This work subsequently was verified by Gurvich et al. (1959).

Popularization of the Method

The technique afforded the potential of preparing immunochemically pure proteins, but saw little research use until Axen et al. (1967) developed the cyanogen bromide (CNBr) method of activating polysaccharide matrices and Cuatrecasas et al. (1968) popularized affinity chromatography. As of 1982, immunosorbent technology had been applied to 52 different antigens in research applications (Table 9.1, from Wilchek et al., 1984).

Table 9.1. Proteins Purified on Immunosorbents[a]

Protein	Eluent
Adenosine deaminase	Urea
beta-Adrenergic receptor	Propranolol
Alkaline phosphatase	High pH
alpha-Fetoprotein	Urea
$alpha_2$-macroglobulin	Low pH
Aminopeptidase	Tris-HCl
Angiotensin I converting enzyme	$MgCl_2$
Astroglial protein	Acetic acid-urea
C1g, C1r, C1s subcomponents	EDTA, ethylene glycol
C5	KBr
Carcinoembryonic antigen	Urea
Cardiotoxin	Acetic acid-NaCl
Choriogonadotropin	$MgCl_2$
Estrogen receptor	NaSCN
Factor VIII	NH_4SCN
Factor VIII	Ethylene glycol
Factor V	NaCl
Factor IX	NaCl, KSCN
Glucocorticoid receptor	Acetic acid
Hepatitis A virus	NaI
HLA-B and B	Diethylamine pH 11.5
HCG-receptor complex	HCG[b]

Continued on next page

Continued from previous page

Protein	Eluent
Insulin receptor	$MgCl_2$
Interferon	Ethylene glycol + citrate
Kallikreins	Guanidine
Monoamine oxidase	KSCN
Myosin	Guanidine
Neutrophil migration inhibition factor	Urea
Phosphotyrosine proteins	Phenyl phosphate[b]
Poliovirus	KSCN
Properdin	NaCl
Somatostatin	Acetic acid
Terminal deoxynucleotidyl-transferase	NH_4OH
Urokinase	Glycine-HCl
von Willibrand's factor	100 m*M* Triethanolamine 0.1 *M* NaCl pH 9.4

[a]Modified from Wilchek et al. (1984)

Oligosaccharides hold some promise as immunomodulators, both as immunostimulants such as the fraction PS1 from BCG (Lou 1993) and as immunosuppressants such as the lipoarabinomannan found in all mycobacteria (Brennan 1990; Chan et al. 1991; Sibley et al. 1988). Purification usually is based on poorly discriminating methods (size-exclusion chromatography (SEC) or salting-out behavior) or lectin affinity chromatography. However, immobilized monoclonal antibodies (MAbs) can be used in the column purification of oligosaccharides under isocratic conditions (Ohlson et al. 1988; Zopf et al. 1987). Because the K_ds of MAbs for oligosaccharides are of the order of 10^{-2} to 10^{-4}, which usually is too high for affinity chromatography, such separations must be done on column and with modest loading of sample—probably no more than 5 percent of the column volume. This is not practical for large-scale separations, but may suffice for research purposes. The K_ds of MAbs for peptides and proteins usually are of the order of 10^{-6} and lower; Abs with such K_ds are suitable for use in immunosorbents.

Evaluation of the Product

Before a purification scheme can be devised, test methods for the solute of interest must be in hand and, preferably, validated. The routine test(s) for process usually are both qualitative and quantitative. Such tests provide key data for estimating product yield and column longevity. In the case of factor VIII (F-VIII), product usually is eluted in a background of about 1 percent human serum albumin (HSA), hence A_{280} is not useful. Possibilities include SDS–PAGE, which resolves product and HSA. However, SDS–PAGE takes at least 30 minutes to run. Capillary electrophoresis is rapid, quantitative and is micro-scaled. High-performance size-exclusion separations (HP–SEC) also is rapid and sufficiently sensitive for applications like these. An enzyme-linked immunosorbent assay (ELISA) also is useful because it is both qualitative and quantitative. My choices for process monitoring would be HP–SEC and ELISA. The Laurell rocket (Laurell 1966) also is a good assay, but is less precise and accurate than an ELISA test. Monitoring of process entails pH and ionic strength of the eluate as it exits the column, and the temperature of the column jacket.

IMMUNOSORBENTS FOR THE PURIFICATION OF HUMAN CLOTTING FACTOR VIII FOR INJECTION

Economic Incentives for Dramatic Changes in Preparative Methods

In 1981 the first CDC–reported cases of what we now refer to as AIDS (acquired immunodeficiency syndrome) were reported in New York, Los Angeles, and San Francisco (Stine 1993). In the absence of a clinical test for the causal agent, the disease was found increasingly in hemophiliacs receiving clotting factors VIII or IX from human plasma. Seventy-five cases had been reported in hemophiliacs by 1987. By 1992, 1,391 hemophiliacs were seropositive for human immunodeficiency virus (HIV). However, it is unlikely that any new AIDS cases have been reported as a result of infection post-1985, when the F-VIII administered in the U.S. was heat treated. The Hyland division of Baxter marketed heated, freeze-dried F-VIII in 1984 and pioneered other levels of screening to reduce the hazard (Olson 1987).

In the early 1980s it had become increasingly clear that the clotting factors recovered from human plasma required the inactivation or removal of HIV and, hopefully, other viruses (Bechtel et al. 1988; Griffith et al. 1991; Olson 1987; Prince et al. 1986). Hemophiliacs in the early 1980s were faced with an impossible decision—avoid use of factors VIII or IX preparations during bleeding episodes and run the risk of death by hemorrhage, or inject the human-source material and run the equally odious risk of a lingering death from AIDS.

Nonimmunosorbent approaches to the elimination or destruction of viruses are discussed below because MAbs are generated by hybridomas in culture, or from mouse ascites; because murine MAbs play an important role in the processing of human-source factor VIII; and because non-HIV human viruses, murine viruses, and the viruses that may be present in cell lines must be eliminated from injectable bioproducts.

The immunosorbent method pioneered by Campbell et al. (1951) poses several problems when PAbs are immobilized on the column. The nature of polyclonals is that the various clones of Abs may be, and usually are, directed against the many antigenic segments of particular domains of the Ag. A number of proteins share domains with a high degree of homology. As a result, cross-reactions are common with PAbs; experienced immunologists add powdered preparations of cross-reacting Ags to antisera with a view to precipitating the Ag-Ab complexes for cross-reacting proteins. This strategy often is satisfactory for the generation of clinical diagnostic reagents utilizing PAbs.

Why Monoclonal Antibodies

The other problem is that—since an Ag, bound by multiple Abs directed against a variety of epitopes on the various domains of the protein, is trapped on the Ab column by a variety of Ag-Ab reactions varying in affinity—elution from the column requires conditions that will break the strongest Ag-Ab bond (i.e., the lowest K_d). Usually, the result is that extreme conditions are used for elution (e.g., the glycine-HCl method of Campbell et al. at a pH of 2.3–3.0), which may denature the protein of interest.

Unpurified PAb preparations as intravenous immune globulins (IVIGs) have been in routine therapeutic use for the treatment of hepatitis and other indications for many years. The IVIGs from donors specifically immunized with a vaccine are designated as specific polyclonal Abs (SpAs). SpAs are marketed for several indications (e.g., cytomegalovirus infection). Although SpAs constitute a growth market (Harkonen and Gregg 1993), they are unsuitable for use in the purification of most biologics for injections, since SpAs are not directed against other human proteins.

Monoclonal Abs have ameliorated the problems of polyclonals. The technique of recovering individual spleen cells and fusing each with a myeloma cell made possible the immortalization of an individual B cell and its daughters, hence the secretion of a single ideotype by the clone (Köhler and Milstein 1975). In the course of preparing a MAb directed against one protein of many domains, such as F-VIII, many different clones are developed, each directed against an unique epitope of the protein. Where a given clone is directed primarily against an hydrophobic segment of the Ag, the interactions will, of necessity, be largely hydrophobic in nature, and so forth. Therefore, it is possible not only to select a MAb that binds to an unique segment of the protein (hence shows no or little cross-reactivity with contaminating proteins), but also to select a MAb that can be dissociated from the Ag under mild conditions, (e.g., elution with salt and polyols). For a large protein such as F-VIII, mild conditions for elution are important in avoiding significant levels of denaturation of the product. Also, when a MAb is used, the K_d can be determined and the pH of column operation selected with a view to obtaining a higher or lower K_d. Mejan et al. (1988) show K_a ($K_d = 1/K_a$) as a function of pH for an immobilized anti-vWF MAb. Generally, $K_d > 10^{-4}$ is too high for affinity interactions. The exception seems to be with immobilized antioligosaccharide Abs, which can be used with some success although they do not bind nearly so strongly as most antiprotein Abs (Zopf et al. 1987).

The Scripps Clinic Approach

The initial MAb approach to the purification of F-VIII from human plasma (the Scripps Clinic patent, Zimmerman and Fulcher 1982) was very clever. The F-VIII in mammalian plasma is associated, noncovalently, with von Willibrand's factor (vWF), a large protein that is generated by normal people and by F-VIII-deficient hemophiliacs; hence it is not usually required for therapy. This patent has been licensed by Scripps Clinic to the Armour Pharmaceutical division of Rhone-Poulenc-Rorer, Kankakee, IL. The F-VIII produced by this process is tradenamed Monoclate™ and the Monoclate process has been described by Hrinda et al. (1989). The process is shown in Figure 9.1.

MAbs directed against vWF are immobilized on a support and as vWF is sorbed onto the Ab, F-VIII also is recovered because it forms a noncovalent complex with vWF. The crude F-VIII-containing material is approximately equilibrated with what Zimmerman and Fulcher refer to as the VIII:C buffer (0.02 *M* imidazole, 0.15 *M* NaCl, 0.10 *M* lysine, 0.02 percent NaN_3, pH adjusted to 6.8 with conc. HCl). The

Figure 9.1. The Monoclate (Scripps/Armour) process for F-VIII purification.

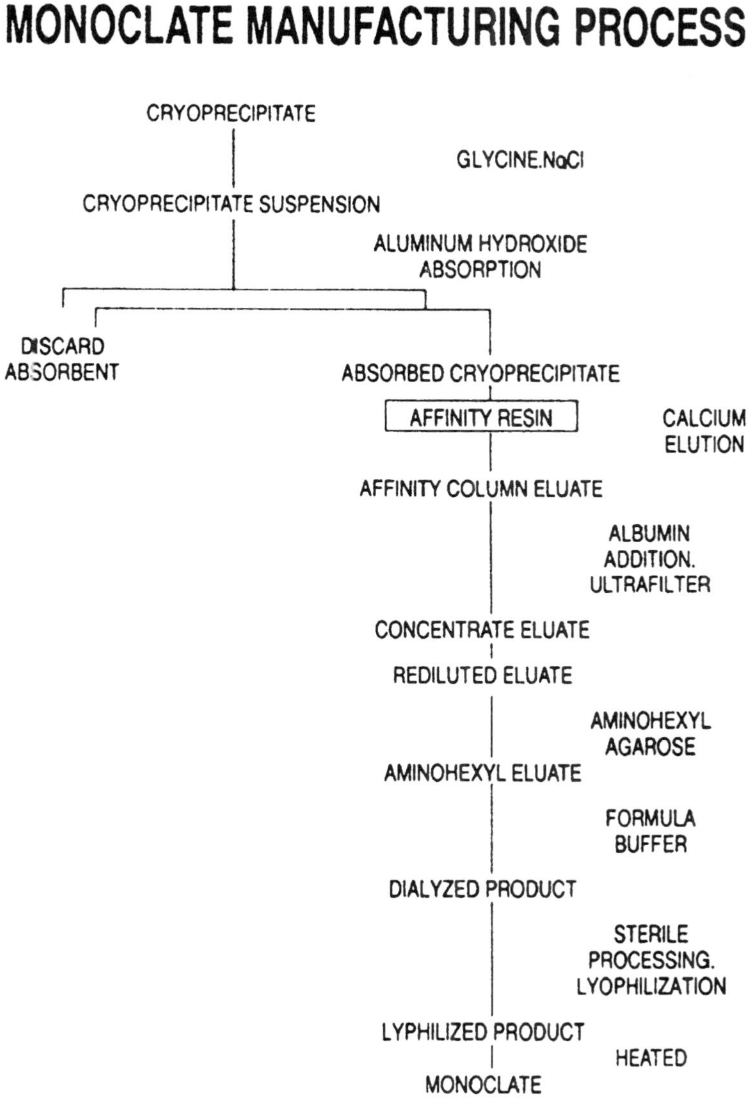

column is equilibrated in the same buffer. After the adsorption of the F-VIII by the anti-vWF MAb, the column is washed with the VIII:C buffer to remove unreacted proteins. Elution of the F-VIII from the

Ab-bound vWF is with a 0.3–0.35 *M* solution of $CaCl_2$ in the VIII:C buffer.

The elution profile for F-VIII from immobilized vWF is exquisitely pH-sensitive (Delaage et al. 1989). At alkaline pHs, F-VIII and vWF elute as a complex. At pH 9.4–9.8 the complex is relatively pure; whereas at higher pH, considerable contaminants co-elute. This likely is attributable to the ion-exchange effects of CNBr-activated matrices.

According to the patent, F-VIII then is concentrated on a column of aminohexyl agarose (AH). Although AH is an anion exchanger and doubtless retains F-VIII for subsequent elution as a concentrate, I believe that the primary function of the AH is to retain F-VIII while any murine anti-vWF antibodies (which presumably are uncharged or slightly cationic at physiologic pH) that leach from the immunosorbent pass through the AH (which also is cationic). Regeneration of the immunosorbent bed is with 3 *M* sodium thiocyanate, which interdicts the Ag-Ab interaction and causes the vWF to elute. Downstream, the purified, formulated, and freeze-dried F-VIII is pasteurized; heating of the dried material kills many viruses, including HIV (Olson 1987).

Hyland's Method M

The process (Method M; Griffith et al. 1991) developed at the Hyland division of Baxter Healthcare (now Baxter Biotech), Los Angeles, CA, incorporates three elements:

1. Virus kill by means of the solvent/detergent systems developed at NYBC (see below)
2. Factor VIII purification on a MAb column directed against the F-VIII molecule
3. Ion-exchange downstream of the monoclonal column.

Some details are provided in Table 9.2.

For the human-source material the cryoprecipitate (cold-insoluble globulin fraction) is collected at 95 percent thaw of frozen plasma. The resuspended cryoprecipitate then is rechilled so as to reprecipitate the cold-insoluble proteins and reduce the co-precipitants recovered at thaw (Liu et al. 1989). Tri-*N*-butyl phosphate (TNBP) and Triton X-100™ then are added to 0.3 percent and 1.0 percent (w/v), respectively, and the solution is warmed to 20°C. This solution is applied to the immunoaffinity column that binds F-VIII. The column is washed with buffer which washes out non-F-VIII proteins and viruses (Figure 9.2). Factor VIII is eluted with buffer containing 40 percent ethylene glycol. The eluted F-VIII and ethylene glycol then is applied to a quaternary aminoethyl (QAE)-Sepharose column under conditions that

Table 9.2. The Method M Process

Purification Step	Description
Plasma	Human plasma, frozen at -80°C, is thawed (95 percent; some ice remains) and pooled.
Cryoprecipitate suspension	The cold plasma, containing about 5 percent ice, is mixed to form a uniform dispersion because, although the bulk of the protein (albumin, peptides of low molecular weight) remains in solution, much of the protein is insoluble.
Cold precipitation	The 95 percent thawed plasma is centrifuged to precipitate the euglobulin fraction which, at 0°C, contains most of the F-VIII, alpha$_2$-macroglobulin, IgM, and other proteins of high molecular weight at low solubility in cold saline. The precipitate is resuspended at 20–25°C so as to form a true solution.
TNBP/Triton X-100™	Tri-*N*-butyl phosphate and Triton X-100™ are added to the resuspended, resolubilized euglobulin fraction and mixed for 15 minutes at 20–25°C. This is the solvent/detergent step that destroys the integrity of the envelopes of most or all enveloped viruses.
Immunoaffinity column	The solvent/detergent treated euglobulin fraction is filtered to remove particles and then reacted for 10 hours with monoclonal anti-human F-VIII antibody immobilized on a solid support, in this instance porous beaded agarose in a column. Contaminating proteins and viruses and the residues of these are washed from the column and the antibody-captured F-VIII is eluted under conditions that break the Ag-Ab bond (40 percent ethylene glycol for this Ag-Ab system).

Continued on next page

Continued from previous page

QAE ion-exchange	Both the MAb and the F-VIII have well-chromatography defined isoionic points (pIs). Therefore, the column eluate is passed through a strong anion exchanger (quaternary amino ethyl or QAE is the Pharmacia terminology) under conditions that bind the F-VIII but not the MAb that may have dissociated from the matrix. The ethylene glycol, TNBP and Triton X-100™ pass through the QAE exchange matrix prior to elution of the F-VIII.

Figure 9.2. Washing of viruses from immunosorbent columns.

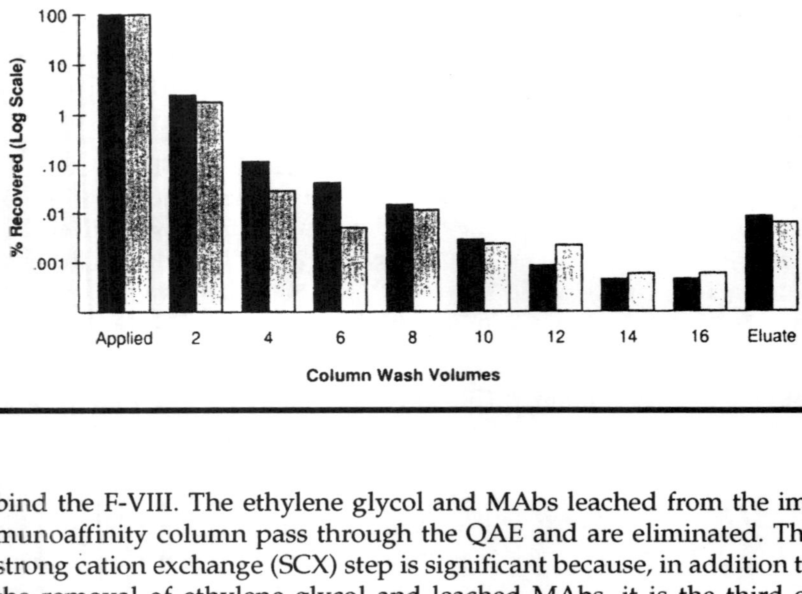

bind the F-VIII. The ethylene glycol and MAbs leached from the immunoaffinity column pass through the QAE and are eliminated. The strong cation exchange (SCX) step is significant because, in addition to the removal of ethylene glycol and leached MAbs, it is the third of three steps in process that eliminate or denature viruses. Representative data shown in Figure 9.3 (Griffith et al. 1991) are for the washout of EMC virus and murine MAb from the QAE column. Note that the ordinate is in $\log_{10}$ units.

Alternative processes to purify F-VIII invariably would have resulted in dramatically decreased yields. Consider that, even if each step in process results in a 95 percent yield (an exceptionally optimistic

Figure 9.3. Removal of ectromelia virus and murine MAbs with a strong anion exchanger step in the Method M process.

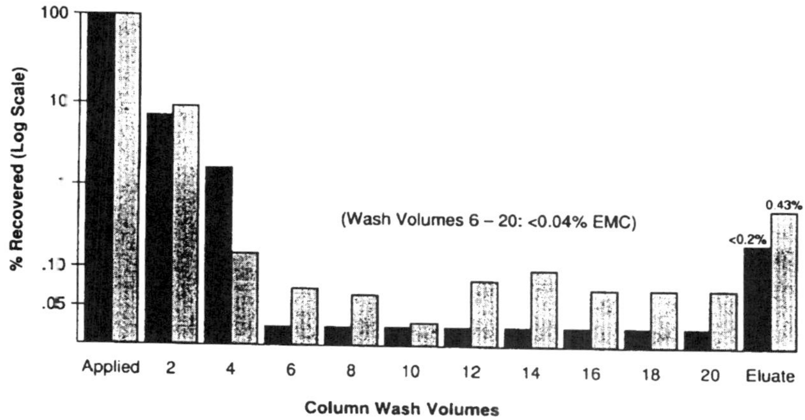

result), eight steps in process will result in an overall yield of 66.3 percent. Consequently, utilization of a one-step purification process almost invariably improves yield dramatically.

The differences in the immunoaffinity-purified F-VIII preparations of Hyland (human- or recombinant-sourced) and Rhone-Poulenc-Rorer (human-sourced) appear to be negligible as a crossover, single-dose pharmacokinetic study indicated that clearance, mean residence time, and the volume of distribution were alike (Morfini et al. 1991).

Mantovaara et al. (1991) played upon the interaction of F-VIII with vWF in another way. They loaded carboxymethylated aspartic acid agarose with Ca^{2+} and sorbed VIII:C. After washing, F-VIII was desorbed with a Ca^{2+}-containing buffer. This system *may* prove adequate for the purification of F-VIII from plasma-source proteins that are otherwise processed for virus inactivation and removal, but would be insufficiently specific to be of use with spent medium from recombinant Chinese hamster ovary (CHO) or other suitable cells.

THE ISSUE OF VIRUS ELIMINATION AND/OR DESTRUCTION

Viruses in Human-Source F-VIII

In addition to HIV, human-source factor VIII might also contain hepatitis A virus (HAV), hepatitis virus B (HBV) and HCV (HCV formerly was known as nonA-nonB virus), together with a variety of other, less virulent species. For HIV the routine testing of each blood donation (which commenced in the U.S. with the Abbott Diagnostics test in the spring of 1984) is for anti-gp120 Ab, and such Abs may be produced three months or longer after HIV infection. That is to say, there is a lag phase, following infection, before a clearly detectable Ab response can be shown. HIV may be present in the plasma during this period; hence the plasma, which tests negative for the virus, would be thawed and included in a plasma pool from which F-VIII is prepared (Olson 1987). This also is true of other viruses. Therefore, in 1984, the development of strategies for the elimination and/or destruction of HIV and other viruses became an *A* priority for the blood community.

The first F-VIII products in the marketplace that had a reduced risk of HIV transmission were the heat-treated fractions prepared by the Hyland division of Baxter. Apparently because the anti-gp120 test and the heat-treated F-VIII were expensive U.S. products, the National Center for Blood Transfusion (CNTS) in France at first would not use the U.S. test while a French test was in preparation. Nor would CNTS use the more expensive heat-treated F-VIII until existing stocks had been used. Therefore, during 1985 French hemophiliacs received F-VIII that was suspect of being HIV contaminated (Maddox 1993). The result was that about half of France's hemophiliacs seroconverted to HIV positive.

Among those brought to trial in France for this debacle were Dr. Michel Garretta, director of CNTS as of October 1984 (four-year prison sentence); Dr. Jean-Pierre Allain, former director of research at CNTS (four years imprisonment, two years suspended); Dr. Jacques Roux, director-general of health in the National Ministry of Health in 1985 (four-year suspended sentence); and Dr. Robert Netta, director-general of health in the National Ministry of Health (four-year suspended sentence). There was considerable scapegoating in the CNTS affair (Dickson 1993). The lessons to *all* of us involved in the manufacture and the regulation of biotech drugs should be that

- Product safety takes priority over budgets.
- Blame trickles downward.
- Issues involving ethics require rigorous discussion, despite the discomfort of executives, and require careful documentation.

Viruses and Other Materials in Recombinant Factor VIII

Most mammalian cell lines are assumed to contain viral material and oncogenes in the genome. Murine MAbs are assumed to contain any of a number of murine viruses unless testing precludes their presence. Host proteins or DNA or both are assumed to be present in the human r-protein unless they can be shown to be absent. The concern over animal viruses is not far-fetched. HIV likely arose in monkeys and was transmitted to man by a mechanism not yet delineated. Influenza viruses pass through birds enroute to man. Although the risks may be modest, regulatory agencies have chosen, rightly, to act on the side of caution. Where recombinant human protein is produced in *E. coli*, viral considerations relate to the exclusion of T phages from the manufacturing environment.

Similarly, since host cell protein, whether from CHO cells or *E. coli*, may trigger anaphylactic shock, they cannot be tolerated in the final product. In this chapter the focus is on viral inactivation or removal as the highest priority.

The New York Blood Center Method for Destruction of Enveloped Viruses

Concerted efforts at the New York Blood Center (NYBC) in Manhattan resulted in the solvent/detergent (SD) strategy, wherein tri-*N*-butyl phosphate (TNBP) and sodium cholate were added to F-VIII preparations containing various model viruses (Horowitz et al. 1985; Prince et al. 1986). The final composition was TNBP/polysorbate 80, and this SD process has been in use by NYBC, Miles (Cutter; Clayton, NC) and Alpha Therapeutics (Los Angeles). The Hyland group use TNBP/oxtoxynol (Fricke 1994). Oxtoxynol is otherwise known as Triton X-100™. There is substantial confirmation of the NYBC work (Table 9.3). At Hyland the reductions of HIV by tri-*N*-butyl phosphate and Triton X-100™ mixtures were > 10^{11} (Piszkiewicz et al. 1989). Mitra et al. (1994) at Miles, Inc. (Berkeley, CA) have shown that at least 9 logs of HAV is destroyed by the SD process. The Blood Products Advisory Committee of FDA has concluded that the risk of infection with HAV, or parvovirus (or any other virus) from clotting factor concentrates is extremely low.

Viruses can be excluded from solutions of proteins by ultrafiltration (UF). This approach can work, but high losses of F-VIII are possible. The problem is in the casting and evaluation of the UF membrane; many of the UF membranes available commercially do not perform within advertised parameters (McGregor 1986). In fact, UF membranes usually show exclusion limits that range over an order of magnitude of molecular weights. Major increases in molecular weight usually are accompanied by modest increases in radius because of the extent of folding of most proteins. Consequently, UF membranes with very fine definition by molecular weight probably are not possible. However, the Millipore group (Bedford, MA) have developed an UF line,

Table 9.3. Solvent-Detergent Kill of Viruses[a]

Treatment Conditions	Virus[b]	Log_{10} Inactivation
New York Blood Center		
0.3 percent TNBP/	Sindbis	>5.0
0.2 percent Na cholate	Vesicular	>4.5
4 hours, 20°C	stomatitis	
	Sendai	>5.5
	FMC	0
Baxter Biotech (Hyland)		
0.3 percent TNBP/	Sindbis	>5.0
1.0 percent	Vesicular	
Triton X-100™	stomatitis	>4.7
3 minutes minimum,	PRC	>4.0
20°C	HIV-1	>11.0[c]
	EMC	0

[a]After Griffith et al. (1991)

[b]EMC and FMC are myocarditis viruses that contain RNA in naked icosahedral nucleocapsids (i.e., they are not encapsulated in protein envelopes). Unenveloped viruses are resistant to solvent/detergent inactivation.

[c]From Piszkiewicz et al. (1989)

Virasorb, that appears to be quite reproducible in behavior, and which passes most proteins while retaining most viruses (see Bechtel et al. 1988). Asahi Chemical Co. (Osaka, Japan) now produces a 15 nm and 35 nm deadend filters, which appear to remove over 6 logs of the smallest animal viruses, including HIV.

Elution of Viruses from Factor VIII Immunosorbent Columns

Other than by UF, viruses also can be eliminated during the purification of the Ab-immobilized F-VIII. Clearly, when the immunosorbent column is washed, virus should be lost to waste during the washing of the column, prior to development. Data for the washout of model viruses is given in Figure 9.3 from the Hyland data. Method M and the Scripps Clinic immunosorbent steps most likely are equivalent in this regard.

A method that appeared attractive was the use of a hydrophobe-substituted matrix for the retention of hepatitis B (Iwarson et al. 1987) and other viruses. The concept was good, but the efficiency and reliability were poor in practice.

Dry Heating

Viruses in solution are killed by heat (Kaplan 1958). HIV and some other viruses in freeze-dried F-VIII also are killed by heat, but dry heat is not especially effective in the kill of HCV (Daenen et al. 1987; see review by Olson 1987). The Armour data for heat treatment show 2.36 logs of vesicular stomatitis virus (VSV) kill and 4.8 logs of kill for HIV (Hrinda et al. 1989). Armour has been as successful with human-source F-VIII that is prepared on an immunosorbent column and then dry heated for 30 hours at 60°C as Hyland has been with solvent/detergent-treated, immunosorbent purified F-VIII, in reducing the HIV and hepatitis C virus (HCV) risks to near nil (Mannucci and Colombo 1988). Immuno have shown 10^6 reduction in HIV titers to zero detectable virus at 10 hr at 60°C; the Immuno process utilizes water vapor at temperature, rather than dry heat (Barrett et al. 1989).

Current Status of Human-Source and Recombinant F-VIII

At this writing, Method M is applied to the purification of F-VIII prepared in mammalian cell culture (Recombinate™) as well as to human-source material, and the Scripps method is applied to human source-material only (Monoclate™). The Cutter division of Miles (Berkeley, CA) terms rF-VIII *KoGENate* which almost certainly is purified by immunosorbent methods. The human-source material likely is very safe for the seroconversion rate appears to be nil in the U.S. since the shift to immunosorbent-purified F-VIII.

PREFERRED METHODS FOR ANTIBODY IMMOBILIZATION

The Matrix

The first consideration for preparative-scale immunoaffinity separations is the choice of matrix which, traditionally, was 2 percent and later 6 percent beaded agarose. Polysaccharide matrices are preferred because they pose the fewest problems in adsorption. Hydrophobic matrices (e.g., polystyrene) and generally sorptive materials, (e.g., glass) tend to be unacceptable unless particular care is taken to mask sorptive sites. Glass and silica also tend to dissolve at pH > 8–8.5, restricting the utility of these matrices. Dissolved silica might be seen as a very small bit of turbidity in the reconstituted final container and a cause for lot rejection. The issues with polysaccharide supports *as matrices* usually are flow rate and compaction. Six percent agarose generally provides acceptable flow rates (Tharakan et al. 1992). Rapid development at high flow rates is done with radially-configured sheets of cellulose (Huang et al. 1988; Menozzi et al. 1992) because the resistance of the medium is minimal. However, product eluted from the radial cartridge system almost always is more dilute than that from an agarose column (Olson unpublished).

Beaded polysaccharides in columns pose some problems of loading and washing. The bulk of the immobilized Ab is beneath the bead surface, hence diffusion distances in the bead are of consequence as the Ag penetrates to Abs in the interior of each bead. For example, a 200 ml bed volume of beaded agarose with a mean diffusion path length of 10 μm should be loaded and developed at a linear flow rate of 0.02–0.03 ml/sec, whereas a membrane filter sorbent with a mean diffusion path length of ca. 0.3 μm can be loaded and developed at a linear flow rate of ca. 100 ml/sec (Klein 1991). The speed of the loading and development of immunosorbent membranes would be most attractive were it not for the tendency of such filters to clog, rendering them unsuitable for use. The gel beds can be resuspended (a highly successful strategy) or, at worst, the clogged layer can be scooped away and replaced.

The support matrix does not impact K_d appreciably, but does affect the maximum amount of Ab that can be immobilized, the rates of adsorption and desorption of a protein to the immobilized Ab, and the specific activity of the immunosorbent. Data are shown in Table 9.4 (Fowell and Chase 1986) for rat anti–*E. coli* beta-galactosidase MAb immobilized on the various matrices. Note that different coupling chemistries (each of them more or less random) obtained with the different matrices, a reflection of the activated material available in the marketplace. For example, CDI-Trisacryl is *N*-acryloyl-2-amino-2-hydroxymethyl-1,3-propanediol polymer. Activated Ultrogel was glutaraldehyde-activated polyacrylamide-agarose, and CDI-Fractogel FSK was the activated vinyl polymer. From the data one may conclude that the CNBr has a very high capacity for the immobilization of Ab and, for reasons that are unclear, beads of polymerized vinyl (Fractogel) appear in some way to inactivate the immobilized Ab. The Ultrogel product was available from LKB (Bromma, Sweden), which was purchased by Pharmacia in the 1980s.

Table 9.4. Some Matrix Choices for Immunosorbents[a]

Matrix	q_m	K_d	Specific activity
CNBr-Sepharose	5.0	2	0.15
CNBr-Sepharose	9.5	3	0.12
CNBr-Sepharose	0.2	2	0.10
CDI-Trisacryl	3.1	1	0.25
Activated Ultrogel	3.0	9	0.14
Activated Ultrogel	3.8	10	0.23
CDI-Fractogel FSK	1.8	9	0.01
CDI-Fractogel FSK	0.15	1	0.05

[a]See text for significance of the data

q_m: Maximum capacity of the immunosorbent in mol/ml $\times 10^9$

K_d: Antigen-antibody dissociation constant in $M \times 10^{10}$

Specific activity is moles Ag bound per mole Ab (moles/mole)

Membranes likely would be most useful as a preparative immunosorbent matrix when operated in the crossflow mode, with particle-immobilized MAbs on the upstream side of the membrane. The Millipore system (Millipore Corp., Bedford, MA) of binding membranes to plastic supports, wherein woven or nonwoven upstream supports are not used, would seem to be ideal. Coupling to cellulosic membrane filters can be done in most of the ways applicable to agarose and other polysaccharides.

The Cyanogen Bromide Method: Issues and Improvements

The diazo coupling utilized in the 1950s was in small-scale commercial use for research in 1965 (Olson unpublished), but by far the most popular coupling technique has been the CNBr technique developed by Axen et al. (1967) and popularized by Cuatrecasas et al. (1968). A schematic of the steps in activation and coupling by means of CNBr is given in Figure 9.4. Most of the cyanate ester rapidly converts to poorly reactive or inert moieties that behave like weak anion exchangers. The various coupled species also vary in their stability, hence there tends to be significant bleed of Abs coupled to CNBr–activated matrices. Bleed is addressed later in this chapter. Another problem, other than the lability of the coupling, is coupling via the antigen-binding sites of the Ab. Clearly, if coupling is via any portion of the Ab other than the Fc segment, Ag-binding will be reduced or eliminated. Theoretically, one should be able to bind two moles of Ag per mole of immobilized Ab or, if the Ag is sufficiently large that the binding of one molecule impedes binding of a second Ag molecule to the second binding site of the Ab, one would expect at least one mole of Ag bound per mole Ab. In fact, with the CNBr coupling method, one is fortunate to have an Ag:Ab binding ratio of 0.2 (Olson unpublished). This means that a substantial proportion of the murine MAb that bleeds from the column has been unavailable for binding the Ag. Additionally, the CNBr activation method poses hazards to the workers preparing the matrix; the acetone solution used in activation often is concentrated and hazardous. However, when the Zimmerman-Fulcher and the Hyland work were in process, the CNBr process was the best known and most thoroughly tested method, hence was incorporated into those processes. Once the product has been characterized and is in clinical trials, manufacturers are loathe to change process. If the IND has been submitted, regulatory personnel in the industry avoid changing any part of the submissions in process.

Figure 9.4. CNBr activation of polysaccharide matrices.

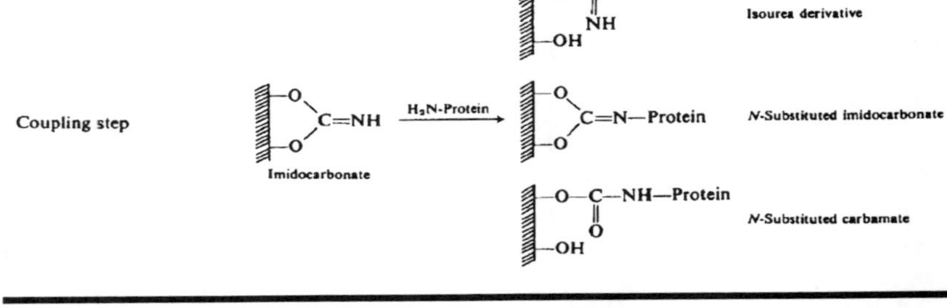

The FDA approves some changes in process by letter, but for substantive changes to process, a "baby" clinical of 10–20 patients may be required. Even though clinical trials cost $10,000–15,000/patient, there may be a real payback if a change in Ab coupling reduces column costs and improves murine MAb retention (reducing potential risks of anaphylaxis in patients). This is worth an enquiry.

If there remains some obscure reason to prefer CNBr activation over other activation/coupling methods, triethylamine should be added with CNBr at neutral pH. The result is a substantial increase in the amount of cyanate ester and high coupling capacities (Kohn and Wilchek 1982; Wilchek et al. 1984).

Periodate Oxidation of the Fc Saccharide

The preferred method is to couple the Ab via some portion of the molecule remote from the Ag binding sites. This approach ensures a high degree of Ag binding (i.e., an Ag:Ab ratio that approaches or is greater than one) and reduces the amount of Ab required for immobilization, perhaps as much as fivefold. One approach is to use a mild oxidizing reagent (e.g., $NaIO_4$), to generate aldehydes in the carbohydrate of the Fc portion of the Ab. The Ab with aldehydes then is reacted with an amine or hydrazide-substituted matrix and, for the reaction with the amine, reduced to form a stable covalent couple. This method has been used often and works well (Fleminger et al. 1990; O'Shannessy and Wilchek 1990). The couplings are no less stable than the CNBr method and the ratio of Ag bound:immobilized Ab is increased significantly.

Kontes (Vineland, NJ) offers Affinochrom™ matrices, which are avidin-coated nonporous glass beads. The concept is that the Fc carbohydrate is $NaIO_4$-oxidized to –C(O)H and reacted with hydrazine biotin. Because K_d for the avidin-biotin complex is of the order of 10^{-15}, immobilization is remarkably stable, provided the avidin (an egg white protein) is well anchored. If the bleed of avidin from the matrix is satisfactorily low and if the adsorption of the antigen to the matrix does not unduly complicate antigen purification or column life, this matrix and approach may merit consideration.

Sulfhydryl Coupling of Fab Fragments

A second approach to binding via regions of the Ab remote from the binding site(s) utilizes sulfhydryl coupling of "half-antibody" or of reduced $F(ab)_2$ fragments (produced by papain digestion; see Andrew and Titus 1992) to an iodoacetyl-activated matrix. The "half-antibody" approach is particularly useful as the Ag:Ab binding ratio is improved since one free site is not sterically hindered by Ag bound to the sister site. These techniques are reviewed in Hermanson et al. (1992), which provides protocols for the techniques (p. 230).

The Nickel Affinity Approach

A most interesting approach that may have considerable merit is the immobilization of the MAb or PAbs by means of Ni^{2+} affinity. For this system the support matrix is substituted with iminodiacetic acid, which then is complexed with Ni^{2+} (Porath et al. 1975). Nickel in such a complex has a very high affinity for histidine. Consequently, if one couples polyhistidine, usually hexahistidine, to the MAb or PAb, the Ab will bind to the matrix but, under defined conditions, can be eluted (Loetscher et al. 1992). The advantages of this approach are considerable. For example, the immobilized Ab or Ag can be desorbed and put aside so that the column matrix can be cleaned rigorously and then steam autoclaved. If desired, the Ab (or Ag) can be repurified by ion-exchange chromatography (IXC) and/or UF or SEC, followed by sterilizing-grade deadend membrane filtration. The Ab then can be formulated with pasteurized human serum albumin or Ficoll or nonreducing saccharides and freeze-dried for long-term storage.

There also are possibilities for prophylactic filtration of F-VIII eluted from the immunosorbent column by means of a short column or membrane containing chelated Ni^{2+}. Provided one can show that the bleed of the histidine-substituted Ab is less than for CNBr–coupled material, and that it can be removed fully from eluates, the method has much to commend it.

These three approaches (direct coupling via carbohydrate on the Ab, coupling via sulfhydryls on Ab fragments, and Ni^{2+} affinity coupling) are, in my view, methods of choice for Ab immobilization as immunosorbents.

Antigen binding is about tenfold higher when the Ab is coupled via the Fc portion of the Ab (Zopf et al. 1987). Spitznagel and Clark (1993) have examined the effect of whole Ab and Fab′ fragment loading on the ratio of Ag to Ab and K_d. As loading of the resin with the whole or fragment of Ab increases, the Ag:Ab ratio declines by as much as 50 percent (presumably as neighboring binding sites compete for the desired Ag). Capacity for large Ags can decrease as Fab′ loading increases. However, the Ab loading of the resin increases considerably when the Fc fragment is eliminated. The authors used Weetall's (1976) silanization method with controlled-pore glass for all studies. It would appear that 75–100 $\times$ 10^{-6} mole Ab or Ab fragment/liter matrix is optimal for resin loading.

Other Methods

Fc binding of Ab should make Protein A immobilization of Ab attractive. The Protein A approach is acceptable if one is preparing a diagnostic; but Protein A, or antigenic fragments generated by proteases in the feedstream, clearly is unacceptable in a therapeutic. Nonetheless, the very inventive coupling method of Ramirez et al. (1993) deserves a special mention. They constructed a fusion protein between Protein A and the cellulose-binding domain of an exoglucanase from *Cellulomonas fimi*. It was shown that the fusion protein binds to cellulose without impeding the Ig-binding of the Protein A portion, and that the cellulose-fusion protein system can be used in immunoassays, affinity chromatography, and immunoprecipitation. However, it is highly questionable for the preparation of injectables.

Nakamura et al. (1990) proposed the use of TSKgel Tresyl-5PW (TOSOH, Tokyo, Japan), a (2,2,2-trifluoroethanesulfonyl) chloride-activated matrix. The activated matrix reacts with amino, thiol, phenol, or imidazole groups under mild conditions. The authors found (on the basis of A_{280} values) that Ab coupling was approximately 100 percent efficient at pHs between 7 and 8.5. They also found (based on UV data) that Ag binding ranged from about 70–80 percent at pHs between 6.5 and 8.0. They seem to have assumed that the Abs coupled via the Fc portion of the molecule, based on the Ag binding efficiency. That conclusion bears investigation. If it is substantiated, the tresyl-activated matrix may have general utility as an immunosorbent.

SCALING FROM THE LIMITING STEP IN PROCESS

In part because of the considerable expense of the MAbs, a critical consideration is the size of the immunosorbent column and the MAb loading. Let us say that the immobilized MAb resin has been characterized and binds one mole of the desired protein (DP) per 5 moles of MAb (this would be common for CNBr-activated agarose). Let us say that the M_r of the Ab is 160 kDa and the M_r of the DP is 25 kDa. If the batch size of DP is to be 25 g, we will require 25 g/25,000 g/mole = 0.001 mole of binding sites per batch. One millimole of the MAb is 160 g, but with the CNBr procedure as originally described, you will require $5 \times 160 = 800$ g of the MAb, simply because the coupling is random and binding is to Fab at or near the binding site as well as to Fc.

If one couples 20 mg of MAb/ml of agarose, a 4.0 ℓ agarose bed will be required for binding all of the DP. The safety factor should be at least 20 percent, hence the finished bed volume is 4.8 ℓ. If a more efficient method of coupling is used, the MAb and gel bed requirement, and accompanying expense, diminish about four- or fivefold.

To verify that this bed volume will be adequate, small-scale experiments in minicolumns are needed, with at least 40–50 cycles per column so as to establish the decline in capacity with use. To this end, the 20 percent oversize is useful—it prolongs the use of the column.

The tendency is to place all of the coupled gel into a single column, but that is not the best strategy. Channeling sometimes occurs, wherein liquid feed bypasses a proportion of the gel bed. In such instances columns in series are a great advantage as problems in the packing of one column seldom are reproduced in another.

CHARGING AND ELUTING THE COLUMN

Avoiding Contamination of the Column

The most serious problem is microbial contamination of the column. The second priority is minimizing the particle burden. Third is minimizing the effect of proteases on the immobilized Abs. The first two are best addressed by filtering all liquids entering the column through a 0.2 μm-rated microporous membrane filter. This requires that the filter is attached to the column and that another filter is in-line with the buffer sources. However, such precautions are moot if the column is poured in an open room; the column should be poured under a HEPA hood by personnel who are gowned, gloved, and masked to reduce the bioburden and particle burden, although most coupling methods are microbicidal and tend to sterilize the resin.

Charging and Rinsing Polyclonal and Monoclonal Columns

The energy of activation of Ag-Ab reactions is sufficiently low that they go to completion even in the cold. However, they are diffusion-limited and are more readily accomplished at 20–30°C than in the cold. Therefore, although MAb columns usually are stored at cold-room temperatures, they are equilibrated to ambient temperatures (22–25°C) for operation. Loading of the column is assumed to be tedious as the epitope of the Ag and the V_H (variable, or Ag-binding region) of the Ab must interact properly if binding is to occur. In solution this reaction seems slow because precipitate formation is slow. The advantage of the column with immobilized Ab is that as the Ag front advances down (or up) the column, Ab consistently is in excess. Because the association constant is so high, there is no need for a long bed provided the appropriate Ag-binding capacity is present. A column with a width-to-height ratio of 1:2 to 1:4 usually suffices and avoids overly long loading or development times (Eketorp 1980).

Elution Schemes for Polyclonal and Monoclonal Columns

Polyclonal columns usually are eluted under the alkaline or acidic conditions (e.g., glycine-HCl buffer, pH 3–4) (especially for immobilized PAbs, Ghatge et al. 1991).

Mild changes in pH may be suitable for useful changes in the Ag-Ab K_a, per Berry et al. (1990). With the MAb that Berry et al. were studying, the K_a was high at pH 5.5 and dropped precipitously at physiologic pH, favoring elution. Among the gentlest of solutes for the elution of MAbs are the polyols such as ethylene glycol (Thompson et al. 1990). Monoclonal antibodies also vary considerably in temperature sensitivity for oligosaccharide binding. For one MAb the K_a doubled with each 8°C decrease in temperature (Zopf et al. 1987). Consequently, one may wish to jacket the column with a view to operation at 30–37°C during elution.

So as to better understand what is involved in elution from immunosorbent columns, the following terminology is borrowed from SEC:

$$V_T = V_O + V_I + V_G$$

where V_T = total column volume, V_O = void volume (liquid volume surrounding the gel beads), V_I = *vida infra* (liquid volume within the gel beads), and V_G = volume of the gel, together with bound water. Because the column beads are microporous, rapid buffer flow promotes flow around the beads, via V_O, to the detriment of flow into and through the beads. All manufacturers of chromatographic resins and equipment recommend optimal flow rates (as face velocities); these will merit testing so as to establish flow rates at which $V_O + V_I$ elutes in a volume consistently less than V_T.

Elution curves of both PAb and MAb columns are skewed left, with a long tail following the main peak. Product in the "tail" usually is indistinguishable from that of the main peak. It probably represents product bound deep within the beads, and which was dependent on diffusive flow from the beads for elution into the void volume of the column.

Minimal Denaturation of Product and Antibodies

One of the most useful studies on optimum conditions for immunoaffinity column development was that of Tsang and Wilkins (1991), wherein human IgG (as Ag) was immobilized on Affi-prep-10 and used as a sorbent for the purification of hyperimmune goat anti-hIgG (PAbs). The optimum dissociation reagent was 3.0 M $MgCl_2$-$6H_2O$ in 0.075 M HEPES/NaOH with 25 percent ethylene glycol, pH 7.20. The PAbs recovered with this dissociation reagent showed a specific activity 1.87 times greater than that for PAbs eluted with 1.0 *M* glycine/HCl, pH 2.00; the total recovery was 6.33 times greater with the milder reagent.

ISSUES WITH IMMOBILIZED ANTIBODIES

Column Half-Life

The practical half-life of the column must be estimated at bench level before scaleup is commenced. Half-life is the number of cycles at which the amount of Ag bound is half the amount at t_0. Matrices fragment, Ab is lost during the cycle or denatures, and Ab sites may be obscured by particles. From the half-life one can estimate the amount of Ag that can be purified over the practical life of the column. Depending on the antibody, the coupling method and other variables, the decline in column capacity may be first-order or may be virtually nil (Hamman and Calton 1985).

Column Sanitization and Storage

Immunosorbent columns for purifications of the scale of F-VIII injections are very expensive. A large column might easily cost $100,000 in MAbs, let alone the resin, modifications to the resin, chromatographic reagents, the chromatographic system, and automation of it. Therefore, stability and longevity of the columns is of considerable interest.

Cleaning can be done with a mild detergent (e.g., a solution of Tween 80), and transient variations in pH (e.g., 3 to 9 or 10). These conditions may not destroy all bacteria or viruses lodged in the column. However, in the case of the Ni^{2+}-modified column for Ab immobilization, after Ab removal the matrix can be cleaned with harsh detergents, extremes of pH, and could be steam autoclaved. However, the less rigorous practices in current use clearly produce safe product, and in the absence of a compelling reason, the existing methods will not be changed.

Column Clogging and Repacking

Column cleanliness, both in bacteria and particles, is maintained to a considerable degree by filtering all column influent through 0.2 μm-rated membrane filters. Even buffered distilled water contains a considerable particle burden and must be filtered. For best results, the filtration must be in-line with the column. However, column clogging may yet occur, as shown by a reduction in flow rate at a constant head pressure, or increasing back pressure on the column at a constant flow rate provided with a pump. The consequence of column clogging is that portions of the immunosorbent matrix become unavailable for the immobilization of F-VIII (i.e., column capacity is effectively reduced). The clog of a column is countered in part by development in the direction opposite to the direction of loading and wash. That strategy alone promotes physical column longevity.

Bleed of the Antibody into the Product

A column stability issue and an issue in product safety is the presence of the Abs from the column in the product. The anti-F-VIII Abs used in the preparation of the immunosorbent usually are murine and, if present in sufficient concentration and injected repeatedly into a hemophiliac, generate anti-murine Ig Abs. The potential for anaphylactic shock exists. Minimally, one must screen product for mIgG and, in the best circumstances, one utilizes strategies for the removal from product of any immunosorbent Abs, or fragments thereof, that might coelute with the product. To this end, the Method M strategy includes an ion-exchange step downstream of the immunosorbent column.

Potential Advantage of the Nickel Affinity Method

It would be of interest to know whether, for intact hexahistidine-modified MAbs, scavenging by chelated Ni^{+} is as efficient on the downstream side as on the upstream. That likely is the case, but does not salvage antibody fragments that lack hexahistidine from the product.

Phage Antibodies

One of the major considerations with MAbs, aside from the issue of sensitization of the patient to mouse Ags, is cost. McCafferty et al. (1990) displayed Ag-binding fragments of Abs on the surface of a bacteriophage. They call these "coliclonal" Abs, which appear to represent an orders-of-magnitude increase in speed of manufacture and diversity of the Abs (Chiswell and McCafferty 1992), together with probable reductions in production costs. The concept is that every immune cell in any mammal, including humans, contains DNA encoding the specific Ab displayed on the surface of each cell. Repertoires of Ab genes (Winter and Milstein 1991) can be amplified using the polymerase chain reaction (PCR). The Ab variable-region genes are cloned into the gene encoding the third gene of the filamentous bacteriophage fd. The fusion protein thus created consists of the encoded Ab fragment at the N-terminus of the phage coat protein. The genome of each recombinant fd phage contains the DNA of the specific Ab, and the Ab is expressed on the surface of the phage. One of the beauties of this approach is that a phage-Ab library can be derived from *unimmunized* donors (Marks et al. 1991), and the desired phage-Ab with a K_d of ca. 100 n*M* can be isolated from populations of the order of 10^7–10^8 of other, weakly or non sorbing phage Abs. Particulars of the method are provided in some detail in the references cited in Chiswell and McCafferty. Issues regarding optimal methods for fd immobilization with the Abs unrestricted, and the exclusion of phage components from the product, require resolution. However, the method definitely merits investigation for the purification of recombinant products for injections.

Yet another twist on the coliclonal Ab approach is the preparation of genetically-engineered *E. coli* that expresses, on the outer surface membrane protein LamB, the *antigen* to which the coliclonal Ab is directed (Bradbury et al. 1993). My inclination would be to sorb the recombinant *E. coli* onto 3 mm diameter glass beads and to fix the cells with 1.5 percent formaldehyde *after* they had sorbed onto the beads, then to wash the beads, in the column, with three column volumes of PBS containing 0.05 percent Tween 20, followed by two column volumes of PBS. Specifically absorbing phage are eluted in a small volume of 100 m*M* triethylamine in PBS, passed very slowly through the column. This approach for the purification of the high-affinity recombinant fd phages, prior to phage immobilization for the purification of recombinant product, need not be so rigorous in the elimination of *E. coli* byproducts since the high-affinity fd phages subsequently are cloned in the host for expansion. Bradbury et al. claim that as few as 10 phage in a background of 10^{10} can be purified on these "living columns."

Stratgene Cloning Systems and Scripps Clinic (La Jolla, CA) have set procedures, techniques, and reagents for the creation of immunoglobulin expression libraries in *E. coli*. This enables one to clone Ig heavy chains and light chains separately into what they term the ImmunoZAP expression vectors, which are derivatives of the Lambda phage. This enables a chemist or biologist with basic microbiological skills to clone Abs. Stratagene also offers the coliclonal fd system as a cookbook procedure. As with most successful biotech companies, their technical support is quite good.

AUTOMATION OF THE CYCLE

A system for automating a column immunoaffinity system is shown by Eveleigh (1982). Equipment is available from many sources, including Pharmacia (Uppsala, Sweden and Piscataway, NJ) and Bio-Rad (Richmond, CA). The controllers are simple as is the valving.

Peristaltic pumps are preferred as there is no prospect of product exposure to lubricants and non-sterile surfaces.

The advantages of automation are as follows:

- Uniformity of conditions for product purification, hence less lot-to-lot variability.
- Assurance of permanent records of purification arameters, allowing for retrospectives on system performance
- Reduction in labor hours required for product purification.

Since one worker year costs at least $30,000 in most venues, reduction of personnel required from two to one likely provides the payback to justify the correct decision to automate.

CELL PURIFICATION WITH IMMOBILIZED ANTIBODIES

Lymphocyte purifications based on affinity chromatography were first published by Wigzell and Andersson (1969). Braun and Kumel (1986) coupled lymphocyte-specific MAbs to CNBr-activated Sepharose beads of 200–300 μm diameter or, more cleverly, reacted the cells with the specific MAb and immobilized Protein A on the column matrix. Particulars are:

> *T cells were incubated for 30 min in PBS (pH 7.2, 2 percent fetal calf serum, 0.01 percent NaN_3) at 4°C, run into a 5 ml column (sufficient to separate 10^8 cells), and the system incubated for 30 min at 4°C followed by up to 50 ml of washes with cold PBS to wash out nonadherent cells. Adherent cells were released with an excess of at least 5 μg MAb/10^7 cells, which also incubated in the column for 30 min at 4°C. Recovery is 10–20 percent. Stirring or vortexing increased the yield but decreased cell viability (Ghete et al. 1978). Dead cells were removed by Ficoll/metrizoate gradient centrifugation.*

Although this approach works, the system employed by Applied Immune Sciences (Santa Clara, CA) seems superior and is cited below.

Cell purification by means of immobilized Abs has not been applied on what we usually consider an industrial scale. However, there are therapeutic applications of the method and it presently is in use. The application of greatest interest is in the purification of $CD34^+$ cells (progenitor or stem cells from bone marrow). Stem cells are recovered from a candidate for gene therapy; transfected with a viral, plasmid, or other construct containing the required gene; and returned to the patient by intravenous injection.

This method for the purification of cells, followed by the insertion of a particular gene(s), eventually will replace at least certain of the therapies that involve intravenous delivery of purified proteins. The sterile device in which the anti–$CD34^+$ MAbs are immobilized will be used routinely in gene therapy applications. An appropriate MAb is coupled covalently to the plastic wall of a Roux bottle or to a large, perhaps magnetic, bead. Applied Immune Sciences (Santa Clara, CA) presently prepare such surfaces with coupled MAbs (anti–$CD34^+$, for example).

Cells (e.g., bone marrow cells), suspended in appropriate medium, are added (in the AIS system) to a flask containing the immobilized $CD34^+$ MAbs, swirled and incubated. Following several rinses to remove nonadherent cells, the immobilized cells are incubated with medium and, over time, the immobilized cells are resuspended in the medium and can be recovered by pipeting. Alternatively, cells attached to MAbs (which, in turn are coupled to large beads) can be rinsed free of unreacted cells and then agitated to release the $CD34^+$. Presumably, the mass and velocity of the immobilized cell overcome the anchorage of the $CD34^+$ Ag in the cell membrane, the cell is freed and the membrane (by virtue of its fluidity) reanneals.

Immobilized CD8 cells are used to recover tumor infiltrating lymphocytes (TILs) from kidney tumors removed surgically. The TILs are IL-2 activated, multiplied 100-fold in vitro, and reinfused into the patient (in phase-I trial as of August, 1993). CD8 cells, activated 100-fold in vitro with IL-2, also are in experimental therapy with AIDS patients. This extension of immunosorbent technology is likely to undergo considerable expansion in the near future.

Columns present problems with cells, but they can be overcome. Biotinylated MAb specific to a cell line can be added to a mixture of cells which then are passed through Sepharose 6B (Pharmacia, Piscataway, NJ) or Biogel P-30 (Bio-Rad, Hercules, CA) to which avidin has been conjugated by one of several methods. The K_d for the avidin-biotin reaction is of the order of 10^{-13} M, thus the cells are effectively immobilized. Cell release is achieved by recovering the gel from the column and developing shear with the gel beads and buffer by means of a Pasteur pipette. The method can be used for cell depletion and the removal of neoplastic cells from bone marrow.

Hydrazide-substituted polystyrene beads are available from Pierce Chemical Co. (Rockford, IL) and the –C(O)–NH–NH_2 group on the beads is reacted with bis-glutaraldehyde. The glutaraldehyde-activated bead then is reacted with the Ab of interest and $NaCNBH_3$ (cyanoborohydride), which results in a secondary amine link between the bead and Ab. Hexanediamine-substituted polystyrene beads can be reacted with sulfosuccinimidyl-4-(*N*-maleimidomethyl)cyclohexane-1-carboxylate (Sulfo-SMCC, Pierce Chemical Co.) which then reacts with the cysteine–SH of reduced IgG (half-Ab, incorporating one binding site and half of the Fc). Hermanson et al. (1992) have published coupling protocols for both methods; this author's preference is in the latter.

COMPARISON OF IMMUNOSORBENT WITH OTHER METHODS

Simplicity is at the heart of efficient, cost-effective manufacture of anything, including proteins for injections. Consider that a process involving eight steps, for each of which the yield is 95 percent of theoretical, then the overall yield of the process is 66.34 percent. If the number of steps can be decreased to 4, the overall yield is 81.45 percent. Consequently, an immunosorbent step will tend to result in high yields. The issues are the dollar and time costs of the immunosorbent and the risk potential for the inclusion of murine or other animal Abs in the product.

By way of contrast, IXC utilizes relatively inexpensive matrices that are amenable to cleaning with harsh reagents and sterilization by steam autoclaving, if desired. Ion exchangers have high capacity and the strong exchangers are capable of subtle separations. Although IXC also is a powerful tool for concentration of the product, purification remains very difficult if the product is dilute. Where the protein of interest is in low concentration, and there is a considerable background of contaminating proteins, affinity methods are necessary. In the absence of a highly specific interaction of the desired protein with one of the Procion dyes, immunosorbent methods are preferred. As a rule-of-thumb, immunosorbents are the last resort as a purification method on the basis of cost and the efforts required to ensure that the product is not inadvertently contaminated.

REFERENCES

Andrew, S. M., and J. A. Titus. 1992. Large-scale fragmentation of IgG to Fab using papain. In *Current protocols in immunology*, edited by J. E. Coligan, A. M. Kruisbeek, D. H. Margulies, E. M. Shevach, and W. Strober. NY: Green Publishing Assoc. and Wiley-Interscience.

Axen, R., J. Porath, and S. Ernback. 1967. Chemical coupling of peptides and proteins to polysaccharides by means of cyanogen halides. *Nature* 214:1302–1304.

Barrett, P. N., J. Eibl, and G. Wober. 1989. Virus inactivation of plasma proteins by vapour heating. In *Biotechnologie des proteines du plasma,* edited by J. F. Stoltz and C. Rivat. Colloque INSERM Vol. 175.

Bechtel, M. K., A. Bagdasarian, W. P. Olson, and T. N. Estep. 1988. Virus removal or inactivation in hemoglobin solutions by ultrafiltration or detergent/solvent treatment. *Biomater. Artif. Cell Organs* 16 (1–3):123–128.

Berry, M. J., J. Piron, M. M. Gani, and P. Porter. 1990. Immunoaffinity purification of guar alpha-galactosidase using mild elution conditions. In *Separations for biotechnology, Vol. 2,* edited by D. L. Pyle, London: Elsevier Applied Science.

Bradbury, A., L. Persic, T. Werge, and A. Cattaneo. 1993. Use of living columns to select specific phage antibodies. *Bio/Technology* 11:1565–1569.

Braun, R. W., and G. Kumel. 1986. Separation of T cell subpopulations by monoclonal antibodies and affinity chromatography. *Methods Enzymol.* 121:737–748.

Brennan, P. J., S. W. Hunter, M. McNeil, D. Chatterjee, and M. Daffe. 1990. Reappraisal of the chemistry of mycobacterial cell walls, with a view to understanding the roles of individual entities in disease processes. In *Microbial determinants of virulence and host processes,* edited by E. M. Ayoub, G. H. Cassell, W. C. Branche, and T. J. Henry. Washington, DC: American Society of Microbiology.

Campbell, D. H., L. Leuscher, and L. S. Lerman. 1951. The use of an immunologically specific adsorbent for purification of antibody. *Proc. Natl Acad. Sci. USA* 37:575–578.

Chan, J., X. Fan, S. W. Hunter, P. J. Brennan, and B. R. Bloom. 1991. Lipoarabinomannan, a possible virulence factor involved in persistence of *Mycobacterium tuberculosis* within macrophages. *Inf. Immun.* 59:1755–1761.

Chiswell, D. J., and J. McCafferty. 1992. Phage antibodies: will new 'coliclonal' antibodies replace monclonal antibodies? *Trends Biotechnol.* 10:80–84.

Cuatrecasas, P., M. Wilchek, and C. B. Anifinsen. 1968. Selective enzyme purification by affinity chromatography. *Proc. Natl Acad. Sci. USA* 61:636–643.

Daenen, S., Y. Hoogeveen, J. W. Smit, G. W. van Imhoff, J. van der Meer, J. T. M. de Wolf, C. T. Smit Sibinga, and R. M. Halie. 1987.

Risk of transmission of human immunodeficiency virus (HIV) by heat-treated factor VIII concentrates in patients with severe hemophilia A. *Transfusion* 27:482–484.

Davis, R.V., R. M. Blanken, and R. J. Beagle. 1969. Diazotized *m*-aminobenzyloxymethylcellulose as the insoluble matrix for an immunosorbent used in the purification of antigens and antibodies. *Biochemistry* 8:2706–2711.

Delaage, M., O. Mejan, and A. Bourgois. 1989. Affinity chromatography and ion exchange effects, the case of factor VIII-von Willebrandt factor complex. In *Biotechnologie des proteines du plasma*, edited by J. F. Stoltz, and C. Rivat. Colloque INSERM Vol. 175.

Dickson, D. 1993. British group backs French scientist. *Nature* 363:663.

Eketorp, R. 1980. Affinity chromatography in industrial ethanol fractionation of human plasma. In *Methods of plasma protein fractionation*, edited by J. M. Curling. London: Academic Press.

Eveleigh, J. W. 1982. Practical considerations in the use of immunosorbents and associated instrumentation. In *Affinity chromatography and related techniques*, edited by T. C. J. Gribnau, J. Visser and R. J. F. Nivard. Amsterdam: Elsevier.

Fleminger, G., E. Hadas, T. Wolf, and B. Solomon. 1990. Oriented immobilization of periodate-oxidized monoclonal antibodies on amino and hydrazide derivatives of Eupergit C. *Appl. Biochem. Biotechnol.* 23:123–137.

Fowell, S. L., and H. A. Chase. 1986. A comparison of some activated matrices for preparation of immunosorbents. *J. Biotechnol.* 4:355–368.

Fricke, W. 1994. FDA perspective on and response to the risk of hepatitis A from blood products. *Vox Sang.* 67(suppl 4):16–18.

Geisow, M. J. 1992. Improved selection systems for man-made antibodies. *Trends Biotechnol.* 10:75–76.

Ghatge, M., Y. Mawal, S. Gaikwad, and V. Deshpande. 1991. Immunoaffinity purification of glucose/xylose isomerase from *Streptomyces. Appl. Biochem. Biotechnol.* 31:11–20.

Ghete, V., G. Mota, and J. Sjoquist. 1978. Separation of cells by affinity chromatography on SpA-Sepharose 6MB. *J. Immunol. Methods* 21:133–141.

Griffith, M. J., S.-L. Liu, G. Neslund, and H. S. Kingdon. 1991. Reducing the risk of viral transmission in biotechnology processes.

In *Sterile pharmaceutical manufacturing: Applications for the 1990s, Vol. 1,* edited by M. J. Groves, W. P. Olson, and M. H. Anisfeld. Buffalo Grove, IL: Interpharm Press.

Gurvich, A. E., R. B. Kapner, and R. B. Nezlin. 1959. Isolation and examination of pure antibodies prepared by the use of antigens fixed on cellulose. *Biokhimiya* 24(1):129–140.

Hamman, J. P., and G. J. Calton. 1985. Immunosorbent chromatography for recovery of protein products. In *Purification of fermentation products,* edited by D. LeRoith, J. Shiloach, and T. J. Leahy. ACS Symposium Series No. 271. Washington, DC: American Chemical Society.

Harkonen, W. S., and K. A. Gregg. 1993. Back to the future with polyclonal antibodies. *Bio/Technology* 11:1518.

Hermanson, G. T., A. K. Mallia, and P. K. Smith. 1992. *Immobilized affinity ligand techniques.* San Diego: Academic Press.

Horowitz, B., M. E. Wiebe, A. Lippin, and M. H. Stryker. 1985. Inactivation of viruses in labile blood derivatives. I. Disruption of lipid-enveloped viruses by tri(n-butyl)phosphate detergent combinations. *Transfusion* 25:516–522.

Hrinda, M. E., C. Tarr, W. Curry, J. Newman, A. B. Schreiber, and R. D'Alisa. 1989. Monoclonal antibody affinity purification of plasma proetins removes viral contaminants. In *Biotechnologie des proteines du plasma,* edited by J. F. Stoltz, and C. Rivat. Colloque INSERM, Vol. 175.

Huang, S. H., S. Roy, K. C. Hou, and G. T. Tsao. 1988. Scaling-up of affinity chromatography by purification. *Biotechnol. Progress* 4:159–165.

Iwarson, S., M. Einarsson, L. Smallwood, P. Snoy, and R. J. Gerety. 1987. Chromatographic removal of hepatitis B virus from a factor IX concentrate. *Transfusion* 27:171–173.

Kaplan, C. 1958. The heat inactivation of vaccinia virus. *J. Gen. Microbiol.* 18:58–63.

Klein, E. 1991. *Affinity membranes.* New York: Wiley-Interscience.

Köhler, G. and C. Milstein. 1975. Continuous cultures of fused cells secreting antibody of predefined specificity. *Nature* 256:495–497.

Kohn, J., and M. Wilchek. 1982. A new approach (cyano-transfer) for cyanogen bromide activation of Sepharose at neutral pH, which yields activated resins, free of interfering nitrogen derivatives. *Biochem. Biophys. Res. Commun.* 107:878–884.

Laurell, C. B. 1966. Quantitative estimation of proteins by electrophoresis in agarose gel containing antibodies. *Anal. Biochem.* 15:45–52.

Liu, S., J. Addiego, E. Gomperts, C. Kessler, L. Garanchon, G. Neslund, V. Foster, R. Berkebile, S. Courter, M. Lee, H. Kingdon, and M. Griffith. 1989. A plasma factor VIII:C concentrate purified by immunoaffinity. In *Biotechnologie des proteines du plasma*, edited by J. F. Stoltz, and C. Rivat. Colloque INSERM, Vol. 175.

Loetscher, P., L. Mottlau, and E. Hochuli. 1992. Immobilization of monoclonal antibodies for affinity chromatography using a chelating peptide. *J. Chromatog.* 595:113–119.

Lou, Y. 1993. The isolation, properties and formulation of an antineoplastic agent derived from bacillus Calmette-Guerin vaccine. Ph.D. dissertation, University of Illinois at Chicago.

Maddox, J. 1993. Contaminated blood case nears end. *Nature* 363: 491.

Mannucci, P. M., and M. Colombo. 1988. Virucidal treatment of clotting factor concentrates. *Lancet* ii:782–785.

Mantovaara, T., H. Pertoft, and J. Porath. 1991. Purification of factor VIII:c coagulant activity from rat liver nonparenchymal cell culture medium by immobilized metal ion affinity chromatography. *Biotechnol. Appl. Biochem.* 13:120–126.

Marks, J. D., H. R. Hoogenboom, T. P. Bonnert, J. McCafferty, A. D. Griffiths, and G. Winter. 1991. Human antibodies from V-gene libraries displayed on phage. *J. Molec. Biol.* 222:581–597.

McCafferty, J., A. D. Griffiths, G. Winter, and D. J. Chiswell. 1990. Phage antibodies: filamentous phage displaying antibody variable domains. *Nature* 348:552–554.

McGregor, W. C. 1986. Selection and use of ultrafiltration membranes, In *Membrane separations in biotechnology*, edited by W. C. McGregor. New York: Marcel Dekker.

Mejan, O., V. Fert, M. Delezay, M. Delaage, R. Cheballah, and A. Bourgois. 1988. Immunopurification of human factor VIII/vWF complex from plasma. *Thromb. Haemost.* 59:364–371.

Menozzi, F. D., K. C. Hou, and A. O. A. Miller. 1992. Development of recombinant protein A cart ridge for radial affinity chromatography. *Chimica oggi* March: 17–21.

Mitra, G., M. Dobkin, M. Dumas, P. Ng, G. Roldan, and C. Galloway. 1994. Hepatitis A viral safety of plasma-derived factor VIII concentrate Koate™-HP. *Vox Sang.* 67(suppl 1):80–82.

Morfini, M., P. M. Mannucci, G. Longo, S. Cinotti, and A. Messori. 1991. Comparative evaluation of the pharmacokinetics of three monoclonal factor VIII concentrates. *Thromb. Res.* 61:285–290.

Nakamura, K., T. Hashimoto, Y. Kato, K. Shimura, and K.-I. Kasai. 1990. Tresyl-activated support for high-performance affinity chromatography. *J. Chromatog.* 510:101–113.

Ohlson, S., A. Lundblad, and D. Zopf. 1988. Novel approach to affinity chromatography using "weak" monoclonal antibodies. *Anal. Biochem.* 169:204–208.

Olson, W. P. 1987. Viruses in therapeutics from human plasma: upstream and downstream strategies. In *Aseptic pharmaceutical manufacturing: Technology for the 1990s,* edited by W. P. Olson and M. J. Groves. Buffalo Grove: Interpharm Press.

O'Shannessy, D. J., and M. Wilchek, M. 1990. Immobilization of glycoconjugates by their oligosaccharides: Use of hydrazido-derived matrices. *Anal. Biochem.* 191:1–8.

Piszkiewicz, D., C.-S. Sun, and S. C. Tondreau. 1989. Inactivation and removal of human immunodeficiency virus in monoclonal purified antihemophilic factor (human) (Hemofil M). *Thromb. Res.* 55:627–634.

Porath, J., J. Carlsson, I. Olsson, and G. Belfrase. 1975. Metal chelate affinity chromatography, a new approach to protein fractionation. *Nature* 258:598–599.

Prince, A. M., B. Horowitz, and B. Brotman. 1986. Sterilization of hepatitis and HTLV-III viruses by exposure to tri(n-butyl) phosphate and sodium cholate. *Lancet* i:706–710.

Ramirez, C., J. Fung, R. C. Miller Jr., R. Antony, J. Warren, and D. G. Kilburn. 1993. A bifunctional affinity linker to couple antibodies to cellulose. *Bio/Technology* 11:1570–1573.

Sibley, L. D., S. W. Hunter, P. J. Brennan, and J. L. Krahenbuhl. 1988. Mycobacterial lipoarabinomannan inhibits gamma interferon-mediated activation of macrophages. *Inf. Immun.* 56:1232–1236.

Spitznagel, T. M., and D. S. Clark. 1993. Surface-density and orientation effects on immobilized antibodies and antibody fragments. *Bio/Technology* 11:825–829.

Stine, G. J. (1993) *Acquired immune deficiency syndrome. Biological, medical, social and legal issues*. Englewood Cliffs, N.J.: Prentice Hall.

Tharakan, J., F. Highsmith, D. Clark, and W. Drohan. 1992. Physical and biochemical characterization of five commercial resins for immunoaffinity purification of factor IX. *J. Chromatog.* 595:113–119.

Thompson, N. E., D. B. Aronson, and R. R. Burgess. 1990. Purification of eukaryotic RNA polymerase II by immunoaffinity chromatography. *J. Biol. Chem.* 265:7069–7077.

Tsang, V. C. W., and P. P. Wilkins. 1991. Optimum dissociating condition for immunoaffinity and preferential isolation of antibodies with high specific activity. *J. Immunol. Methods* 138: 291–299.

Weetall, H. H. 1976. Covalent coupling methods for inorganic support materials. *Methods Enzymol.* 44:134–148.

Wigzell, H., and B. Andersson. 1969. Cell separation on antigen-coated columns. Elimination of high rate antibody-forming cells and immunological memory cells. *J. Exp. Med.* 129:23–36.

Wilchek, M., T. Miron, and J. Kohn. 1984. Affinity chromatography. *Methods Enzymol.* 104 (Part C):3–55.

Winter, G., and C. Milstein. 1991. Man-made antibodies. *Nature* 349: 293–299.

Zimmerman, T. S., and C. A. Fulcher. 1982. Ultrapurification of factor VIII using monoclonal antibodies. U.S. Patent 4,361,509. 30 November, 1982. Assignee Scripps Clinic and Research Foundation, La Jolla, CA.

Zopf, D., K. Schroer, G. Hansson, J. Dakour, and A. Lundblad. 1987. Affinity purification of oligosaccharides using monoclonal antibodies. *Methods Enzymol.* 138:307–313.

10

Process Validation of Separation Systems

Steven S. Kuwahara
Jane H. Chuan

Process validation is the procedure that allows one to establish the operating parameters of a separation process. These operating parameters extend beyond simple questions of the efficiency of a particular separation step and involve matters related to the quality of the product. For the experimenter new to the concepts of industrial quality assurance, the idea of considering several "irrelevant issues," which appear to be only incidentally related to the separation step, is often difficult to grasp when validating a process. This chapter is designed to provide the worker with the basic concepts for performing a validation study on separation processes. Discussions will provide for specific separation systems, but the reader must realize that these examples will be limited by the experience of the authors and the available amount of space and by no means represent all of the potential scenarios that may confront a process developer.

TYPES OF VALIDATION

All of the following validation studies should be performed by the user in facilities and under conditions that are controlled by the user and that are as close as possible to the normal production situation.

Note that there are no provisions made for adopting studies from the vendors of instruments or consumables. Many vendors protect themselves by adding a statement in fine print that says, "The user of this product is ultimately responsible for determining its suitability for its intended use." This is a way of saying that users must conduct their own validation studies and not rely on the vendor's studies. The user should be wary of uncritically accepting studies that were performed by others.

Prospective Process Validation

This chapter will deal primarily with prospective validations. Prospective validations result from executing a validation protocol and obtaining data before a process is placed into commercial production. Ideally, all validations should be prospective to avoid surprises during commercial work; however, cost and other considerations may create a need to bypass the prospective studies, at least in part. Some studies must be done as prospective validations. For example, the ability to produce a sterile product must be validated before attempting even limited commercial or clinical production.

Concurrent Process Validation

Concurrent validation can only be done with a process that can be shown to be in a state of control. One cannot simply decide to conduct process validation studies concurrently with commercial production, if there are no data to indicate that the process is already well defined and under control. An example of an acceptable concurrent validation study would be the one validating the installation of a new membrane filtration device into an already validated and operating process.

Retrospective Process Validation

Retrospective process validation is basically a method of using historical data to establish the parameters of a process. The advantage to this method is that there is usually a large amount of data available on a process that is already running and the costs will already have been absorbed. This type of validation can be especially useful for reestablishing or resetting operational parameters that may have been too broad or uncertain during the early stages of manufacturing or for introducing a new control parameter whose utility was not recognized in the early stages of manufacturing. The disadvantages are that it cannot be used for initial licensing; its retrospective nature means that measurements of current interest may not have been performed or the work may have been performed with controls or equipment that are no longer in use.

WHY VALIDATE?

Practical Considerations

It is important to know if your system works as you thought it would. The validation study is not only useful for providing assurance of reproducible processing to the regulatory agencies, but should serve to provide project management with assurances that the processing method is reproducible and works as expected. It is very important to demonstrate that the system operates in a predictable and controllable manner and that important processing criteria can be met.

Process Control

Process validation data may be employed in the initiation of a system for statistical process control (Gopal 1989; Gershorn 1991). The data can be used to set the initial ranges for the control parameters and provide early assurance that the process is a capable process. Even in the absence of a formal system for statistical process control, the process developer will want to establish the operational parameters for the system. Once these parameters are established, it will be necessary to check the system to see if it is capable of producing a good product when running at the extremes of the ranges of operational parameters. This knowledge will allow the worker to validate the operational ranges that will be employed in setting the control and capability parameters of the process. It is not necessary to go to the extremes that the process or machinery is capable of attaining, if the extremes of the operational parameters do not reach these levels.

Regulatory Requirements

There are regulatory requirements that require a producer to show that the process is controlled and is capable of producing a uniform product over time. The U.S. (Food and Drug Administration, 1987; 21CFR 58; 21CFR 211; 21CFR 610; 21CFR 820), EEC (Comm. Euro. Commun. 1989) and Japan (Kazuyuki 1990) have similar regulations that require the demonstration of a uniform product produced by a controlled process. These requirements grow out of the principles of Good Manufacturing Practices (GMPs). These principles are based on the following ideas:

- Testing and inspection alone cannot assure quality or safety in a product.
- The manufacturing process must be designed to build quality, safety and efficacy into a product.
- Each step of the manufacturing process must be controlled to maximize the probability that the finished product will meet quality and safety requirements.

If a separation step is mentioned in a license application or a manufacturing procedure, a request to examine the validation study should be expected from the regulatory agencies.

WHO, WHAT, AND WHERE TO CONDUCT VALIDATIONS

Where

The process validation study is often conducted under conditions that have little to do with the actual conditions that will be employed for manufacturing. To the extent that these conditions differ, the study must be considered to be flawed. Having said this, it must be realized that it is often unrealistic to expect that a process validation can be done under exact operating conditions. In many cases the process validation study will be conducted at a time when the manufacturing facility has not yet been constructed. In the case where the ability of the system to remove viruses or other microbes is being validated, simple prudence dictates the need for conducting these studies in an isolated location away from normal production areas.

Problems of Scale

In many situations the full manufacturing process may require the separation of material on a kilogram scale, while the process validation study may only need a few grams of product. In these cases financial considerations will dictate a scaled-down study. With microbial challenge studies, the use of large quantities of infectious agents may not be acceptable and a scaledown may be required.

The problem here is that one must be careful in considering the scaledown. While a tank may be reduced in size, the amount of time that the material is held and stirred in that tank should not be reduced. If a transfer from one tank to another normally takes 10 hr, that rate of transfer should be retained even if the volume of the tank has been reduced to a tenth of the production size. One of the major problems that arises when the process is validated before the construction of the final production facility is that the times involved during the full production activities may be quite different from the times employed in the validation study. This, in turn, can lead to questions about the validity of the validation study itself, product stability, aging and crystal growth, cooling or heating of intermediates, and so on.

Safety Issues

Some of the issues that must be considered during a validation study may create situations that are unsafe if performed in the usual production facility. The nature of the product or process could lead to a need to challenge the ability of a purification step to remove toxic substances, biologically active substances, or pathogens. Rather than expose the entire production area and personnel to these materials, it would be wise to conduct these validations in an isolated area with specially trained personnel.

Employee Qualifications and Realistic Situations

One of the major problems with validation studies is that they are often conducted by specialized teams of highly trained and educated employees (product development team) whose abilities and dedication may be significantly different from the employees who will perform the regular production operations. These employees may also be working during normal hours and will not be suffering from the stresses caused by off-hours work and minimal supervision. It is also known that employees can be expected to be more meticulous about their work when they realize that they are under observation and that the results will be under close scrutiny. The end result of these effects may be to produce product with high yields and little variation in the process parameters. This, in turn, may cause the setting of process parameters with a narrower range than can be maintained under normal operating conditions. For this reason it is usually not wise to maintain a special set of instruments, equipment, and employees for validation studies.

Selection of Models

Choosing the proper model for a study often requires a very careful and informed management decision. Ideally, the process will be validated with the same equipment, personnel, facilities, and raw material that will be employed in regular production runs. However, for reasons that were given above and elsewhere in this chapter, it is not always possible to comply with these requirements. In these cases it is very important to construct a model system that is as close as possible to the actual operational system. Simply taking a 10 ℓ system and reducing it to a 10 ml system and claiming a 1,000-fold scaledown is often unrealistic as incubation times, heat transfer rates, surface-to-volume ratios, and pumping rates often cannot be so simply scaled down.

One question that arises with models where there are multiple steps involved concerns the use of fresh preparations for validating each step. An example of this arises in studies where the ability of separation steps to remove viruses is investigated. Because of the high level of clearance that is often seen and the practical limits to the density of viral cultures, it is quite possible for a separation step to reduce a count from 10^9 to 10^4 plaque-forming units (PFUs) per ml. If the next step can be expected to be equally effective, the number of PFUs that remain are not sufficient to demonstrate the full potential of the step. The usual procedure here would be to start over with a fresh preparation that contains 10^9 PFUs/ml and if a reduction to 10^6 PFUs/ml were seen, a statement would be made saying that an overall reduction of 10^8 was attained. The problem is that it is quite possible that the virus preparation contains virions with differing binding properties and some will be removed by nonspecific binding on the first column. The virions removed will not be available for nonspecific binding to the second column and, therefore, the second column should not be expected to give the same clearance factor as it will when a fresh preparation is used. This results in a misleading value for the clearance factor if the addition procedure is employed. The actual purification procedure used a heterogeneous mixture that had already been exposed to a column before loading the second column. The fresh preparation has properties that differ from the actual process solutions. The interpretation of clearance factors for heterogeneous preparations can be difficult (Hageman, 1991; Parenteral Drug Association, 1992) and a careful review of the study must be conducted.

EXTENT OF VALIDATIONS

Cost Considerations

The cost of validation studies is usually quite high as a large amount of product and personnel time must be consumed in their performance. For a validation study to be useful, the amount of sampling, staff involvement, and supervision will exceed the amount that will be employed during the course of normal manufacturing operations. The number of studies that must be performed is often quite large, although good technical management can keep these numbers to a minimum. Although this chapter mainly concentrates on the validation of separation processes, these separations will be a part of a general overall process; the management of the validation of this overall process can make a great deal of difference to the costs involved. If properly planned and executed, a single processing run can provide the process developer with a great deal of information and greatly reduce the need for the costly repetition of processing runs just to gather individual sets of data.

Statistical Needs

The allowable extremes of a process must be tested with enough repetition to provide statistically valid results. It is very important to consult a competent statistician during the early, planning stages of a process validation study. A poorly managed study is performed first, then the data are passed to a statistician to "massage" the data for statistical validity. This practice is guaranteed to produce useless information at additional cost.

To properly test a system, validation studies must be carried out at the allowable extremes of the parameters for each step. For instance, if a process specification for a particular separation requires a pH of 7.5 ±0.5, at 5° ±2°C, with a salt concentration of 1.0 ±0.1 *M*, the validation study will require nine runs with combinations starting at pH 7.0, 3°C, and 0.9 *M* salt; and going to pH 8.0, 7°C, and 1.1 *M* salt with all of the other combinations in between. Ideally, each parameter of each run will be measured with six replicates as this number of replicates should provide precision estimates that are within a reasonable range of the population values (Anderson 1987, 298). Clearly, this practice can lead to astronomical numbers of tests and samples. A statistician familiar with the method of Box and Hunter and other experimental design techniques (Montgomery 1991) can devise methods for obtaining valid study results while greatly reducing the number of tests that must be performed.

It is important to have statistically valid estimates of the variance and mean for the critical parameters of a separation step. If one follows the procedure of taking actual process data and using them to set acceptable ranges for process and other parameters, then a two standard deviation range about the mean creates a situation where 5 percent of the test results will fail by chance alone. This rate of failure and the associated need for retesting may not be economically acceptable to management. A 1 percent failure rate may be obtained by going to a three standard deviation range, but this may create an unacceptably wide range for the parameter specifications. A large variance that results in a wide parameter range is usually an indication that the process is not controlled and indicates the need for further process development before the process parameters can be established.

GENERAL CONSIDERATIONS

Differences Between Synthetic Drugs and Secondary Metabolites Versus Biologics

With respect to process validations, a major difference between synthetic drugs, secondary metabolites, and biologics is the level of purity that one may expect to encounter in the preparations that one handles. A synthetic drug or secondary metabolite is usually a relatively small molecule that can be obtained in a high degree of purity. Biologicals are often complex, high-molecular-weight substances that are a minor component of the mixture in which they are found. This lack of purity is often due to the lability of their biological activity and the prohibitive cost of isolating the substance in pure form. The instability of the product also means that a biological must be administered parenterally and stringent standards of purity and safety may need to be met. The practical effect of this difference is that a biological usually requires more extensive process validation studies.

What Needs to Be Validated

Process validation subjects may be divided into three main groups. The first group consists of product characteristics such as purity and activity. A second group consists of process parameters such as yield, the number of theoretical plates, flow rate, and the like. The third group would be those items that represent the removal of contaminants. Studies on the removal of contaminants may require investigations into potential contaminants that are not necessarily present in the raw materials or intermediates, such as viruses or carcinogens. Contaminants also may appear in the product through accidental or other means. Under current rules it is not enough to validate only process parameters. For maximum utility and cost efficiency, one needs to combine many studies into one. This is a major challenge for the technical manager who needs to be aware of all of the factors that may impact the process. The following describes some of the points that need to be considered.

Sterility or Bioburden

Sterility and *bioburden* usually are applied to bacterial or fungal contamination of products and process intermediates. A preparation that will be administered parenterally needs to be sterile. Drugs for other methods of administration need to meet some standard for maximum permissible levels of microbial contamination. While a drug or biological may not need to be sterile, practical and safety considerations usually dictate that some bioburden standard needs to be maintained, especially in re label or other product claims for the sterility or safety of a product.

It is impossible to give absolute assurances of product sterility or low bioburden without consuming large amounts of the product during the QC testing of the finished material. To avoid extensive testing, it is best to perform in-process testing to demonstrate that the process produces a product that is sterile or has a low bioburden.

Sterility and bioburden testing are usually performed according to procedures that may be found in the appropriate Pharmacopeia (USP Convention 1994, 1976–1981; 21CFR 211; Convention on the Elaboration of a EP 1983a, 1986; Society of the Japanese Pharmacopeia 1987). One important consideration here is the validation of the bioassays. Most pharmacopeias will list a set of indicator organisms whose growth in the presence of the test substance is crucial to the validity of the assay. These organisms are usually bacteria and fungi with at least one yeast included among the fungi. If the test substance inhibits the growth of one or more of the indicator organisms, the worker must consider strategies for removing the inhibitor. One of the most common methods is to pass the sample through a sterilizing filter and wash the filter with a growth medium to remove any inhibitory material. The filter is then placed on a layer of nutrient agar to allow the growth of the organisms.

Pyrogenicity

The pyrogenicity of a product is of great concern if the product will be administered parenterally. Testing on separated fractions is usually conducted by animal testing in rabbits or by the use of the *Limulus* amebocyte lysate (LAL) test. The LAL test offers the advantages of speed, low cost, and at least partial quantitation. The rabbit test, however, may be more representative of how a human will react toward a product. The LAL test is fairly specific for Gram-negative bacterial endotoxins, while the rabbit test will also detect other pyrogenic (fever-inducing) substances. In the great majority of cases, the pyrogen will be a bacterial endotoxin, but pyrogens other than Gram-negative endotoxins are known to exist (Kuwahara et al. 1985). The validation of the pyrogen or LAL test is a matter of concern. The LAL test, being enzyme based, is more likely to suffer the effects of inhibitors than the pyrogen test; rabbits, however, can be susceptible to antipyretic substances present in the preparation.

Viruses

The ability of a processing step to remove viruses from a preparation is important for products that are produced from material of biological origins, especially mammalian cell culture. Overall, it is usually desirable to demonstrate that the purification processes can reduce levels of indicator viruses by factors of 10^8 to 10^{13}. The indicator viruses are chosen to be representative of classes that are of concern for the particular product. Specific viruses that are to be tested are often selected using the European Note for Guidance (Commission of the European Communities 1991), but one must realize that different regulatory authorities may have different ideas concerning the importance of different viruses.

When working with viruses, especially human or animal pathogens, the potential for contaminating a facility or employees is high, especially employees who are not trained in sterile techniques. Consequently, the work is often sent to a contract laboratory that can provide an isolated facility where the proper level of containment may be attained. The contractor should have experienced employees who possess well-developed techniques for the handling of viruses. The group submitting the work must audit the contractor to verify their ability to conduct the study and provide valid results. The ultimate responsibility lies with the group that seeks the license. That group must verify the contractor's ability to perform the separations procedure and assay the level of viral contamination. Since the manufacturer has expertise in the separation technique while the contractor has the virologists, a common practice is to have the manufacturer's personnel travel to the contractor's containment facility and perform a scaled down process with the contractor's personnel in attendance, then to provide the contractor with samples for virus enumeration. Other studies should be carried out at this time to reduce the number of replicate runs that will be needed and to make maximum use of the travel funds and the material that will be generated during the study.

Mycoplasmas

Mycoplasmas are of concern with products derived from mammalian cell cultures. Because mycoplasmas can be introduced from human sources, it is not possible to perform a validation study and thereby eliminate all further testing. The testing protocols are specified by specific regulations (21CFR 610; Convention on the Elaboration of a EP 1983b; Association of Biologicals Manufacturers of Japan 1986, 334–336) although other "equivalent" methods may be employed (Barile and Kern 1971; Chen 1977; Stanbridge 1981). The normal validation practice is to test either the finished product or to test the product at the first intermediate step after the cells have been removed.

Transmissible Spongiform Encephalopathies

The transmissible spongiform encephalopathies are neurological diseases such as Scrapie, Bovine Spongiform Encephalopathy (BSE), Kuru, and Creutzfeldt-Jakob Disease (CJD). They are extremely slow diseases with long latent periods and are apparently transmitted by direct exposure or consumption of infected tissue. As many as two years may be required for a bioassay to reach an end point. The infectious agents are not well understood and appear to be resistant to heat and several of the usual disinfecting procedures. The concern here is primarily with products that are produced with poorly or partially purified material from human, bovine, or ovine tissues. The European Guidelines (Committee for Proprietary Medicinal Products 1992) provide useful information for this situation. The preferred procedure is to ensure that the human material is obtained from individuals in a donor monitoring program where the donors can be shown to be free from CJD and that a program for retrospective notifications are in place. With bovine and ovine material the simplest procedure is to show that the substances came from countries like Australia and New Zealand that are free from Scrapie and BSE. The manufacturer has the responsibility for conducting audits to show that the suppliers can verify and validate the sources of their material.

Residues

DNA and Other Nucleic Acids. There is concern about the potential for a DNA fragment to be incorporated into the human genome. Several scenarios based on the incorporation of an oncogene into cellular DNA have been developed. If a separation is being developed for a process that will be applied to preparations derived from cell cultures or other sources that may contain DNA or RNA, it is important to determine the ability of the separation step to remove nucleic acids from the preparation. It may not be necessary to test the nucleic acid content of each lot of finished product if the removal of nucleic acids can be established by the process validation studies.

Proteins and Allergens. Biologics and secondary metabolites are often obtained in cell-culture fluids that contain proteins or other substances that may act as antigens. The ability of a process to remove these substances is of great interest; this information should be developed for the separation steps in a process validation study. This is particularly important when the cells are derived from animals or other organisms for which the general human population is known to display a high rate of allergic reactions.

Other Substances. Other biologically active substances, such as growth factors and hormones, known to be present in culture media or other source preparations, need to be removed during the purification process. It is important to demonstrate the ability of individual separation steps to remove these substances. Other substances of concern are toxins or carcinogens that may be present in culture media or process chemicals, and the process chemicals themselves. Even if a contaminant is believed to be nontoxic, it is desirable to reduce its level to the lowest point possible. The removal of these substances must be validated for each separation step.

Process Parameters

Those who are new to process validations will assume that process parameters are the only parameters that require validation, but we have already seen that a process validation goes beyond these items. The process parameters listed below are the most common ones for consideration and those that are more specific to a type of separation are given later with the method.

Yields. Step yields are of great interest for the process developer. While the yields are more a product of the system rather than a controllable parameter, it is important to establish expected ranges for the yields. During routine production a sudden departure from an expected range for the step yield may signal the advent of difficulties. A yield greater than 100 percent for a product whose activity is normally monitored may indicate an unwanted activation of the material.

Purification Factors and Purity. As with step yields, purification factors are process parameters that are not necessarily controllable. Their importance, especially when derived from changes in the specific activity, is that they serve to monitor the operational effect of the separation step.

Time Intervals. Time intervals between processing steps can be a source of problems and also symptoms of processing problems. The monitoring and observance of time intervals can be a useful tool for process management. In fact, within the context of the Good Manufacturing Practices (GMPs) Act, large variations in processing times are often cited as points requiring explanation by the manufacturers (21CFR 211.111). A process that is under control and operating within defined parameters should possess stable processing times for individual steps and for the overall process. In a process validation study it is important to execute the separations within the time frame that will be allowed for the step during the processing operation.

Problems with rates of transfer and flow rates will be noticeable if the times are monitored. In addition, there is a stability question. If column A normally requires 6 hr for elution and the eluate is held at 5°C for 2 hr before loading onto column B, an event that causes the elution of column A to take 10 hr and prolong the holding period to 5 hr should raise questions about the stability of the product over this period. If the manufacturer chooses to accept the prolonged process, they should conduct a stability study as part of their process validation. In this manner the manufacturer will be able to provide assurance that the quality of the product was not compromised as a result of the extended processing times.

Matrix Regeneration. As an economic factor, column matrices, especially those containing covalently bound monoclonal antibodies, need to be regenerated. Other matrices, such as asymmetric membranes and ultrafiltration membranes, may also require regeneration. It is important to demonstrate the removal of contaminating substances such as proteins, microorganisms, and lipids that may be nonspecifically bound to these structures; it is important to show that the regenerated matrix really is equivalent to the original material. The validation study should be designed to demonstrate decontamination and maintenance of the original properties. The point to remember is that a validation study cannot cover all possible routes for matrix degradation or contamination, so it is unrealistic to expect that a matrix will be useful indefinitely. The process developer who has a matrix that is apparently capable of indefinite regeneration should, nonetheless, set an economically sensible point after which the matrix will be discarded.

It should be remembered that if a procedure calls for using a given matrix 50 times, the validation studies must qualify the matrix for the 50th use as well as after the first use. While it is not necessary to validate every cycle, it would be wise to validate the first, second, third, fifth, tenth, and probably every tenth use after that. During production a careful QA program would probably provide for a requalification of the matrix after every fifth use until the point where there was confidence that the interval could be extended with the approval of the regulatory authorities.

GETTING STARTED: WHAT NEEDS TO BE IN PLACE BEFORE BEGINNING YOUR VALIDATION

Validation Protocol

A validation protocol must be prepared and circulated for approval before initiating the validation study. This will provide an opportunity to obtain the ideas and support of key individuals. By sending the protocol around for review and approvals, you will also have the opportunity to receive input from diverse sources and to have some assurance that the study does not contain any blind spots.

Test Instrument Validations and Calibrations

Before a process validation is conducted, the instruments that will be used for the testing of samples, monitoring of the process, and general data gathering must, themselves, be calibrated and validated. In many cases instruments are calibrated, but their ability to operate properly is not validated. For example, a recording thermometer may be calibrated at specific temperatures, but the thermometer may not respond well to brief, but significant, temperature fluctuations. It makes no sense to conduct a study and draw conclusions about operating parameters if the instruments employed in the control, monitoring, and measuring of the process are not known to be generating valid data. A good process validation protocol will contain provisions for verifying the status of the instruments that will be employed. Variations introduced by a lack of precision in instrument readings will create limitations on the ability to make statements about the level of control.

Assay Validations

The assays that measure the processing parameters and the quality of the product must also be validated before initiating the study. Particular attention should be paid to the levels of variation and sensitivity of the assays. An assay that has a precision of ±10 percent cannot be used to support a contention that a process step is controllable to within ±5 percent. Similarly, if an assay has an incremental sensitivity of ±1.0 mg/ml, it is difficult to support an argument that a run-to-run variation of 0.7 mg/ml is significant.

Environmental Monitoring and Certifications

All processes should be run under defined environmental conditions. Factors such as temperature, humidity, and even light levels may have a significant impact upon operational results. Air flow, particle, and microbe levels all contribute to the results that will be obtained from the study. The process validation study should be conducted in an environment that closely matches the one in which the actual process will be run. As this will require control and monitoring, the instruments used for this purpose will also need to be calibrated and their operations validated.

Equipment Validations

The equipment that will be used in the process validation should also undergo a validation. This is especially important if the process validation is not using the actual process equipment as in the case of a scaled-down or pilot-plant validation study. If the process will be run in 316L stainless steel tanks, it is important to verify that the tanks that will be used in the study are also made of 316L stainless steel and are prepared for use in the same manner as the process tank. If a different pump is to be used, will the nature of its operation affect the process? Is it capable of producing the flow rates and pressures that will be required?

Raw Material Qualifications

As with equipment, the raw material that is to be used should be checked and qualified for use. If the process is geared for the use of technical grade chemicals, the use of analytical reagent grade chemicals in a scaled-down study may produce different results. Similarly, if the process specifications call for the use of USP–grade material, it is important to verify that the raw material used in the validation study can also meet USP specifications.

SEPARATION SYSTEMS

The following comments are intended to cover those points that are more or less specific to the methods. General requirements for validation studies have been previously discussed and will not be repeated here.

Extractions and Partitions

Liquid-solid extraction and liquid-liquid partitions should have partition coefficients added to the list of process parameters that require monitoring. These partition coefficients may be calculated from the data that are used to obtain the yield figures. As the partition coefficient should be a constant, dependent only upon temperature, significant variations in this parameter may indicate serious processing problems.

Selectivity is a related matter. Many extractions, adsorptions, and partitions are conducted in a manner that has all of the product passing into one phase with very little partitioning. While this is acceptable, the process developer should determine the extent to which contaminants also pass into the phase containing the product. In some cases the product is separated from one contaminant while another contaminant is concentrated with the product.

Extractions and partitions also lead into simple chromatography. In many cases chromatography columns are not operated so as to take advantage of the multiple partitions that can occur between the mobile and stationary phases. There are many "chromatographic separations" that are really nothing more than an adsorption desorption step performed in a cylindrical container. In these situations the standard chromatographic process parameters, such as the "height equivalent of a theoretical plate" (HETP), may be meaningless and simple partition coefficients may be more useful. A frequently encountered example is the use of a monoclonal antibody coupled to a support to capture a specific protein. The column is eluted with a chaotropic agent that disrupts the antigen-antibody complexes in such a way that there really is no equilibrium. Elution takes place with the product traveling in the solvent front. This process could have been performed in a stirred reaction tank using a centrifuge or filter to separate the solid matrix from the liquid phase.

Liquid-Solid Separations

The following procedures may be viewed as methods by which solid and liquid phases can be separated.

Evaportation, Distillation, and Lyophilization

Evaporation, flash evaporation, distillation, and lyophilization are basically solvent removal methods, although they do have other useful applications. These methods have the common property of removing the solvent as a gas during so that the amount of residual solvent is low. While sublimation (lyophilization) does not involve a liquid, it is a separation based on the removal of a vapor. The control of environmental factors, such as temperature and pressure, are important as are questions about residual solvents, especially if they are toxic.

Crystallization, Precipitation, and Solubilization

Basically, crystallization, precipitation, and solubilization represent liquid-solid separations where the liquid is removed without passing through the gas phase. As with other equilibrium phenomena, temperature control is critical as are holding times, solution polarity and ionic strength. The ionic strength of the solvent can increase the solubility of some proteins while reducing the solubility of other proteins. The possibility of differential solubilization means that the control of ionic composition as well as ionic strength are important parameters. Controlling pH can be especially critical as variations in the concentration of hydronium ions can produce major changes in the state of ionization of acids and bases. The solubility of acids and bases can change rapidly with pH changes when the pH is in the region around the pK_a.

Chromatography

Biologicals, synthetic drugs, and secondary metabolites may be purified by separations that are based upon molecular size, shape, net charge, charge density, hydrophobicity, isoelectric point, metal binding, antigenicity, and substrate or allosteric binding. These properties can be exploited with specific chromatographic techniques such as ion-exchange, reversed-phase, or hydrophobic interaction chromatography; gel filtration; affinity chromatography; and immobilized metal ion chromatography. The net result can be a very specific separation that produces material with a high degree of purity.

When dealing with chromatographic separations, a common cause of internal controversy is the issue of using new or freshly packed columns for each validation run versus the reuse of a single column. The problem here is that the preparation that is being chromatographed may contain material, such as inert particles, that is capable of slowly plugging the column or substances that bind irreversibly to the column matrix and cause a slow, progressive decline in the column's performance. If the column regeneration procedure is unable to reverse these effects, each successive use of a single column will produce a set of slightly different results. Consequently, five or six runs with the same column will not be truly representative of a manufacturing process where the same column may be reused 25 times.

Irreversible nonspecific binding to the column matrix is a problem as are interactions with the column matrix that arise from unusual properties of the product and create unexpected chromatographic effects. The classic case was the unexpected interaction of aromatic residues with cross-linked dextran gel permeation columns. These columns were often employed in the measurement of molecular weights of proteins; the unexpected interaction resulted in proteins that contained larger than normal amounts of aromatic residues being retarded during flow through the column. These proteins then were assigned an incorrectly low molecular weight. However, those who were aware of this property exploited it to separate proteins of nearly equal molecular weight that differed in content of aromatic amino acids.

Another problem comes from the leaching of column material into the product stream. Some of the stationary phases are unstable or may dissolve into the mobile phase. It is common to see chromatographic processes with a "cleanup column" designed to remove contaminants that have leached into the product stream from the column packings.

Chromatographic separations are often validated through the use of scaled-down models. This can have unfortunate consequences when the work is not conducted in accordance with the basic principles of chromatography. An often heard rule of thumb is that the column diameter should be reduced while retaining the column height. While this is an easy rule for a "cookbook" procedure, the actual parameters for measurement are often difficult to determine and require the ability to give a description of the column in quantitative terms (Janson and Ryden 1989, 76–82). When comparing a scaled-down column with the original, it is very important to measure key operational parameters, such as plate number, pressure differentials across the column length, porosity, and the linear flow rate (cm/hr) of the columns. Regardless of the type of chromatography, if a system produces multiple partitions and generates a Gaussian elution profile, one may determine column performance factors, such as the HETP and the porosity (Snyder and Kirkland 1979, 218–225). If the column height is maintained while the diameter is reduced in proportion to the change in volume, a cursory review will suggest that the HETP will be maintained. The problem is that there will still be a significant change in the surface-to-volume ratio; this may promote conditions where the porosity may not be maintained. As a result, the HETP will change. The importance of measuring the HETP and maintaining the general shape of the elution curve has been discussed (Parenteral Drug Association, 1992). If the diameter of the column (D_c) is a significant multiple of the packing particle diameter (d_p) (i.e., $D_c/d_p > 100$) the porosity will be maintained. If the linear flow rate also is maintained and there are no significant heat flow problems or changes in the interactions between the solutes and the column wall, the scaled-down column will be a good model for the production column.

The following sections provide more specific considerations for validating the common types of chromatography.

Partition Chromatography: Liquid-Liquid, Liquid-Solid, Gas-Liquid, Gas-Solid, Silica Gel

In partition chromatography the maintenance of the stationary phase can be a problem. In most cases a polar or nonpolar solvent layer is coated onto the stationary phase, or the stationary phase may be the polar or nonpolar material itself. A solvent with the opposite polarity is employed as the mobile phase. It is important to control the partition coefficients of the materials that are being separated and demonstrate that temperature and solvent composition controls are good enough to maintain the partition coefficients.

A second problem is in the contamination of the product by the stationary phase. It is common for the components of the solvent layer or the stationary phase to dissolve into the mobile phase and become part of the product stream. While it may not be possible to prevent this, the degree to which this happens should be determined.

Ion-Exchange Chromatography: Cation Exchange, Anion Exchange, Ion Exclusion, Mixed Bed Ion Exchange, Ion Chelation

With ion-exchange chromatography (IXC) the control of the original ionic form of the column and the retention of its stationary ionic groups are important considerations. Raw material testing of newly purchased or regenerated packings should assure the presence of the proper counterion; titration studies should confirm the presence of sufficient numbers of the required ionic groups on the stationary phase. The retention of sufficient numbers of these groups over the lifetime of the matrix should be demonstrated.

Ion-exchange columns, especially those of the mixed bed type accumulate undesirable ions on the matrix. Unless vigorous regeneration procedures are used, the column may accumulate enough contaminating ions (especially macromolecular ions) that performance may degrade. In addition, the rate of "bleed" of these contaminating ions may become significant. Usually it is wise to discard these columns after a set number of cycles, even if data are available showing the potential for a longer life.

Affinity Chromatography: Antibody Coupled; Metal Chelate; Hydroxyapatite; Chiral Ligand; Dye Binding; Cofactor, Inhibitor, or Substrate Binding; and Other Specific Techniques such as Protein A or G Chromatography

Affinity chromatography columns are basically highly selective columns whose specificity is conferred by ligands that have been chosen for their ability to interact in specific ways with substances of interest. These ligands vary in complexity from monoclonal antibodies to simple amino acids or inorganic ions. With all of these preparations, it is important to determine the stability of the linkage that holds the ligand on the stationary phase. Ligand leaching out of the stationary phase and passing into the product stream always is of concern to regulatory agencies. Many coupling methods permit a small, but steady, "bleed" into the product stream. Consequently, even if one has a washing step to remove accumulated material that has been released from the column, it is necessary to study the amount of material that can be transferred to the product stream during production. This study is especially important with columns that contain monoclonal antibodies, polyamino acids, or other potential enzyme substrates where the ligand may be attacked by components of the separation mixture. A related subject is the degradation of the ligand by chemical or microbial action. Dyes, enzyme cofactors, substrate or inhibitor analogs, polysaccharides, cyclodextrins, polyamino acids, and chiral ligands are subject to degradation through hydrolytic or oxidative mechanisms. Their temporal stability, especially during repeated recycling, should be established.

The high specificity or selectivity of affinity systems often leads to a belief that only the material of interest is being bound to the column. This can lead to a false sense of security. In complex mixtures, such as those encountered with biologics, there may be more than one substance that is capable of binding to the ligand. A substrate- or cofactor-based ligand may bind several enzymes. The validation study should establish the purity of the eluate. Even if the impurities are nothing more than isoenzymes, the purity should be demonstrated, so that future changes can be noted.

Gel Filtration: Size Exclusion, Gel Permeation, Controlled Pore

Size exclusion, gel permiation, and controlled pore permeation really refer to a single chromatographic technique. This system is believed to be basically passive with the separation based upon a molecular sieving effect created by the matrix of the packing material. Some of these column packings are mechanically fragile; especially the hollow beads. Excessive stirring damages the beads and allows fragments of the packing material to pass into the product stream. As with affinity packings, the validation study should verify the integrity of the column matrix and examine the possibility that portions of the matrix may pass into the product stream. The pressure drop along the column length, the porosity, the solvent flow rate, and the sample size are critical factors for producing reproducible separations.

Hydrophobic Interaction: Hydrophobic Ligand, Reversed Phase

Hydrophobic interaction chromatography (HIC) separates molecules on the basis of their ability to interact with weakly hydrophobic ligands. By varying the ionic strength of the solvent, conditions are created where hydrophobic regions of macromolecules are able to interact with hydrophobic regions on the packing material. Reversed-phase chromatography (RPC) uses a relatively nonpolar mobile phase and strongly hydrophobic ligands as the stationary phase. The hydrophobic interactions may be so strong that biologics can be denatured and inactivated while being eluted. Both techniques frequently use gradients and validations of the gradient composition are important. The potential for strong binding and denaturations means that the columns can be expected to accumulate irreversibly bound material during prolonged usage. Also the denaturing effect may result in the presence of molecules that behave like normal molecules but do not possess the same biological activity or, worse, exhibit neoantigenic properties. These factors should be studied during the validation studies.

Isoelectric Focusing, Chromatofocusing, Isotachophoresis, Electrophoresis

The focusing methods are separations primarily based on the isoelectric points of macromolecules, while the phoresis procedures are based on mobility due to charge densities. These methods have proven themselves to be extremely useful as analytical techniques in the QC laboratory and as small-scale preparative techniques for the R&D laboratory. The authors believe that these methods will find increasing applicability in industrial separations, especially for molecules of high activity where only small quantities will need to be purified to produce many doses.

Focusing techniques develop a pH gradient through the use of ampholytes. The stability of this gradient and the removal of the ampholytes from the product are important points for validation. In isotachophoresis and electrophoresis the electrical field gradient and removal of the electrolytes from the product are of concern. Because of the electrical currents used in these separations, control of the heat generated and the safety of the workers are also important points for validation.

Filtration Processes

Filtration processes encompass many different techniques that are employed for various reasons. The following discussion uses the concept of filtration as a process where material is separated as a result of a sieving effect across a thin layer of some porous material.

Sterilization and Particle Removal

The removal of infectious organisms and the removal of potentially harmful particles are very important processes in the production of parenteral drugs. The usual procedure for the process validation of filters that are used for these purposes is to challenge them. With sterile filtration procedures it is normal to perform the challenge with a small bacterium such as *Pseudomonas diminuta.* For particle removal filtrations, the challenge can be made with latex or other beads or particles of defined composition and size. As with bacteria, the validator must be able to quantitate the particles upstream and downstream from the filter. The detection system must be sensitive enough to measure the particles at the desirable levels. The advantage of using bacteria or yeasts in the challenge is that these organisms will amplify themselves by growing, and they may prove to be useful even when the interest is only in the removal of particles of defined dimension.

Dialysis and Ultrafiltration

Dialysis and ultrafiltration (UF) are macromolecular filtration processes applied to proteins and polysaccharides. These processes separate small molecules from the macromolecules. Dialysis is a passive process; solutes pass across a thin, microporous membrane under the influence of a concentration gradient. In UF a pressure gradient is applied across the membrane. Because of the semipermeable nature of the membrane, osmotic effects can develop, especially in dialysis. The pressure differentials can cause the stretching of the membranes and the pores within them, changing the exclusion limit of the pores. For these reasons the validation study should include kinetic studies on the rates of transfer of material across the membrane and a validation of the nominal molecular weight limit of the membrane.

Reverse Osmosis and Pervaporation

Reverse osmosis (RO) is driven by a liquid pressure differential while pervaporation is driven by vapor pressure differences. In both cases selectivity is obtained by employing asymmetric membranes. In both the selectivity of the process must be validated along with other factors, such as rejection rates, throughput, pressure differentials, and temperature effects.

REFERENCES

Anderson, R. L. 1987. *Practical statistics for analytical chemists.* New York, NY: Van Nostrand Reinhold Co.

Association of Biologicals Manufacturers of Japan. 1986. Sterility test for mycoplasmas. In *Minimum requirements for biological products*

(Eng. ver.). Tokyo: Association of Biologicals Manufacturers of Japan.

Barile, M. F., and J. Kern. 1971. Isolation of *Mycoplasma arginini* from commercial bovine sera and its implication in contaminated cell cultures. *Proc. Soc. Exptl. Biol. Med.* 138:432–437.

Chen, T. R. 1977. In situ detection of mycoplasma contamination in cell cultures by fluorescent Hoechst 33258 stain. *Exp. Cell Res.* 104:255–262.

Code of Federal Regulations 1994. Good laboratory practice for nonclinical laboratory studies, Title 21, Part 58, sections 61 to 130. Washington, DC: U.S. Government Printing Office.

Code of Federal Regulations 1994. Current good manufacturing practice for finished pharmaceuticals, Title 21, Part 211, sections 100 to 115. Washington, DC.: U.S. Government Printing Office.

Code of Federal Regulations 1994. Current good manufacturing practice for finished pharmaceuticals, Title 21, Part 211, section 111. Washington, DC.: U.S. Government Printing Office.

Code of Federal Regulations 1994. Current good manufacturing practices for blood and blood components, Title 21, Parts 606. Washington, DC.: U.S. Government Printing Office.

Code of Federal Regulations 1994. Good manufacturing practice for medical devices: General. Title 21, Part 820. Washington, DC.: U.S. Government Printing Office.

Code of Federal Regulations 1994. General biological products standards, Title 21, Part 610, section 30. Washington, D.C.: U.S. Government Printing Office.

Commission of the European Communities 1989. Guide to good manufacturing of medicinal products. Vol. IV. *The rules governing medicinal products in the european community.* Luxembourg: Office for Official Publications of the European Communities.

Commission of the European Communities, Ad Hoc Working Party on Biotechnology/Pharmacy 1991. *Validation of virus removal and inactivation procedures. Note for guidance.* Document #111/8115/89–EN Final. Brussels: Commission of the European Communities.

Committee for Proprietary Medicinal Products, Ad Hoc Working Party on Biotechnology/Pharmacy and Working Party on the Safety of Medicines. 1992. Note for guidance: Guidelines for minimizing the risk of transmitting agents causing spongiform encephalopathy via medicinal products. *Biologicals* 20:155–158.

Convention on the Elaboration of a European Pharmacopeia. 1983a. Microbial contamination of products not required to comply with the test for sterility. *European Pharmacopeia, 2nd ed.,* section V.2.1.8., Ste. Ruffine, France: Maisonneuve S.A.

Convention on the Elaboration of a European Pharmacopeia. 1983b. Test for mycoplasmas. *European Pharmacopeia,* 2nd ed., section V.2.1.3.6. Ste. Ruffine, France: Maisonneuve S.A.

Convention on the Elaboration of a European Pharmacopeia. 1986. Test for sterility. *European Pharmacopeia,* 2nd ed., section VIII.3. Ste. Ruffine, France: Maisonneuve S.A.

Food and Drug Administration, Centers for Drugs, Biologics, Devices and Radiologic Health. 1987. *Guidelines on general principles of process validations.* Rockville, MD: Food and Drug Administration.

Gershorn, M. 1991. Statistical process control for the pharmaceutical industry. *J. Paren. Sci. Technol.* 45:41–50.

Gopal, C. 1989. Statistical process control (SPC). *Pharm. Technol.* 13:56–65.

Hageman, T. C. 1991. An analysis of clearance factor measurements performed by spiking experiments. *BioPharm.* July/Aug: 39–41.

Janson, J.-C., and L. Ryden. 1989. *Protein purification: Principles, high resolution methods, and applications.* New York, NY: VCH Publishers.

Kazuyuki, M. 1990. Highlights of good manufacturing practice in Japan. *J. Paren. Sci. Technol.* 44:35–38.

Kuwahara, S. S., J. W. Mehling, D. O. Huggett, and T. C. Yang. 1985. The limulus amebocyte lysate test and its application to human plasma albumin preparations. *Plasma Ther. Trans. Technol.* 6:95–103.

Montgomery, D. C. 1991. *Design and analysis of experiments,* 3rd ed. New York: John Wiley and Sons.

Parenteral Drug Association. 1992. Industry perspective on the validation of column-based separation processes for the purification of proteins. *J. Paren. Sci. Technol.* 46:87–97.

Snyder, L. R., and J. J. Kirkland. 1979. *Introduction to modern liquid chromatography,* 2nd ed. New York: John Wiley and Sons.

Society of Japanese Pharmacopeia. 1987. Section 43. Sterility test. *The Pharmacopeia of Japan,* 11th ed. Tokyo: Yakuji Nippo Ltd.

Stanbridge, E. J. 1981. Mycoplasma detection—An obligation to scientific accuracy. *Israel. J. Med. Sci.* 17:563–568.

United States Pharmacopeial Convention. 1994. Section <1211>. Sterilization and sterility assurance of compendial articles *United States Pharmacopeia* 23. Rockville, MD: United States Pharmacopeial Convention, Inc.

11

A Salute to the Human Plasma Processing Industry and Its Founders, E. J. Cohn et al.

Robert A. Tenold

Immediately prior to U.S. involvement in World War II, Edwin J. Cohn and his associates at the Harvard University Department of Physical Chemistry did their part for the war effort of the 1940s. Their work was heroic in that they lost no time in developing and optimizing an ingenious system of isolation and purification of albumin adaptable to large-volume plasma fractionation (Cohn et al. 1946). One would have difficulty imagining something like this in today's world. In fact, it would probably take a committee to do a feasibility study and this alone would take the better part of a year after some industrial politicians decided who should be on the committee!

Cohn received federal grants to isolate albumin. His first grant amounted to $1,500! The devotion and commitment of these scientists made it possible to fly albumin to Pearl Harbor to treat shock cases a short time after the December 7th attack occurred. A DC-3 flew out of Boston, making several stops to pick up all the available albumin. It took about three days to arrive in Honolulu. By this time most of the shock cases had been resolved, but many were to follow (Tullis 1988). One must remember that at that time, whole blood and plasma presented all sorts of logistical problems. There were no M.A.S.H. (Mobile Army Surgical Hospital) (Bogikes 1968–1993) units and few refrigerators where biologicals could be stored.

Before the arrival of albumin in the field, military corpsmen would restore a 500 ml container of plasma that had been lyophilized (freeze-dried for preservation); it was necessary to reconstitute dried plasma with a bottle of diluent (sterile distilled water)—a slow process. Plasma tended to become hydrophobic during lyophilization and particles of plasma would float, making wetting and restoration all the more difficult. Each bottle of diluent weighed almost two pounds and the plasma flask, itself, about half as much. The plasma flask was under vacuum. Each item was issued in an olive drab or navy gray can to protect its contents from rough handling. The whole system was very bulky. The introduction of albumin to the battlefield medic made a cumbersome task simpler. A 50 ml unit of albumin is osmotically equivalent (as a volume expander in shock) to that bulky 500 ml bottle of reconstituted pooled plasma and represented very little bulk in the corpsman's pack.

The disease problems associated with pooled human plasma were well recognized. Dried plasma was not irradiated until the 1950s through the use of a Dill machine; this offered very little safety (Hart et al. 1993). Quite often while plasma was being processed, the lamps would lose their intensity. This would go unnoticed since lamps were not read continually. In all, the systems were very crude, and physicians would weigh the use of plasma against the odds of the patient contracting hepatitis. In the case of the need of a volume expander with a patient in shock, the physician did not have to be concerned about albumin because it was terminally heated for 10 hours at 60°C (pasteurization), which made it safe with regard to the "agent of hepatitis" (terminology of the time).

Protein characterization was in its infancy at this time and the scientists that Cohn sent into industry were very conservative and willing to accept only those methods of purification that had been demonstrated to be safe and clinically sound. Yet, some of the testing performed at that time was more precise than testing currently required. Tiselius moving boundary electrophoresis, a four-hour test to determine purity, requiring as many hours to set up, gave more information more about the mobility of a molecule (and the purity of a preparation) than today's cellulose acetate 30-minute procedure. Testing was not possible at the time to prove whether viral inactivation was a result of ethanol exposure or pasteurization, but there was no question as to the safety of the preparation when terminally heated in its final container.

Dr. Cohn was of medium height and to compensate for this, he was a meticulous dresser and was often pictured with a boutonniere on his lapel. It was very evident to his coworkers that Cohn was an excellent organizer and had that special talent to bring together very diverse types, who themselves were dedicated scientists in the finest sense of the word and would support Cohn in most of his theories. Many of his coworkers had been with him for five or more years. He also was not afraid to get involved in the process. There is the story that Dr. Tullis tells of the demonstration of an automated blood/plasma collection system that was given a trial on stage in Spain to an auditorium filled with physicians. Dr. Cohn was the donor; and due to some overzealous use of silicone grease, the device sprayed the audience with Cohn's blood (Tullis 1988).

I worked with Jack Weare, Cohn's chief technician at Harvard, when he was at Hyland Laboratories in the mid-1950s. I found that his great knowledge and familiarity with plasma proteins were a direct result of his hands-on "large-scale" experience. Jack and Alex Kohler built the Armour plant in Fort Worth, Texas. I believe this was a common thread observed in all Cohn's coworkers: Their strength was not a result of how much theory they had learned or which texts they had read, but rather the good observations and real scientific evaluations they had made. Many of today's fractionation "experts" have never "stirred the pot at midnight." In wartime the primary concern was the preparation of albumin for use on the battlefield, which surely accounted for a reduction in the number of combat casualties. In peacetime the emphasis shifted to component therapy to economically meet national health needs and the specific requirements of the National Polio Foundation. Dr. Cohn's objective was centered around component therapy to conserve the natural biological resources of the nation. Cohn's specialists were trained in the lab at Harvard and then were sent out to develop knowledgeable people in industry. It was a very efficient technical transfer. Dr. Cohn himself assigned the royalties for his discoveries to the formation of the Protein Foundation. This Foundation was the common meeting place for the world's plasma fractionators. When first formed, annual meetings took place at its headquarters in Boston. The last "fractionation" meeting at the then renamed Blood Research Institute was in the late 1970s. It seems that the industry did not continue to provide financial support to the institute; other scientific interests changed the focus of the organization to permit continued funding.

Each company—Armour, Cutter, Parke-Davis, Squibb, Hyland and Courtland, Merck, Lederle, and Pittman-Moore (animal blood therapeutics)—shared the national healthcare mission. The capacity of Armour's Fort Worth plant was greater than any other facility in the U.S. It was necessary to process Afro-American plasma separately due to prevalent racial attitudes in some parts of the country. As required, Afro-American plasma was processed separately in the same vessel with plasma from people of European extraction. Based on the number of Afro-American donors, containers of albumin would be finished with the distinguishing plasma source noted on the label. The process area in the Armour plant was designed with process vessels located on the mezzanine to gravity feed the centrifuges on the main floor. The supernatant was then pumped back to the mezzanine for further processing. When the "Armour" people came to Hyland in the 1950s, they designed the process area on one level instead of two and used pressure/vacuum systems, eliminating the need for any pumps and thus eliminating the possibility of pump shear of the plasma proteins, cleanliness of closed pathways, and the high cost of maintenance.

In the early days one would often hear an honest and real concern for the plasma side of national healthcare. One company would be more than happy to help another "solve" a production problem. All were in a learning phase, none of the Cohn-trained people would fail to support each other. There were no secrets when it came to national healthcare or the war.

Even after the war it was a common practice for production scientists at one company to phone an acquaintance at a competitor company to resolve a problem in processing. A good example of this would be that during the 1950s, when much of the albumin that was prepared during and shortly after the war was reprocessed with new expiration dates for civil defense organizations. Albumin prepared at that time contained phenylmercuric borate as a preservative. It was highly questionable for its value as a good bacteriostatic agent in albumin; in addition, multiple doses might result in mercury poisoning. During the "rework" process (the sodium reduction step Cohn Fraction V Rework C), it was recommended that the mercury compound be removed (through precipitation and filtration) using neutralized cysteine hydrochloride. Not all manufacturers did this (some used cumbersome dialysis techniques); yet any would have supported another in explaining the technique to eliminate mercury from the reprocessed albumin. During the 1950s the *Code of Federal Regulations* (CFR) ruled against the use of ANY preservative in plasma products. I believe the reason for the illegality of adding preservatives was that toxic substances might be used as a bacteriostatic agent. The nature of albumin is to bind such toxins and this might impair its use as a therapeutic.

In California civil defense agencies stockpiled albumin in case of a national emergency or a natural disaster. The author has been told by reliable sources that in Los Angeles alone there are five "fully equipped" hospitals under the bleachers at Santa Anita Race Track. These hospitals come fully supplied with albumin to treat shock. If a need arose today because of an earthquake or other catastrophe resulting in many people suffering from burns, you would find only outdated (expired) albumin that was reprocessed in the early 1970s. When this albumin was reprocessed, it was estimated that 60 percent of the protein was polymerized. This material should have been discarded and replaced with albumin having a monomer content of at least 90 percent.

The first shipment to the military forces of gamma globulin (IGIM—Intramuscular ISG [Immune Serum Globulin]) was to the Navy in 1944 (Cohn et al. 1946). It was manufactured by Cutter Laboratories in Berkeley, California. Using the methods of Oncley et al. (1949), gamma globulin was prepared in sterile solutions at about 16.5 percent protein preserved with Thimerosal. Although considered "shotgun therapy," gamma globulin was the only prophylaxis indicated for those exposed to poliomyelitis, measles, mumps, and pertussis. Later, tetanus, RhoD hyperimmune, and hepatitis specific ISG came into very general use, especially in Europe.

After the war the interest in the manufacture of plasma source products was on shaky ground. The government sold the facilities and equipment to the respective companies for about ten cents on the dollar.

The National Polio Foundation, with funds gathered through drives of the March of Dimes, contracted the former military plasma processors to furnish IGIM. Plasma was not the only source of IGIM; human placentas were processed using variations of Cohn and Oncley. The Polio Foundation, along with domestic albumin needs and the Office of Civil Defense [both state and federal], kept the fractionators in business in the U.S. Physicians lost interest in component therapy, believing fractionated therapeutics were too expensive and not clearly effective. The public had geared to blood drives during the war; after the war they still supported the efforts of the American Red Cross (ARC) in filling national blood needs. [Even today, the ARC has not made the commitment to do its own plasma processing!] Interestingly, blood donations typically ran very high after the Christmas holidays, which offset the demand on the supply brought on by elective surgeries that occur in the first quarter of the year!

In the postwar slump some manufacturers processed bovine plasma for reagent albumin for in vitro testing, a step-by-step process using the standard Cohn methodology. They processed bovine for six months and then returned to processing human plasma. Rigorous cleaning took place after bovine fractionation. A carryover from those days is the precipitin identity test that is part of the release criteria to assure the absence of bovine and other animal proteins in albumin. In the early 1960s there was much interest in using bovine albumin in cosmetic preparations and the industry geared to meet the orders for bovine albumin. It was the greatest wrinkle remover since plastic surgery! Literally, trainloads of bovine albumin were in transit to the leading cosmetic manufacturers when the Food and Drug Administration (FDA) stepped in and stopped all shipments. There was a fear that splashing bovine albumin all over one's body might result in some antigenic responses. That was the end of possible windfall profits that would have kept much of the industry afloat.

Most contracts after the war were either military or civil defense and were generally based on cost plus 10 percent; yields were on a best effort basis with some minimum standard.

The Korean War again raised the questions of using ISG to offer prophylaxis to servicemen sent to the Orient. Multiple studies used gamma globulin from the U.S. to protect GIs from the U.S. Antibodies developed in the Orient were lacking in U.S. preparations.

Immune serum globulin lost much of its efficacy as a result of the "Korean experience." These studies might have had quite different results if ISG had been processed from those having active immunity for the diseases specific for that area of the world. In fact, Europeans have led the U.S. in this area; for example, a resident of France who was planning to vacation in Greece, might see his physician to receive an injection of "Greek" gamma globulin (gamma globulin processed from Greek donors). U.S. medics were weaned from ISG preparations because the Korean war effort depleted the U.S. supply. It has not been until the early 1980s that interest in the therapeutic value of gamma globulin has regained consciousness in the U.S. The driving force this time was due to overseas markets in Europe and Japan. These areas of the world have been using intravenous gamma globulin (IGIV) much longer than the U.S.

U.S. laboratories have undergone many changes in the last 40 years. They have changed ownership, many into the hands of foreign, multinational chemical giants. Jonas Salk eliminated the demand for large volumes of PIG (polio immune globulin). On the path of fighting poliomyelitis, Cutter Laboratories in Berkeley, California, worked with the National Institutes of Health (NIH) to prepare an attenuated virus vaccine for polio. The processing was done under the supervision of the government with government experts on-site. Inadvertently, some of the product was not completely inactivated and this resulted in several cases of infection.

Cutter was held responsible in the matter because of a label statement claiming that the product was "free of live virus." Although Cutter carried 1 million dollars in liability insurance, suits totalled more than 5 million dollars (over 25 million 1993 dollars). It was not until the early 1970s that all claims were settled. Some years after settling the claims, Cutter faced a problem with their solutions. This after they assisted Abbott Laboratories in preserving Abbott's market share in solutions when Abbott had a plant pathogen contamination in the overseal of their solution containers. The solution problem occurred after Cutter had planned an expansion in North Carolina. In 1974 Cutter was acquired by Bayer AG and today is a wholly owned subsidiary (Miles Inc.) of the German chemical company.

Armour was sold by the meat packers to Greyhound Bus Lines. Still in financial difficulties, it was purchased by Revlon. Revlon sold its interests in Armour Pharmaceutical to Rorer, which later became Rhone-Poulenc-Rohrer, a French chemical company.

In 1968 Abbott Laboratories purchased Courtland Laboratories, Los Angeles, California. They became known as Abbott Scientific Products. Due to problems in profitability (and after terminating some of those who had built the company), they were spun off to the Green Cross of Japan, becoming Alpha Therapeutics.

Lederle decided to heat albumin in bulk instead of the final container. After repercussions following a review with the Center for Biologics Evaluation and Research (CBER), they decided to cease fractionating.

Squibb processing areas could not meet the new cGMPs and decided to leave the fractionation business in about 1975. Merck decided to drop their fractionation lines shortly after Squibb.

Dow (Pittman-Moore) was unable to settle a union dispute and closed down their brand new fractionation facilities in the early 1970s. Only Baxter (Hyland/Travenol) remains as a U.S.–owned company. Some small regional processors are still about. Among these are Melville Biologics (New York Blood Center [NYBC]), orignally a nonprofit corporation, but seeking to become a for-profit company; North American Biologics, Miami, Florida, and Immuno (Parke-Davis facility) in Detroit, Michigan. State-run operations in Massachusetts and Michigan (as do nonprofit operations) continue to process to meet intrastate requirements.

Anyone who has fractionated plasma into acceptable final containers meeting all the minimum requirements of the CBER (formerly the NIH) cannot help but admire the thought and wisdom of these plasma pioneers. From a textbook perspective one might tend to assume that the adjustment parameters to bring proteins in and out of solution somehow perform in the same fashion from step to step and are fairly straightforward. Intimate fractionation experience will very quickly show that all steps are not the same. The Cohn fractionation system deals with proteins in a variety of ways to achieve the goals set by the original investigators who were very aware of what occurred when the parameters were met. Within the Cohn process, we see proteins washed (extracted), denatured (altered sometimes irreversibly), and harvested (precipitated). This progression of steps yields products of demonstrated safety.

Today, equipment may change and techniques may vary, but the work of Dr. E. J. Cohn stands essentially unchanged for over 50 years. One must give him his due and respect his concepts for having withstood the advancements made in science since his original work, the safety associated with Cohn albumin and ISG, and his process philosophy that resulted in IGIV and plasma protein preparations.

FRACTIONATION AUTOMATION

In 1972 Dr. John Watt, D.V.M., of the Royal Scottish Transfusion Service traveled extensively, visiting all the fractionation facilities in the world to determine their interest in his development of a system to continuously process plasma to albumin. I spent an afternoon with Dr. Watt and found his ideas somewhat suspect. He faulted the industry for their methods and equipment.

Cutter Laboratories bought the invention and rights from Dr. Watt. George Mitra and John Lundblad worked long and hard on trying to get the Watt Trolley to perform. It had a gear system that would have confused a watchmaker. Precipitated fractions would not separate; because they were so rapidly formed, they became colloidal. To reduce the heat sink of the system, it was moved into the +2 refrigerator. This did not help. In reviewing the problems with the inventors, it was felt that the trolley would work better if the holding stages of the process would be treated by a Gaant ultrasound device. The thought was that a better aggregation of the formed precipitates would happen at different ultrasound frequencies. The whole arrangement ended as a total fiasco and for years the Watt Trolley stood as a monument in the pilot plant in Berkeley.

CENTRIFUGES

The Sharples™ Centrifuge

The Sharples™ centrifuge was the universally recognized tool of the industry. It adapted well to the volumes being processed and the parameters needed to safely handle the plasma stream. Hybrids of this centrifuge have been made in about every country in the world. Originally designed for the petroleum industry, its use carried over to oils in the citrus industry and the separation of water from fuel and lubricating oils on submarines and other vessels. The engineering expressed in this centrifuge permitted reasonably good maintenance and care and above all SAFETY. The bowl in the Sharples™ model

AS-16 and AS-26 should rotate at 15,000 rpm to achieve the g force necessary for the routine processing of Cohn separations. As with all centrifuges of tubular design, as the bowl (rotor) collects the precipitate, the cross dimension diminishes, resulting in a reduction in g force. There is evidence, too, of a zonal effect within the bowl. Lighter precipitates are collected in the uppermost region of the rotor or where the product is about to exit. This zonal effect has never been capitalized on and might be under certain process conditions. Bowl liners have taken the drudgery out of "bowl scrapping." These sheets of Teflon® are cut to size, dipped in cold ethanol, and inserted into the bowl.

Centrifuges have made harvesting precipitate much easier than the methods of using multiple stainless steel tools to reach in and remove the precipitate. Sharples centrifuges are supplied through an inlet port at the base of the equipment. Feeding banks of centrifuges, whether 16s or 26s, is best accomplished through a manifold supply line with minimum pressure. Gravity will generally supply sufficient pressure and the appropriate diameter tubing sufficient volume. Adjusting the orifice or feed nozzle opening will fine-tune the desired flow rate. Feed nozzles come in a variety of sizes; if the size needed is not available, a stainless hex screw may be threaded into the delivery port of the feed nozzle with the needed opening. Plasma processors used the Sharples centrifuge not only for collecting precipitates, but also for separating cells from whole blood. The Model 16 Sharples™ centrifuge was installed on killing floors in slaughter houses for hemoglobin-poor bovine plasma (Moura 1965). The little T-1 air turbine/electric drive centrifuge is very good in separating cells from outdated whole blood. Very low levels of hemoglobin are also seen here using #3 discharge screws or a #30 ring dam on the Sharples™ AS-16 bowl; a drive pulley that reduces the bowl speed to 12,000 rpm. With the T-1 Sharples™ the three way (stainless steel flow director) must be inserted no more or less than $^1/_2$ inch into the mouth of the bowl. In all cases, if cells are the product separated, the temperature of the blood must be elevated to about 18°C to obtain the ideal viscosity for separation.

The industry adapted equipment for its use, but there was no limit to its inventiveness. One obstacle in the use of the Sharples™ was the need for cooling. This is always a problem with high-speed centrifugals. Coiling a centrifuge and using freon or "brine" to carry away the generated heat was used by most processors. Coil temperatures needed to be at least -35°C. When the coils were held at this temperature, the supernatant/effluent/centrifugate* temperatures would match the temperature of the process stream entering the

*The author uses these different terms so that the product is understandable by all manufacturers.

machine. They were also equal to the temperature of the room. Because of air being drawn in and out of the Sharples™ bowl case, it was necessary to be sure that the room temperature was the same as the product.

Without fail, the first sign of effluent temperatures moving out of range was when the room had reached a defrost cycle. Experience taught that the Sharples™ ran best if it was running all the time. Those who fought slippery drive belts at midnight on a Sunday realized that a little care and patience made the job a lot easier. Starting the machine after a weekend shutdown always went better if it was spun for a short while at a low speed; after the lubricants had an opportunity to move to their place of need, the centrifuge would then be brought to full speed. Sharples™ separation of Cohn precipitates is labor intensive and equipment limited.

The Sharples™ centrifuge is very expensive; as companies have grown, alternatives to the centrifuge have been sought. Some companies even manufactured their own Sharples™ "type" centrifuge in stainless steel at a fraction of the cost of used machines.

Maintenance of the Sharples™ centrifuge is directly proportional to the training of the people who operate them. Failure points are well known. Tighten the spindle locking nut with the appropriate wrenches spread as far apart as possible, and a spindle will be destroyed. Spindles are about $75.00 each. Try to start/run the machine with a leaking bowl bottom gasket, and there is a good chance you will burn out the motor. Ignore critical speeds (speeds at which vibration is excessive) when bringing a centrifuge up to speed and you risk destroying the drag internals. Remove the bowl from the machine without a thread protector, or mishandle the bowl to effect its balance and lost time and money will result.

Westphalia™ Centrifuges

Some manufacturers have tried to solve the centrifuge "governor" on their process with Westphalia™ centrifuges. These machines were very popular in the eyes of management because of their great capacity, quiet operation, and closed system possibilities. They also offered greater flow rates with greater g forces. In the early 1970s management was so impressed with the capacity of the large Westphalias™ that they failed to consider the thermal effect of the machine on the product.

Various methods have been used to solve the thermal impact while processing with Westphalias™, but eventually their only real value was in process steps that did not mind a 7°C rise in the process stream or as an anchor for very large vessels. The primary concern to remember when working with Westphalias™ is to wear your safety shoes if you are prone to dropping wrenches.

Ultracentrifuges

Ultracentrifuges, such as those manufactured by Electronucleonics in Oak Ridge, Tennessee, have been used preparatory to sterile filtrations. These instruments require dedicated technicians in their operation. They eliminate the need for acres of depth media to handle very dear product streams. Abbott Laboratories installed a K-II ultracentrifuge for the clarification of antihaemophilic factor (AHF) in the late 1960s. Attention was given to the dedicated and well-trained operators, but due to some poor installation engineering, a catastrophe followed that resulted in a product recall. The upper and lower bearing seals were cooled with water on a recirculation circuit. Product interchanged with the cooling water in turn began to grow. The reuse of this coolant could not be positively proven not to have interchanged with the product stream. There was a fear that contaminated product was on its way to the clinic. The company immediately recalled all product. Fresh, pyrogen-free distilled water to these seals will eliminate the possibility of contamination through this mode.

Other Centrifugals

In the early days all plasma was pooled from centrifuged units of whole blood that were either bled for that specific purpose or became outdated if drawn for transfusion. Large ten-place refrigerated swinging bucket centrifuges were used. The distance across the yoke was about 36 inches and they were spun at about 1900 rpm. It took about 45 minutes to obtain a good (firm) cell pack to permit the pooling people to manually aspirate the plasma. Until the advent of Haemonetics, Inc., techniques, the refrigerated Sorval centrifuge was used in most donor centers. Here the whole blood drawings would be spun, the plasma pressed into satellite bags in a closed system, and the cells resuspended in physiological saline for reinfusion. The Sorval centrifuges were expensive to maintain and generated a lot of heat in the donor center. This promoted the use of membrane separation of cells from plasma where peristaltic pumps do the work of circulation.

FILTERS AND PROCESS FILTER MEDIA

Plate and Frame Filter

Cohn and his coworkers were not satisfied with the separations they could effect with the Sharples centrifuge. To achieve a higher quality process stream, they used nephelometry to determine the extent of the separation. Standards were established that would reject an improperly treated stream and cause it to be refiltered to obtain a better turbidity reading.

Certain process steps became areas of concern, such as the Fraction IV-4 Filtrate and the Fraction V Rework C Filtrate. The early workers determined that the future quality of the albumin (especially its stability) was relative to low nephelometer readings. Until about 18 years ago, all fractionation filter media was made of asbestos. Sheets of asbestos were graded as to their ability to retain solids of different submicron sizes. The electrostatic action of the asbestos fibrous media did much to contribute to the quality of products produced with it. It was not only used in general processing, but it was also sterilized in place and was used to sterile filter solutions. Since asbestos is the source of so many inhalation problems, the pharmaceutical industry was asked to replace it with other fibers as soon as possible. Since 1968, I have used terminal nonsluffing filter media on all sterile filtrations to prevent the migration of asbestos fibers into bulk receivers and then to final containers. Some large-volume manufacturers did not eliminate asbestos filtrations until 1975. Large filters of the vertical Sparkler or Alsop type or horizontal Hormann, Ertel, and Schenk plate and frame presses were used to hold inexpensive asbestos flat stock.

Disposable Filter Cartridges

The advent of the filter cartridge with great sludge capacity sparked the imagination of some who, for the lack of centrifuge availability and aggressive schedulers, looked at filtration separation. Initially, the filtration of waste precipitates seemed to do well on a system like the Cuno or Alsop, using multiples of four-stack filters. Abbott Scientific routinely used filters alone to separate Cohn Fraction I, IV-1, and IV-4. While at Alpha Laboratories, Randall Kent developed the science of filtration separation to reduce the use of centrifuges (Kent 1990–1993). Several U.S. manufacturers do almost all their separations on filters. Those who utilize filtration separation have experienced some yield improvements and certainly a reduction in the cost of maintenance.

There are some obvious pitfalls in using filtration separation. Primarily, the problem is created when a filter becomes "plugged." It is imperative to recover the trapped plasma process stream materials. Pressurizing or "blowing" the system down with high-quality compressed air may help, but the ideal is a postwash that will not resolubilize fixed solids collected on the filter media.

Filter and Basket Fuges

Filters that incorporate centrifugal action and regenerate their surfaces automatically are interesting and are currently being evaluated by development groups associated with some manufacturers. Tolhurst (Ketema Corp., El Cajon, CA) manufactures a very large basket type centrifuge. This equipment uses filter cloth and a spinning basket to collect resins; resins may even be washed while held in place on this type of instrument.

Another concept in filtration separation is washing in place of the cake. Several questions must be answered: How is channeling prevented? What is the best solubilizing or nonsolubilizing agent to use? Some researchers have looked at scintered stainless steel screens that offer submicron passageways. These would permit aggressive cleaning methods and could replace depth cartridge type filters that are used only once and are very expensive.

Filtration technology has grown through the years with the variety of products requiring filtration and the characteristics of protein solutions. Plate and frame type systems are still in use by plasma processors. These are difficult to clean and, depending on usage, should be electropolished on a regular basis (probably once a year).

Funda Filters

In 1964 at Cutter Labs, Jack Smiley demonstrated the capabilities of the Funda filter using Teflon® cloth in place of standard scintered stainless steel. He was able to separate all fractions routinely. Application of the Funda was relegated to acetone washing of albumin and plasma protein fraction (PPF) as per Hink et al. (1957); previous work went unnoticed (Smiley 1978–1993). Many innovations were a result of Jack's interest in doing things better in fractionation.

WIPED OR THIN FILM EVAPORATION SYSTEMS

Rodney Hunt—Luwa Systems

Some years back Alex Kohler developed and used the wiped film evaporator for the removal of ethanol from Cohn Fraction V suspensions. The principles are quite simple and straightforward. An alcohol solution of protein is introduced into a spinning rotor or tube at such a rate that a wiper lays the product on the surface of the chamber to a fractional depth. The chamber walls are heated to offset the evaporative action of the water and alcohol that is being removed under a vacuum, which is just less than the boiling pressure, permitting the transfer of azeotropic vapors to the Sarco™ pump for removal from the system. Every pass through the system concentrates the protein about sevenfold, while reducing the ethanol by a factor of seven. Hence, if a 5 percent solution of albumin containing a diluted amount of less than 0.4 percent ethanol would be run through the system, it could be finished to meet standard final release criteria. (The final ethanol concentration would be about 0.006 percent.) Abbott Scientific attempted to process PPF through thin film evaporators in the early 1970s, but was unsuccessful. They did not train a dedicated group to handle such fragile instrumentation. Today, it is a natural adjunct to concentrate protein that has been washed on ultrafilters. Prior to large-scale ultrafiltration, thin film evaporators would concentrate salts; today this would not be a problem because we have diafiltration capabilities.

SCREEN SYSTEMS

Sweco Shaker

When World War II ended, Cohn fractionation principles were used to purify gamma globulin from the rich placental source for the prophylactic treatment of persons exposed to poliomyelitis. The major problem in dealing with tissue source extracted proteins was handled with very large meat grinding equipment. The ground tissue was collected in a suspension of saline, paper pulp, and cold distilled water. This mixture was then pitchered onto a sanitary Sweco shaker that held back most of the large pieces of tissue. The liquid portion of the mixture was harvested and centrifuged on an AS-16 for clarification. The very hemolyzed protein-rich stream was then introduced to several alcohol precipitations to reduce the heme and eliminate unwanted contaminating proteins. Although Sweco shakers worked well in this application, they were not as sanitary as one might wish. If the need presents itself in the future, a shaker could be made sanitary.

A major problem facing Cohn fractionators is the reduction of salts. Originally, albumin was "salt poor," which meant that the sodium level of albumin was about 100 meq/ℓ. The rationale for salt "poorness" was probably due to the instability of the contaminating alpha globulins (the major contaminant of Cohn albumin).

In the early processing days turbidity was the primary test to anticipate the stability of albumin solutions. Religiously, turbidity measurements were taken on all filtrates, and albumin final containers had federal minimum standards for turbidity (not to exceed 16 national turbidity units [NTU] after heating a sample at 50 hours at 57°C). When the sodium level was increased to isotonicity in the early 1970s, turbidity levels were difficult to meet and at that time there were no instruments available for turbidity measurement. In fact, most companies built their own equipment with varying success.

Today, this is not a problem. The Hach Co. of Loveland, Colorado, manufactures a very good ratio turbidimeter. Priced in the neighborhood of $1200, it is an excellent device. At some companies plasma processing again is under the control of nephelometry to assure the quality of in-process filtration and to make determinations when proteinaceous solutions are teetering between stability and instability or having some atypical appearance. Some companies also have in-house turbidity limits for final container release.

Today, we no longer need to resuspend Fraction V paste and go through several process steps simply to remove salt. We have learned that through ultra- and diafiltration, salt removal or reduction on very large volumes is accomplished with ease.

Many original workers thought Cohn's Fraction V Rework C was supposed to improve purity as well as remove salt. Experiments since have shown that reprecipitation of the albumin does very little if anything to improve purity. Reprecipitation using Cohn process methodology would probably compare to the crystallization of albumin; crystallization yields a diminishing level of contaminants as one recrystallizes over and over.

ULTRAFILTRATION/DIAFILTRATION

Hollow-Fiber Media

Protein fractionation is very feasible on ultrafiltration membranes and the world of purification has not really looked closely at these tools. The elimination of ethanol used in the fractionation process is also easily dealt with through diafiltration (although thin film evaporation makes much more sense because it causes less shear). Five to six volume replacements will generally be a satisfactory wash; a combination of saline and distilled water washes also works very well. It is possible to calculate the rate of elimination of any solute permeable through the use of simple arithmetic. Each volume exchange reduces the solute by $66^2/_3$ percent. Gel polarization has been a problem when using membranes for the removal of ethanol, water, salts, unwanted proteins, and other low molecular weight species. When a membrane gel polarizes, a layer of concentrated protein will lie adjacent to the membrane. This formation acts as a barrier to interfere with the rate of flux or permeation. The only way to disrupt this barrier is to offer turbulent flow. This makes it necessary to determine the flow volume to achieve an appropriate Reynolds number necessary to disrupt the gel layer (Hein and Pappenhagen 1979). When maximum turbulence is achieved, the ideal flux or permeation rate is observed.

Spiral and Thin Channel Media

Spiral and thin channel media are also available, but if they are used, it is unlikely that a gel layer can be avoided. Spiral media are constructed from flat stock and various types of separators, as opposed to hollow-fiber media that is potted and fixed in epoxy. The sheets are welded with an adhesive that cannot withstand the normal cleaning conditions used for hollow-fiber media. In one particular instance a product processed on a large ultrafiltration system was tending toward pyrogenicity. The ultrafilter was normally cleaned in 0.05 *M* sodium hydroxide at 60°C. The production manager increased the temperature of the sodium hydroxide by 10° to about 70°C, and the pyrogen trend returned to baseline.

John Hink and coworkers (1957) patented the use of acetone for ethanol removal, and this process has been used for years. It is much less expensive than lyophilization, but it does strip the natural lipids from the albumin. To stabilize the albumin or make it possible to heat the final containers for 10 hours at 60°C, vegetable fat source compounds such as sodium caprylate are added back. In a 25 percent protein solution of albumin, one will find about 0.02 *M* acetyl-DL-tryptophan and about 0.02 *M* sodium caprylate.

There has been no change in the stabilizing excipient in all the years that albumin has been manufactured in the U.S. Originally, the stabilization of albumin was being evaluated by two researchers. One was working with acetyl-DL-tryptophan; the other used fatty acids. In a short dispute it was decided to add equal amounts of each, and this is the original formulation.

Ultra- and diafiltration of proteins has opened an entire world of options for separation, fractionation, and even purification. The behavior of various protein solutions during the course of diafiltration is interesting. While diafiltering ISG with cold distilled water, a hydrogen ion pump action occurs so that hydrogen ions (HCl) must be added to the replacement distilled water to maintain the pH in equilibrium. We have also learned the necessity for sodium ions on the removal of low molecular weights across ultrafiltration membranes. Reports have confirmed the work where contaminating metals such as aluminum are removed through the use of salt rich washes.

GEL CHROMATOGRAPHY

Exclusion and Gel Columns

The large-scale processing of plasma protein solutions and the fractionation of these systems is well documented by the Pharmacia

(Piscataway, NJ) and Bio Sepra Inc. (Marlborough, MA) folks. These applications are readily available by contacting these specific vendors.

Ion-Exchange Resins

Ion-exchange resins are interesting for their use with plasma proteins. The impact of rapid pH variances limits their usefulness. Regeneration of the resin is simple and repeatable. A particular application comes to mind where a liquid plasma preparation required the use of resin to exchange calcium chloride with sodium chloride in a defibrination process. Essentially, an amount of calcium chloride was added to fresh pooled plasma. The containers were "rocked" for 4 hours after the addition. Then the containers were solidly frozen for at least 24 hours at -30°C, were removed from the freezer, and thawed in a +30° bath. The formed clot was then hung and drained for at least 12 hours. The pooled plasma was then metered through ion-exchange resin at no more than 250 ml/min (pools were about 100 liters). The pool was then clarified and sterile filtered for filling. In all, an exhausting process! Across town, a competitor collected the 100 liter pool, lowered the pool pH to 4.9, and held the pool in the +5° overnight. The pool was then filtered to remove denatured fibrinogen. The pool pH was raised to neutrality and stabilized with 5 percent dextrose. The pool was then sterile filtered and filled into final containers (method of Paul Savedra 1964)! Many times there is a much simpler and easier way than we think! Resinous exchanges and exclusions may not always be the best and cleanest way to process. Today they are simply in vogue.

LYOPHILIZATION

Freeze-Drying

Courtland Barr (founder of Courtland Laboratories), the laboratory superintendent at Sharpe and Dohme, shipped the first order of freeze-dried plasma to the U.K. during the Battle of Britain in 1940. He solved the problems associated with plasma lyophilization (Barr 1965–1993). Dr. Drew, medical director of the ARC, supervised plasma drawings in Philadelphia. He died after an auto accident in North Carolina when he was refused a transfusion because of his race.

The major production of albumin manufactured through the 1960s was freeze-dried to remove the alcohol. This meant that the industry had to develop methods and build equipment to handle this large volume of temperature critical protein. Hyland used chambers with electric "toasterlike" elements to drive off the vapor to be collected on the condenser. Lou Jerg, a German surgical tool maker, was the "expert" with this equipment (Jerg 1954–1961). In my opinion his knowledge still exceeds most of the experts in the field. The best example of this is his placement of thermocouples (TCs). Today, everyone places the TC in the ice cake in about the last place to dry. During the drying cycle heat input has no response in the initial phases of drying. Lou, on the other hand, wanted to know the warmest part of the container as he drove off the vapor. He would place a thermocouple in an area that would be bombarded by the molecules of water as they were driven to the condenser.

Using this approach, his heat input was variable and depended on the warmest part of the product, not the coldest part! He had calculated the amount of wattage in any given circuit that would be needed to drive off a given amount of water. His calculations were that each watt of electricity given to the resistance heaters would drive of 1 ml of moisture in one hour.

This made all freeze-drying very predictable. After a certain number of watt hours, the TC would cross from reading a negative temperature to reading a positive temperature. It is well known that if heat is introduced at too rapid a rate, second stage drying is uncontrollable. At this time radiant heat, which in theory is very minuscule, was really very effective in vapor formation. All the refrigeration systems at that time were cascade multistage freon. They very effectively held baffled condensers at -60°C. These systems would not operate at condenser temperatures warmer than -55°C. Powerstats were utilized to feed the circuits with a percent wattage that could be calculated and accumulated on the drying records. Product could not be frozen in the drying chambers. This meant that drying trays had to be loaded in the -30°C freezer. It was possible to almost get frostbite if you spent more than about an hour loading a dryer surgically gowned, using surgical gloves, masked under UV lighting. It was a very eerie scene!

PUTTING OUT FIRES

Pyrogens and Other Disasters

Throughout its history, plasma "businesses" have been in or out of trouble. The major problem is probably due to pyrogenic product. Some products are rarely made (without undue effort) pyrogenic. Plasma protein fraction is such a product.

The loss of integrity of a sterile bulk container of PPF is a situation that almost always will result in a pyrogenic lot of PPF—a rare occurrence. Different methods have been used through the years to resolve such a pyrogenic problem. Typically, Armour Laboratories, discovered after the World War II that if albumin is heated for some time at 50°C, the pyrogenic response could be reduced or eliminated when the product was tested on rabbits. I speak here of a pyrogen that at worst would cause fever and chills in the person receiving the product. Days, weeks, months were spent heating questionable plasma protein products. Once heated for a time and passing animal testing, product was released for use. This worked for a period of time, but on some occasions field reactions would result in a recall by the manufacturer. One manufacturer decided to heat and then refractionate before releasing the product. This also worked, but only until field reports came back to the manufacturer. Now there was a quandary: If it gave passing results on the rabbits and was refractionated, how could there still be a problem?

A particular manufacturer sent product samples to its donor center and with appropriate consent forms performed trials of questionable product. They found that while at the same time they were developing test methods (RCA [red cell agglutination] . . . RIA [radioimmune assay]) for hepatitis, that infected product source donors would stimulate antibody activity during this pyrogen testing for hepatitis! What had happened was that PPF, which contains a good quantity of Fraction IV-4 (heat stable), also contains the antigens that were inactivated by pasteurization. What an ideal way to sensitize and raise hepatitis antibody titers! They also learned that pyrogen testing on healthy humans did not predict reactions at point of use. Arrangements were made with Dr. Bob Pennell at the Blood Research Institute in Boston, where trials with appropriate consent and compensation were performed on the terminally ill to verify whether or not pyrogenic reactions could be predicted. This test method, although rarely used, was extremely accurate in predictability.

The ideal is, of course, not to have the problem in the first place. It never ceases to amaze me that breweries are so clean and sparkling just to produce a beverage. Fractionation facilities are generally dark and dank. Companies purchase stainless steel equipment; because it is stainless steel, they believe it will last forever. However, this is not the case. Metals stress with temperature change and become porous with time. Equipment that shows the stress of use should be replaced with new equipment.

An alert and careful inspection of all processing areas at frequent times will prevent the problem. Many times it can be related to wet product contaminations. These carry the Gram-negative rods or residue to create the pyrogenic response. Common sense cleanliness is the best protection. Electropolishing equipment on a preventive maintenance basis prevents the buildup of contamination. Processing expediently does make a difference. Initial pool bacteriology counts will not predict the outcome of pyrogenic product. The process is too long for this type of a simplistic answer, but if a pool of Fraction I effluent sits around for too many hours, do not be surprised when it finally tests pyrogenic. Fraction I is only 8 percent ethanol and this is insufficient ethanol to prevent growth at 0°C. Rabbit testing on half the animals in the countryside and trying to play the "odds" will not give an answer either.

In my mind today's problems are related to rapid growth and the inexperience of technicians, whether in quality assurance or production. While they are gaining experience, costly decisions may result. Questions that relate to the very heart of the production effort can stymie and exasperate. I think a classic example of such a situation would look like this: While reading cultures on a bulk under sterility testing, the new technician observes a proteinaceous formation. Wolf is immediately shouted! The bulk must be contaminated. In this scenario you are safe since this observation is purely a test anomaly, but what if a bulk is truly contaminated?

A few years ago, production would immediately refilter and retest. Today, a committee must meet to develop an action plan! Common sense goes out the window and millions of dollars in a natural resource are threatened with destruction! The ability to act wisely and quickly is no longer an option for a production manager.

ACKNOWLEDGMENTS

This chapter is especially dedicated to those Cohn workers who came to the industry from Harvard University and elsewhere: Jack Weare, Alex Kohler, John Tullis, Bob Pennell, Fred Johnson, Tillman Gerelough, Dwight Mulford, J. Ashworth, and Bud Mealey to list a few; to those in industry: J. Moody, Jim Horn, Clay Bloom, Joe Noe, Jack Smiley, Conrad Turner, John Hink, Al Pappenhagen, Syl Moura, Randy Kent, Ed Potere, Ron Hein, Johann Vandersaande, Bill Saxton, John Lundblad, Clancy Richter; and to those who gave us literature, especially H. E. Schultze and J. F. Heremans who prepared the masterpiece, *The Molecular Biology of Human Plasma Proteins.* These people have inspired us to remain in the "plasma business" for almost 40 years.

Special thanks also to Clyde Zalut, C. Hilyard Barr, Paul Bogikes, and Marlane Turner.

REFERENCES

Barr, C. H. 1965–1993. Correspondence.

Bogikes, P. M. 1968–1995. Correspondence.

Cohn, E. J., W. L. Hughes, D. J. Mulford, J. N. Ashworth, M. Melin, and H. L. Taylor. 1946. *J. Am. Chem. Soc.* 68:459.

Hart, H., K. Reid, and W. Hart. 1993. *Vox Sang.* 64(2):82–88.

Hein, R. H., and A. R. Pappenhagen. 1979. Unpublished memo.

Hink, J. H., and coworkers. 1957. *Vox Sang.* 2:174.

Jerg, L. 1954–1961. Hyland Laboratories Correspondence.

Kent, R. H. 1990–1993. Correspondence, National Red Cross, Baxter Biotech.

Moura, S. A. 1965. Correspondence at California Transfusion Service.

Oncley, J. L., and coworkers. 1949. *J. Am. Chem. Soc.* 71:541.

Smiley, J. C. 1978–1993. Cutter Laboratories Correspondence.

Tiselius, A. 1937. *Trans. Faraday Soc.* 33:524.

Tullis, J. L. 1988. Cutter/Miles Lecture.

Subject Index